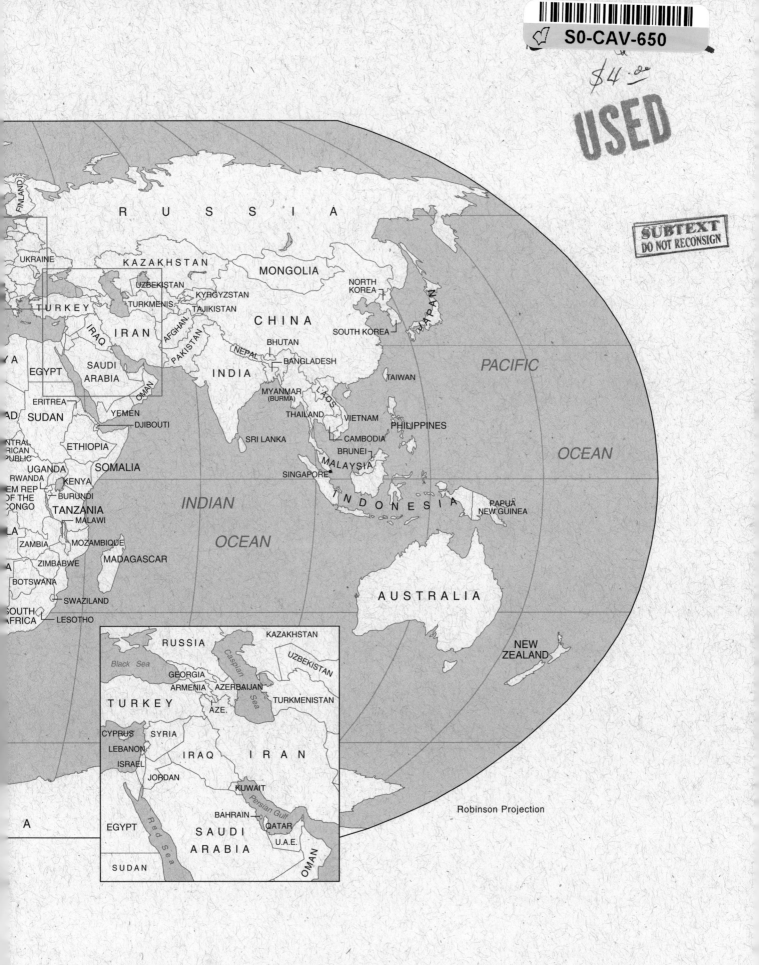

FINLAND

R U S S I A

UKRAINE

KAZAKHSTAN

MONGOLIA

NORTH KOREA

UZBEKISTAN

KYRGYZSTAN

TURKEY

TURKMENIS.

TAJIKISTAN

C H I N A

JAPAN

IRAQ

IRAN

AFGHAN.

SOUTH KOREA

PAKISTAN

NEPAL

BHUTAN

EGYPT

SAUDI ARABIA

OMAN

I N D I A

BANGLADESH

TAIWAN

PACIFIC

YA

ERITREA

YEMEN

MYANMAR (BURMA)

LAOS

SUDAN

DJIBOUTI

THAILAND

VIETNAM

PHILIPPINES

OCEAN

AD

NTRAL
RICAN
PUBLIC

ETHIOPIA

SRI LANKA

CAMBODIA

UGANDA

SOMALIA

BRUNEI

RWANDA

KENYA

MALAYSIA

EM REP
OF THE
CONGO

BURUNDI

SINGAPORE

TANZANIA

MALAWI

I N D O N E S I A

PAPUA
NEW GUINEA

LA

ZAMBIA

MOZAMBIQUE

INDIAN

ZIMBABWE

MADAGASCAR

OCEAN

BOTSWANA

SWAZILAND

AUSTRALIA

SOUTH
AFRICA

LESOTHO

NEW
ZEALAND

A

Robinson Projection

Inset map

RUSSIA

KAZAKHSTAN

Black Sea

Caspian Sea

UZBEKISTAN

GEORGIA

ARMENIA

AZERBAIJAN

TURKEY

AZE.

TURKMENISTAN

CYPRUS

SYRIA

LEBANON

I R A N

ISRAEL

IRAQ

JORDAN

KUWAIT

Persian Gulf

BAHRAIN

EGYPT

QATAR

Red Sea

SAUDI
ARABIA

U.A.E.

OMAN

SUDAN

Second Edition

THE GEOGRAPHY OF ECONOMIC DEVELOPMENT

Regional Changes, Global Challenges

Timothy J. Fik
University of Florida

Boston   Burr Ridge, IL   Dubuque, IA   Madison, WI   New York   San Francisco   St. Louis
Bangkok   Bogotá   Caracas   Lisbon   London   Madrid
Mexico City   Milan   New Delhi   Seoul   Singapore   Sydney   Taipei   Toronto

McGraw-Hill Higher Education 𝒳𝒬
*A Division of The **McGraw-Hill** Companies*

THE GEOGRAPHY OF ECONOMIC DEVELOPMENT: REGIONAL CHANGES, GLOBAL CHALLENGES, SECOND EDITION

2 3 4 5 6 7 8 9 0 KGP/KGP 0 9 8 7 6 5 4 3 2 1 0

ISBN 0–07–365948–7

Vice president and editorial director: *Kevin T. Kane*
Publisher: *Edward E. Bartell*
Sponsoring editor: *Daryl Bruflodt*
Senior marketing manager: *Lisa L. Gottschalk*
Project manager: *Renee C. Russian*
Senior production supervisor: *Mary E. Haas*
Designer: *K. Wayne Harms*
Photo research coordinator: *John C. Leland*
Compositor: *GAC—Indianapolis*
Typeface: *10/12 Times Roman*
Printer: *Quebecor Printing Book Group/Kingsport*

Cover designer: *Nicole Dean*
Cover image: *Marc Yankus/SIS*
Photo research: *Connie Gardner Picture Research*

Library of Congress Cataloging-in-Publication Data

Fik, T. J. (Timothy Joseph), 1957–
  The geography of economic development : regional changes, global
challenges / Timothy J. Fik. — 2ⁿᵈ ed.
      p.   cm.
  Includes bibliographical references and index.
    ISBN 0–07–365948–7
  1. Economic geography.   2. Space in economics.   3. Economic
development.   4. Regional economics.   5. Populations—Economic
aspects. I. Title.
HF1025.F5   2000
338.9—dc21   99–28131
                                                     CIP

www.mhhe.com

To my parents...

> *Edward A. Fik (1913–1990)*
> *Genevieve Fik (1917–1997)*

—the wind beneath my wings.

BRIEF CONTENTS

CONTENTS

CHAPTER **10**

Development Prospects and Obstacles   *270*

PREFACE

WHY I WROTE THE BOOK

As instructor at the University of Florida, I have been in constant pursuit of the "perfect" in-class textbook to use in my course entitled World and Regional Economies. This was no easy task given my desire to find a book that students would enjoy reading and one that was appropriate in terms of topics and subject matter. Moreover, the book had to be multidisciplinary in its emphasis, drawing on the fields of both geography and economics. I envisioned that book to be regional and global in its orientation, highlighting current economic and demographic patterns and trends with discussions and analyses couched from a variety of perspectives and viewpoints. Needless to say, I never found that book.

I was convinced, nonetheless, that a balanced book could be constructed, one that covered a wide range of topics and issues related to the geography of economic development. Many of the so-called standard world economies textbooks seemed to be lacking in one way or another as far as coverage or emphasis. Particularly disturbing was the fact that the majority of these books were far more advanced than what was required at an introductory level. Over the years, students would regularly comment that these books were difficult to read. They seemed to want a book that gives them more of the bottom line.

After roughly ten years of searching and evaluating what was out there, I decided to draft a book that would both appeal to undergraduates and cover the important material in my course. With the help of the McGraw-Hill College Custom Series staff, I constructed a "quick and dirty" first edition. It was released in 1997, partly as an experiment, so I could gain feedback as to the market potential of the book. To my surprise, the book gained a loyal and sincere following, despite its shortcomings. With many changes and additions, that rawhide-rough version has now been crafted into a new and much-improved second edition. I look forward to comments and suggestions on how this edition, yet another work in progress, can be further improved. As it stands, the current edition offers much in the way of discussions on regional and global economic development, and it is my hope that it will be well received.

STRUCTURE OF THE BOOK

The layout of this book is unique in terms of design and content. Unlike conventional world economy texts, which seem to isolate topics by chapter, each chapter in this book can be viewed as a stepping stone which rests across a pool of interrelated subject matter. Each chapter subsequently builds upon the concepts and arguments presented earlier. Key arguments in each chapter are also reinforced with real world data, including forecasts based upon recent trends. It is the sequential integration of theory and discussion followed by empirical evidence and analysis that give the book a unique signature and a "vibe" all its own.

The work begins with an overview of geography as a field of study. Chapter 1 provides a brief sketch of the historical underpinnings of economic geography as a subdiscipline of geography. This provides the reader with an understanding of why geographers have an interest in matters of world and regional economy. It is here that a research agenda is advanced from a "limits to growth" perspective. Note that the notion of limitations (economically speaking) is a central theme that can be found throughout the book.

In Chapter 2, the focus turns to classification of regional economies and economic activities, with emphasis on growth and development and barriers to development. Criteria are established for distinguishing between "developed" and "developing" regions. A geography of economic development is introduced by highlighting the current development status of regions in relation to economic performance and production capacity. Chapter 3 provides in-depth discussions on external growth theory and the importance of exports and trade in the regional development process. Trade policies and development strategies are also examined. Chapter 4 offers essential background on various development perspectives and the impact of political economy. Also included are discussions of stages of development, modernization, and the link between global competitive advantage and uneven development in a capitalist world economy.

Chapter 5 presents an overview of the geography of population. This includes a review of world cities and the limitations imposed by hyper-urbanization. The dualistic nature of production economies are also discussed. Chapter 6 highlights various food and economic security issues and the global food crisis, shedding light on the so-called myths of world hunger and food production ceilings. Chapter 7 introduces the geography of demographic change, with attention to regional variations in population growth as tied to the forces of natural increase and interregional migration. An overview of the world refugee problem is also included.

In light of the world population explosion, Chapter 8 explores issues of economic sustainability in relation to increasing demand for extractive resources. This chapter also provides an overview of the major agricultural systems in developing regions; discussing the future of those systems and their export potential. Chapter 9 devotes serious attention to the globalization of production and industry, underscoring the importance of international investment and the proliferation of global production networks. The controversial role of multinationals in the regional development process is also discussed. This sets the stage for discussions in Chapters 9 and 10, which suggest that many economic development challenges are byproducts of conditions and forces that are beyond the control of individual regions or nations. Population explosion issues notwithstanding, limitations to regional growth and development are linked to global economic and production cycles.

DISCLAIMER

Viewed in its entirety, this book represents an attempt to explain regional variations in economic development status from a Western perspective. It is by no means complete in its coverage or subject matter, and is not intended to be the definitive world economy text (as one does not, nor ever will, exist). Note that a forthcoming edition is planned for the year 2002. That book will include additional sections or chapters, with expanded discussions on energy and world resources, global services, information and technology, international politics, human capital, and women in development. Until then, it is my hope that this edition will serve as a useful alternative to what is currently out there. I trust you will find it to be more than adequate as either a main or supplemental textbook.

T.J. Fik, Ph.D.
Associate Professor
Department of Geography
University of Florida

ACKNOWLEDGMENTS

I am extremely grateful to Daryl Bruflodt and McGraw-Hill for believing in me and this project. Their support and guidance has made all the difference. In addition, I would like to thank my good friend Judy Ice for helping me realize a vision.

I would like to extend my sincerest appreciation to my loving wife Denise for sacrificing moments lost while I worked on revisions.

Many thanks go out to Jim Sloan (cartographer extraordinaire) for the wonderful series of maps he so painstakingly produced and his attention to details. I am also grateful to the external reviewers for their insightful comments and suggestions.

List of reviewers:

Dr. A-M d'Hauteserre
Department of Geography, Southern Connecticut State University

Günter Krumme
University of Washington

Mary P. Lavine
University of Pittsburgh at Johnstown

Neil Reid
The University of Toledo

I would also like to thank so many of my students for their kind words and feedback in regard to this endeavor. Last, but not least, I wish to thank my former advisors and mentors, Professors Ross D. MacKinnon and Gordon F. Mulligan, for their support and encouragement over the years.

ECONOMIC GEOGRAPHY

WHAT IS GEOGRAPHY?

Geography may be defined as *the study or analysis of the locational attributes and spatial variation of phenomena on the earth's surface.* It is a discipline focused on the study of human and physical phenomena—the study of where things are, how things are organized in space, and how and why things vary or are distributed over space. While geography has been associated with the identification and characterization of "regions," name-place recognition, and **cartography**—*the science and art of map making and interpretation,* it is a field of study that encompasses a much broader subject matter and emphasis. In fact, any inquiry that incorporates a spatial viewpoint, or investigates something which implies, pertains to, or involves location or space, may be considered as a geographical inquiry. Viewed in this manner, it is virtually impossible to think of anything that does not involve "space" or geography.

From a geographical perspective, space may be conceptualized in an absolute sense as a distinct or physical entity in and of itself. The distribution of real world phenomena implies that things are separated from one another in physical space. Separation in geographical space can be measured in terms of distance; consider the distance between any two locations on the earth's surface. For instance, consider the distance between Miami, Florida and Dallas, Texas; something that can be easily calculated or measured. The answer may differ depending

upon how one defines separation and distance. For example, the "as-the-crow-flies" distance may certainly be less (in terms of mileage) than, say, the shortest-roadway distance between these two places. Note that separation may also be measured in temporal terms; consider the time it takes to travel between these two locations. Although absolute measurement of separation in space may vary dramatically depending on the measurement units or transportation mode in question, these measures give an indication of the degree to which locations are separated from one another in the physical world.

Alternatively, geographical space may be thought of as an area within which human activities or physical processes occur. Space may also be represented in an abstract manner. More specifically, geographic inquiries may be couched in terms of **relational space**—*the ordering, ranking, arrangement, or positioning of objects or places in relative space, as suggested by evaluations and comparisons of separation as measured in nonphysical terms.* For example, consider a comparison of the economic status of all nations of the world. The United States and Germany, two highly advanced economies (although far apart in terms of physical distance), can be viewed as closer or more similar to one another in relational space than the United States is to, say, the less-advanced and bordering nation of Mexico when compared in economic terms. Note that discussions in this book are primarily couched in terms of relational space with emphasis on economic and demographic variables.

ORIGINS AND TRADITION

[handwritten: Geography = earth depiction]

The word *geography* comes from the Greek roots "geo" meaning earth or world, and "graphos" meaning description or depiction. Geography, therefore, was originally cast as a descriptive field of study and was primarily oriented toward describing the physical world. The description of earthly phenomena led to many useful observations and insights, allowing the field of geography to flourish as a science. Early geographic inquiries tended to focus on exploration and discovery of the physical earth and its treasures. These inquiries were responsible for producing a wealth of information, captured in the form of maps and written accounts. Geographers commonly engaged in the art of reflecting on and interpreting that which they observed and documented. This rich tradition is captured in the writings of American geographer Richard Hartshorne in *The Nature of Geography* (1939) and *Perspective on the Nature of Geography* (1959). Hartshorne portrayed geography as a discipline that sought to provide accurate, orderly, and sensible descriptions and interpretations of geographic phenomena. The spatial variability focus served as a rallying cry for geographers and a way in which the discipline distinguished itself from other disciplines. More importantly, the formal organization of knowledge in geographic terms represented a new and exciting way in which to examine and appreciate the world.

Today, geographers are not alone in incorporating a spatial dimension in their research. Urban and regional economists, planners, ecologists, earth scientists, as well as practitioners in business and government all understand the importance of using geographic information and adopting a geographic perspective. A transportation planner who is investigating traffic counts among major thoroughfares in a large urban area with the intent of forecasting the likely impact of a proposed regional shopping mall on traffic flows, or a chief executive at a large corporation who outlines an innovative marketing strategy for competing with overseas companies in markets abroad, or a regional analyst who monitors changes in international commerce in response to trade policies—all have one thing in common: a thirst for geographic knowledge.

Depending on research objectives and goals of a study, the scope and nature of the problems under review, the type of questions posed, and data availability, geographic analysis of phenomena usually involves considerations of **geographic scale**—*the level at which reality is to be represented or the degree of resolution employed in the observation and measurement of a given phenomenon.* Geographic analyses may involve information gathered from one or more of the following scales: the micro level (also referred to as the local or urban level); the

[handwritten top margin: micro level – local / meso level – regional / macro level – global]

meso level (the regional level); and the *macro level* (the national or global level). Note that discussions in this book will highlight phenomena measured primarily at the macro level.

GEOGRAPHY: A DISCIPLINE DIVIDED, YET ONE

The field of geography can be divided into two separate and distinct subdisciplines: physical geography and human geography. **Physical geography** is *the study of the physical world—systems and processes associated with the earth's atmosphere, hydrosphere, lithosphere, and biosphere as defined in physical terms.* By contrast, human geography is concerned with the human world—focusing on human population, activities, and settlements as defined in cultural, historical, social, political, and/or economic terms.

As a bona fide social science, human geography is a branch of geography that focuses on the activities and behavior of human beings from a spatial perspective. Human geographers, therefore, tend to explore the geography of societies, cultures, and economies. Although different in terms of subject matter and emphasis, both human and physical geographers are very aware of the importance of examining the interdependencies of human and physical systems and the environment. Nonetheless, it is the location and variation of activities over the space in which the human population resides that are of utmost importance to the human geographer.

Human geography may, thus, be defined as *a subfield of geography that seeks to describe and analyze human activity and behavior for the purpose of gaining knowledge and understanding of the factors and forces which affect the human existence* (Figure 1.1).

Note that the spatial viewpoint helps differentiate human geography from other social sciences. While the historian may discuss a phenomenon in terms of a past sequence of events, and a sociologist may examine the possible ramifications of that phenomenon on society at large, or a political scientist may explore its impact on public opinion, it is the human geographer who highlights the spatial variations in the intensity or impact of a phenomenon with the intent to formulate a reasonable interpretation for conditions and changes which are brought about as a result of that phenomenon. Thus, human geographers directly engage in the study of things which affect the human experience as they seek to identify interrelationships operating in or over space. To the human geographer, space is the stage upon which the drama of geography and drama of humanity are simultaneously played out.

Figure 1.1 Contrasting lifestyles: people enjoying lunch overlooking the skyline of a large modern urban area, while a peasant woman prepares food for her family in a small village in a developing country.

THE FIVE UNIFYING THEMES IN GEOGRAPHY

The discipline of geography is unified by five fundamental and unifying themes that allow both human and physical geographers to celebrate the spirit and uniqueness of the geographic perspective: location, place, regions, human-environment relations, and movement.

1. Location

Location may be defined as *a position or site occupied by some distinguishing feature or point on the earth's surface expressed in such a manner that it answers the basic question of "where?" in an absolute or relative sense.*

Absolute (or **mathematical**) **location** describes *an exact position or point on the earth's surface as defined by a set of mapped coordinates obtained from a superimposed grid or measurement system.* For instance, the use of latitude and longitude allows one to measure and locate the exact mathematical position of any given point on the earth's surface. For example, the city of Gainesville, Florida—home of the "fighting Gators"—is known to be located at 29.4° N latitude and 82.2° W longitude. Note that latitude is measured in a north-south direction from the equator (at 0 degrees) along a meridian or great circle where the North and South Poles are positioned at 90° from the equator. Longitude is measured in an east-west direction along a parallel of latitude from the prime meridian (at 0 degrees)—an imaginary line which extends from the North Pole to the South Pole, which runs through the crosshairs of the telescope at the Royal Observatory in Greenwich, England. This measurement system enables a researcher to pinpoint the exact and absolute position of any place or object on the earth's surface.

Relative location refers to *the relational characteristics of a location as described in generalized terms or with respect to other areas or reference points on the earth* (that is, the position of a point or place in relation to the position of other points or places). The relative position of something may be compared using the cardinal directions of the compass (N,S,E and W) or other designators. For example, the Canadian province of Alberta could be described as presiding in or at a location that is west of Saskatchewan, south of the Northwest Territories, east of British Columbia, and on the leeward side of the northerly stretches of the Rocky Mountain chain; a province which shares its entire southern political border with the great state of Montana (a border that is centered approximately 200 miles north of the city of Butte).

The relative location concept is also useful for comparing the status of various economies or nations of the world. For example, it can be argued that the economies of most western European nations are more advanced and diverse, in terms of what they produce and trade, than the economies of, say, Central America. The nations of western Europe could also be thought of as being more highly integrated into the global economic system than their Central American counterparts.

2. Place

Place refers to *a locality with physical and human qualities that help distinguish it from all other localities* (that is, a location having distinctive features which give it meaning and character). In addition, a place can be thought of as having a unique cultural identity—the sum total of the knowledge, beliefs, attitudes, and behavioral patterns shared, acknowledged, and/or transmitted by a group or members of society at a given location. It is an identity embodied in the so-called **cultural landscape**—*the ultimate geographic expression of that culture.* In general, a place may be thought of as a portion of geographical space occupied by a person, group, or object that has at least seven constituent values or components:

1. *a location* (as defined in absolute and/or relative terms);
2. *cohesion* in relation to the workings and the integrated nature of the physical and human systems;
3. *an intensity of social interaction* as defined by the interdependencies and interplay of human agents in the system;
4. *a uniqueness or identity* as defined by physical setting, appearance, topography, site and situation, function, and/or interconnective social, economic, and political structures;
5. *a localized vibe or spirit* which emanates from the heart of the cultural network and which is formed, sustained, and reproduced in geographical space;
6. *an emergence* as defined by its historical and cultural realities and the sequences and changes that have led up to the present and conditions which will influence the future; and
7. *meaning to human agents* who are able to accumulate knowledge and form a deeper understanding of the events, customs, practices, etc., associated with that place through experience, interaction, perceptions, and the processing of information; thereby allowing them to gain a "sense of place."

[handwritten margin note: these make up a place]

If we were to describe a place, say a metropolitan area somewhere in the United States . . . a place that is famous for its frigid winters, heavy lake-effect snow squalls, blue-collar labor force of largely Irish, Italian, and eastern European descent, four-time consecutive Super-Bowl contending AFC championship football team named "the Bills," the birthplace of spicy hot chicken wings and home of the ever-popular "beef-on-weck" sandwich, once-thriving steel and flour-milling industries situated on the eastern shores of Lake Erie in upstate New York, etc. . . . it wouldn't take long before the city of Buffalo came to mind. Certainly the set of attributes noted above are unique to that place and no other, not to mention the meaning that place has to the people living there.

3. Regions

Regions are *areas of the earth's surface which have certain unifying characteristics.* Regions may be physical (take for instance the Andes Mountain Range, which runs along western coast of the continent of South America), political (consider the nation of Mexico as defined by its national boundaries and borders), or economic (consider the so-called "mixed farming" region of the United States—an area centered about the flat, central farmlands of Iowa, including large portions of Illinois, Indiana, and Ohio, where a mix of crops and animals may be found).

Regions imply delineation of an area and areal coverage; that is, regions cover a specific geographic area. Delineation implies that some sort of boundary may be identified. A **boundary** may be defined as *the limit or extent within which a region or system exists or functions.* Boundaries may be either fuzzy (consider zones of agricultural production) or hard and well-defined (consider the Rio Grande river as both a political and physical boundary between Mexico and Texas).

Conceptually, we may identify two general types of regions: uniform regions and functional regions. The **uniform region** may be represented as *a homogenous area that is defined on the basis of some common or unifying characteristic which distinguishes that region from places outside the region (thus implying a degree of internal consistency) as expressed in terms of the presence of one or more activities or attributes.* In other words, a uniform region may be defined by a set of elements that are more or less evenly distributed throughout the area in question, and weakly represented outside of the region.

For example, consider the traditional Dairy Belt region of North America, a region predominantly associated with the rural agricultural landscapes of central Minnesota, Wisconsin, and Michigan (major dairy-producing states). Here a large regional concentration of the black and white Holstein cows graze on the grasses and feed crops grown on the scenic green rolling hills of the northern Midwest; an area noted for its rough topography, glaciated soil types, and relatively short growing season. Portions of this belt also extend across the outer perimeters of the Great Lakes region and extend into southern Ontario, Canada. Although there is no real hard boundary to show where the Dairy Belt starts or ends, and where other agricultural systems take over, the emphasis of traditional dairying activities as the predominant agricultural activity in the areas mentioned above is reason enough to say that such a region does exist (Figure 1.2).

A uniform region may also be defined in terms of physical characteristics. For example, consider the great Saharan Desert—a vast and cruel arid landscape covering a large segment of the northern part of the African continent. It is a region which supports very little life in comparison to many other regions of the world. While the Sahara Desert is homogenous in one regard (climate), it is also very dynamic and ever-changing; with constant shifts in the extent of its boundaries as shaped by the human and physical forces that define and act upon this region. Over time, the Sahara continues to expand its boundaries as land adjacent to the ever-encroaching desert becomes degraded, eroded, and unable to support vegetation.

Figure 1.2 The predominant economic landscape of America's midwestern farming region.

The **functional region** is *an area formed around a central point or group of points in space and delimited by all movement to and/or from one or more points.* In short, a functional region's internal structure is oriented about one or more points in space (where a point defines a concentration of economic or social activity), with a boundary that is defined in terms of the outermost observable interactions. Consider a regional newspaper circulation trade area centered around a major city from which newspapers are distributed (the circulation area for, say, the *Chicago Tribune*), or a downtown Chicago urban commuting zone whose outer boundary is defined by the furthest locations from which people commute to work on a daily basis. In both cases, the activity in question focuses on the same central node: the city of Chicago. Note, however, that the extent of these functional regions may differ dramatically as we would expect the newspaper circulation area to be much greater in terms of its spatial coverage than the functional region defined in terms of the daily commuting patterns of downtown workers. Along similar lines of reasoning, we would expect the functional region formed by the circulation of a major newspaper to be greater in size or areal coverage than the functional region defined by the circulation of a local paper in a small town such as the *Williston Pioneer* in Williston, Florida.

Other examples of functional regions include the designation of metropolitan areas comprised of a central city and surrounding suburban areas or enclaves which are functionally dependent upon that central city as an activity center (for example, consider the many bedroom communities located around the city of Los Angeles, California, from which workers commute on a daily basis). Consider major airline routings (also known as spokes) from a major city (or hub) from which an airline operation is based as part of some larger hub-spoke system; for example, Delta Airline's hub in Atlanta, Georgia, and the functional region defined by its connections throughout the continental United States. A political jurisdiction under control of a governing body that resides in (or acts from) a given location is yet another example of a functional region. Consider the United States as a functional region, and how individuals and components of this region are affected by the many forces and movements which influence legislation on Capitol Hill.

We should make note of two important effects of geographic scale here. First, the same constituent values and components which define a place can also be used to define a region. However, a geographical study of a region will typically focus on one or several elements in the delineation of a region. Note that the level of homogeneity of activities and/or characteristics decreases dramatically as the size of a region increases (the Antarctic and the Sahara being unusual exceptions, perhaps), suggesting that regions are ultimately overlapping. For instance, while the U.S. mixed farming region may run through the states of Illinois and Ohio, it would be very difficult to contend that Chicago and Cleveland (although geographically associated with the mixed farming region) function primarily as agricultural centers. Surely, these urban areas could be thought of as centers of gravity for more complex functional regions which exist within a given agricultural region.

Second, the world economy itself may be viewed as the ultimate functional region. It can be argued that virtually all segments of the global economy are widely influenced and shaped by what happens in a mere handful of nations and leading economic and production centers. The ongoing globalization of local and regional economies, and the outward extension of their boundaries, is a direct outgrowth of changes brought about by political forces, economic reforms, the increasing interdependence of production systems, the internationalization of finance and investment, and the nature of competition in a shrinking and increasingly connecting world. Virtually all local and regional economies have been sensitized to changes within the newly formed and rapidly mutating functional region known as the "global economy."

4. Human-Environment Relations

Human-environment relations refers to *the interdependencies and interrelationships between human beings and their environment and the outcomes and effects of human-environment interactions on the human economic landscape.* In essence, it is the concern of how people alter their environment through various activities and, in turn, how the environment influences behavior. As humans manipulate their surroundings and shape the physical and human economic landscape, environmental change may bring about unanticipated (and often unwelcome) changes that cause humans to modify future behavior. Hence, the human-environment theme typically focuses on the repercussions of human economic activities, and more specifically, how environmental change and degradation impose limitations on future economic activities. From the human geography perspective, it is a recognition that the state of a region and, arguably, the state of the world hinge greatly on the state of the environment.

For example, excessive tree-cutting, overgrazing, and soil erosion from expanded agricultural production in the Himalayan watershed has made Bangladesh's monsoons increasingly more dangerous through time. The loss of vegetation and the inability of the lands to retain moisture has increased the likelihood of soil erosion, which leads to higher run-off rates during the wet season, which in turn increases the likelihood of flooding along the Ganges River. Related shortages in food production systems, which are affected by the loss of soil within this system, cause the region to become more dependent on outside suppliers to meet a growing demand.

Portions of the African continent have also been continuously plagued by drought and erosion from intensified farming efforts, overgrazing of rangelands, and the reduction of natural vegetation from intensified subsistence agriculture and fuelwood gathering efforts in response to localized scarcity. These activities have helped to speed up the process of desertification along the great Saharan Desert. In general, desertification refers to the expansion or encroachment of the desert on fringe areas (usually neighboring grasslands in semiarid environments) (Figure 1.3).

The East African nations of Ethiopia and Sudan, for instance, have experienced significant amounts of soil erosion and dramatic reductions in the productivity of farmland in food production areas as they are overutilized or improperly managed. As productive soils lose their plant-sustaining capabilities due to overutilization or mis-

Figure 1.3 The desertification of cropland in East Africa continues to be a major problem for this deficit food-producing region.

management, water absorption and retention abilities also decline, resulting in dry and more compact soils that are highly prone to topsoil losses during the wet season and wind erosion during the dry season. This also promotes depletion of groundwater in the regional watershed. As the soil hardens, groundwater levels reduce, and nutrient-bearing topsoil is swept away by water or wind, further intensifying desertification. In just a short time, semiproductive agricultural lands on the fringe of a hostile arid environment will be claimed as desert wastelands. Absorbed by the encroaching desert, once productive farmlands will cease to function in a food production capacity; a situation which serves only to worsen localized food and fuelwood shortages.

Another prime example of environmental devastation as a result of human-environment relations is the plight of the dying Aral Sea (located in southern Kazakhstan and northwest Uzbekistan, republics of the former Soviet Union). The diversion of source waters from the sea to support irrigated agricultural projects along the Syr Darya and Amu Darya Rivers is largely responsible for the devastation.

Since the mid-1950s, the production of crops like cotton, rice, and melons throughout the marginal farmlands of southern Turkmenistan has been sustained from water diverted from the Aral Sea via the Kara Kum Canal. This diversion of water has resulted in a dramatic reduction in the water level of the Aral Sea and the receding of its shoreline. The Aral Sea has lost well over 40% of its surface area since 1960—a volume of water that is equivalent to about one-and-one-half times that of Lake Erie. By the year 2000, the Aral Sea is expected to shrink to approximately two-thirds of its present size. If this trend continues, it is likely that this valuable water resource will completely vanish in less than 100 years!

A dying Aral Sea continues to claim many victims. Losses include the devastation of the once flourishing fishing industry in the towns of Muynak and Uchsay, found on the southern shores of the Aral Sea; an area that is now located in the midst of a desert landscape as the sea recedes. Increasing population pressures in nearby areas have increased the demand for water, placing an even greater burden on the sea for water. Increasing salinity of soils in the Aral Sea basin and surrounding irrigated agricultural regions has created problems not only for farmers, but has led to a health crisis in the area. Irrigation has forced the water levels in these areas to rise as naturally saline desert subsoils become waterlogged. Waterlogging has led to heavy salt deposits at the soil surface as the water levels in irrigated farming regions rise and fall. As salts accumulate on the surface and the soil toxicity levels increase, those very soils now poison plant life and render the land unproductive and unable to sustain acceptable levels of agricultural output. Moreover, soil erosion from high winds in this region has led to a high concentration of airborne salts and salt-dust storms. Airborne salts are known to be responsible for the higher incidence of respiratory disease over the last two decades in and around the area. This problem is not unique to this geographic region alone. For instance, the irrigated agricultural lands of the arid southwestern United States have been combating similar increases in the toxicity levels of soils in areas affected by rising groundwater levels.

The examples on the previous page make it is easy to see just how human beings are both able to benefit and suffer from the manipulation of the environment. In some circumstances, economic gains from environmental change come at high cost. The changes human agents impart on the landscape through the manipulation of their environment can indeed come back to haunt them. As geographers seek to understand the very nature of change in human and physical systems, the human-environment relations theme has become increasingly popular.

5. Movement

Movement refers to *the flow of anything transferable or transportable over space; generally between a point of origin and one or more destinations.* In general, this theme focuses on flows in space: the movement of people or migration; the flow of goods and services (trade); the flow of information and ideas (telecommunications); the transfer and exchange of money and commodities (commerce); the spread of innovation and technology (diffusion); and intensity of interaction between two or more locations (spatial interaction).

In theory, spatial flows or movement will most likely occur along the lines of least resistance, in accordance with the so-called **principle of least effort** as advanced by Zipf (1949). The principle of least effort states that *flows are more likely to occur along a shortest distance path or in ways which conform to a least transportation cost solution; that is, along the lines of least resistance.* The path of least resistance is consistent with the solution that minimizes transport cost under the assumption that human beings are rational and act as cost-minimizing agents. The very nature of a transportation cost component implies the existence of a **friction of distance**—*resistance to movement over space as represented in terms of time, money, opportunity costs, or other barriers.* In other words, the friction of distance represents the difficulty of overcoming the effects of separation. Alternatively, the friction of distance may be thought of as the sum of all forces that inhibit the interaction of people and places. In general, the higher the friction of distance, the lower the frequency of flows between two locations.

Movement is influenced by the dependence of economic agents and production processes, and various geographic market forces such as the location of supply and demand points. For example, the production and movement of agricultural commodities amongst the various geographic regions which comprise a given production system are influenced by many things including variations in the commercial viability of agricultural production over space, the location of production and consumption points, and the feasibility of transport under prevailing conditions and cost schedules. As a production system tends toward a "spatial equilibrium," where the market forces are in balance amongst the various geographic regions which define that system, recognizable patterns of commodity production and transport emerge. Price levels are affected by regional variations in product quality and output, changing inventory, surplus and deficit conditions, and transportation cost differentials. As the supply and demand of any commodity vary widely over space, and as supply and demand conditions are typically dissimilar at virtually every point in space, geographic variations in price, production costs, and profit are commonly observed.

Note that demand points may be far removed from the points of production. It is the unequal and uneven spatial distribution of production inputs, resources, technology, output, and demand that make movement and spatial interaction inevitable. The likelihood or intensity of movement or interaction between two points in space, nonetheless, is likely to decline as the distance between those points increases. This regularity is known as the **distance-decay effect**—*the attenuation in the intensity of a process or pattern with increasing distance from a point of reference.* In general, the greater the friction of distance, the stronger the distance-decay effect.

A system in which movement takes place is composed of two fundamental units: places which generate or receive flows (that is, origins and destinations); and the flows themselves. It is important, therefore, to understand the characteristics and conditions which lead to the designation of places as origins and destinations in reference to a particular type of flow and, in turn, how flows change the very nature of the origins and destinations that generate or absorb those flows (Hanink, 1994).

Using the framework developed by Edward Ullman, movement and spatial interaction are known to be guided by three fundamental principles: (i) complementarity; (ii) transferability; and (iii) competing and intervening opportunities.

Complementarity refers to *a condition brought about by the existence of potential supply and demand points in space* where, in theory, the mutually satisfactory exchange of raw materials, agricultural commodities, manufactured products, or information between two regions is possible to fill the needs and wants of each. For example, we may describe a potential commodity flow from a surplus wheat-producing region to a deficit wheat-producing region. Note, however, that a complementary relationship does not necessarily guarantee that a flow will occur between the regions in question. Complementarity is a necessary condition, but certainly not a singularly sufficient condition for spatial interaction. Note that surplus-producing regions are commonly referred to as sources or supply points, while deficit-producing (and potentially commodity-receiving) regions are commonly referred to as sinks or demand points.

Transferability is *a condition which describes the cost-effectiveness of transporting goods, services, and/or information between two regions* (where cost of overcoming the friction of distance is defined in monetary terms or time). It may be alternatively represented as the feasibility of transport as defined and constrained by the implied complementary relationships, the prevailing and competing modes of transportation, the location of transport facilities and the connective properties of the transportation network, acquisition and delivery cost rates and structures, technological and institutional constraints, and existing barriers to movement. If it is not feasible or cost-effective to transfer something between a supply and demand point, even if a complimentary relationship does exist, it is unlikely that a flow will be observed.

Assessment of **competing and intervening opportunities** is a recognition that (a) there are *substitutable and interceding supply, demand, and transshipment points in economic systems,* and (b) flows are influenced by the effects of spatial structure—the way in which space is organized by and implicated in the operation and outcome of human or physical processes. In simpler terms, **spatial structure** refers to *the physical layout and interconnective properties of origins and destinations within a system.* It is well recognized that flows are sensitive to the organizational features of a production system and the relative location of supply and demand points. Given that complementary relationships exist and that a commodity is available and transferable from more than one source, producers from several supply points may simultaneously compete for a share of any given geographic market (Figure 1.4).

These principles are useful in helping us gain an understanding of the direction and magnitude of spatial flows and subsequent economic changes brought about by those flows.

GEOGRAPHY AND ECONOMIC GEOGRAPHY: HISTORICAL UNDERPINNINGS

Historically, geographers adopted the so-called **ideographic approach** in their research—*a descriptive approach which included the creation of detailed, in-depth, and vivid narratives and accounts of observations made in the field or during travels.* The earliest geographers concerned themselves with exploring unknown areas of the earth, describing various places and their features. This typically involved the collection and summary of data on the earth's geographic regions in written form and in the form of maps. This included the earliest recorded observations of the exploration of the Mediterranean

Figure 1.4 The flow of commodities as guided by market forces and the principles of spatial interaction: here cargo is being unloaded from a container ship in a major port as it makes its way to consumers.

basin area (circa 2000 B.C.) and the work of early Roman and Greek geographers who studied the physical earth and physical phenomena (4th century B.C. to 2nd century A.D.). Through their many travels and conquests, the ancient Greeks and Romans provided stimulating accounts of strange new lands at great distances from the Mediterranean region.

Notwithstanding the virtual stagnation of geography during the Middle Ages, early writings of Greek geographers found their way to Muslim scholars. The translation of these works from Greek to Arabic allowed for the rapid dissemination of geographic information which was later tested and verified. Arab geographers such as al-Idrisi and Ibn Batatu also produced detailed maps and accounts of their travels, further adding to the wealth of geographic knowledge. Eventually, Arabic texts and charts were translated to Latin, advancing the study of geography in the Christian world. Translation and scholarly exchange also provided valuable information on distant lands in Africa and southwestern Asia. The world's geographic knowledge base continued to expand with the 13th-century accounts of Marco Polo and his journeys, tales from the Crusades, and the 15th- and 16th-century discoveries associated with the voyages of Spanish and Portuguese explorers. Although isolated during medieval times and unknown to the Western world, the writings of the Mongols and Chinese would later provide important information about the geography of Asia.

The 17th through 19th-century geographers brought about significant changes in the way in which the discipline of geography was characterized. The English

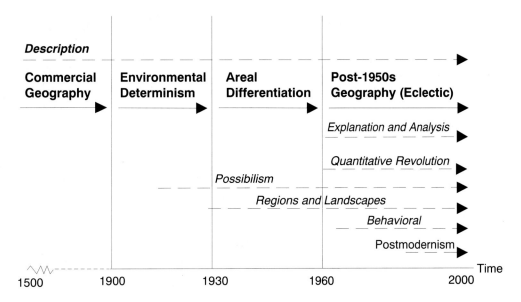

Figure 1.5 Major periods and themes of economic geography.

geographer Nathaniel Carpenter made significant contributions with writings that highlighted spatial interrelationships amongst the earth's physical systems, recognizing the importance of spatial interdependencies. The German geographer Alexander von Humboldt contributed to geographic theory from field observations and newly devised methods for measuring and systematically classifying phenomena. With the publication of *Kosmos* in 1844, a graphic blueprint of the physical geography of the earth, von Humboldt established himself as a key contributing figure in the evolution of the discipline. It was also during this period that the German geographer Bernhardus Varenius discussed geography as a multidimensional field of study. Varenius proposed that geographical inquiries be divided into three separate branches: those which deal with the physical aspects of the earth, those which focus on change and variability in physical systems and climate, and those which emphasize the comparative nature of regions.

In regard to the study of regions, the focus turned to what was unique and particular to each region. The "regional approach" can be traced to the work of the German geographer Carl Ritter. In contrast to von Humboldt, who incorporated a "systematic approach" to geographic analysis—treating physical features as phenomena that were separate from the region itself, it was Ritter who stressed the study of **"regional geography."** The regional approach emphasized how the unique qualities and characteristics of a region could help explain the existence of a given feature or phenomenon at a given location.

Systematic and regional geographic inquiries that described or highlighted human settlement patterns, commerce, and production activities were later labeled as exercises in **"economic geography."** Over time, descrip-

tive analyses were replaced with analyses that sought to establish a body of knowledge for the purpose of formulating theories on the organizational aspects of human settlements and economic activities. In general, economic geography is the study of the spatial variation of economic phenomena. As discussed by Berry et al. (1993), four major periods may be identified in the history of the emergence of economic geography as a specialized field of study (Figure 1.5).

1. The Commercial Geography Period

The birth of economic geography dates back to the age of European exploration and expansion, and the commercial geography period (lasting roughly from the 15th century through the 19th century). In this period there were many explorations over land and sea. Travels and voyages to distant and exotic lands provided geographers with numerous opportunities to produce factual or descriptive accounts of what they observed. Of particular interest to those funding such ventures were the following: the location and availability of land, resources, and raw materials, patterns of colonization and territorial holdings, and, perhaps most importantly, emerging patterns of trade and commerce.

A pioneering work of this period was George Chisholm's *Handbook of Commercial Geography*, first published in 1889. Chisholm's book was essentially a catalog of information on economic activity. Chisholm produced ten editions of this handbook (the last of which was published in 1925), each containing regional narratives of resources, production, and flows. Although Chisholm's work was largely descriptive in its orientation, it offered many insights as to production patterns and trade. Chisholm's work suggested that the world possessed unequal resource endowments and unequal production advantages. He remarked that conditions which were favorable for the production of a commodity at some locations did not exist at other locations.

Chisholm acknowledged that production advantage was not solely responsible for creating regional concentrations of production. Production activities were also influenced by market forces and the characteristics of the transport and distribution systems. Nonetheless, Chisholm viewed trade as "the great equalizer," as trade had the potential to increase the variety of goods available at any given location (as availability was not constrained by what was locally produced) and reduce the difference in price of a commodity sold in any two markets to the cost of transport and duty. In this light, trade was regarded as a mechanism that could eliminate the potential for "spatial arbitrage"—the buying of goods in one geographic market and the movement of those same goods to another market where they could be sold for profit. If uninhibited, trade could reduce the potential for any one region to exploit other regions provided that all regions could compete and actively engage in the exchange of commodities.

In addition, Chisholm elaborated on the eventual deconcentration of industrial activities over space and time, recognizing that there were local production limits in relation to available resources and variabilities in resource depletion rates. Once the highest quality and most conveniently located resources at preeminent production sites neared exhaustion, as production costs began to rise in response to local scarcity, it was envisioned that production would decentralize over the earth's surface. In this way, producers could eventually take advantage of lower quality reserves at other locations. Lower quality reserves at locations closer to market, for example, could become competitive once the premier production sites lost their overall cost advantage in response to scarcity. Hence, changing geographic patterns in production and resource use were largely determined by what later became known as "the law of vanishing assets."

Given the prevailing optimism of this period, there was little or no discussion on resource limitations or limits to growth at the global scale. Most written accounts of production patterns centered around defining commercial opportunities in relation to regions' production, and trade potential in relation to natural resource endowments. In short, trade was viewed as the ultimate vehicle for elevating the economic status of all regions; something which later proved to be untrue given that the distribution of the benefits of trade favored some nations over others.

Note that the later portions of the commercial geography period are associated with modern European colonization (and to a lesser extent American colonization). The formation of colonies, coupled with the rapid industrialization in Europe, the United States, and Japan, set the stage for what we today view as the world of "haves" and the world of "have nots." In particular, the domination of overseas colonies by European interests helped advance European economies at a rate which far exceeded that of their colonial counterparts. Colonial enclaves provided rapidly growing external markets which could easily absorb the surplus production of finished goods from industries "back home," thereby permitting them to operate at a level that was closer to capacity and more economically efficient. In addition, overseas colonies provided the necessary raw materials to fuel industrialization, as well as external geographical markets in which to deploy surplus labor. In reality, trade became a force which promoted uneven growth as the gains were geographically restricted to a handful of dominant and controlling regions.

2. Environmental Determinism and Reactions

By the late 19th century, economic geographers began to elaborate on the importance of the physical environment as an agent of influence and determinant of human economic activity. Environmental determinism was a school of thought rooted in the belief that the environment controls or dictates the course of human action. The environmental determinist movement is, perhaps, best portrayed by the opening statement of Ellen Churchill Semple's book entitled *The Influences of the Geographic Environment* (1911, p.1), where she writes: "Man is a product of the earth's surface." From this standpoint, the environment was viewed as the major determinant in the location of human settlements and economic activity. In essence, the environmental determinist movement sought to establish a link between the physical earth and regional conditions and the locational tendencies of human settlements (as depicted, for example, in Ellsworth Huntington's *Civilization and Climate,* published in 1915).

The origins of the environmental determinist movement can be traced to the writings of the German philosopher Friedrich Ratzel (1844–1904), who discussed the relationships between physical systems and human action and the overarching effect that the natural environment had on people and places in *Anthropogeographie* (1882). Ratzel attempted to show that the spatial distribution of human settlements was determined by the forces of nature and the environment. Depicting geography as the "science of distribution," he called for the study of restrictive areas. Understanding the limitations imposed by these areas would, as he saw it, provide insights into the distribution of activity in other places about the globe.

Environmental determinism, as a philosophy which sought to establish a causal link between the location and quality of human existence in relation to the physical environment, was later abandoned by geographers in favor of the realization that the environment does not control or dictate human activity nor does the environment determine where activities or settlements can or cannot flourish. The physical environment may favor certain types of activities in certain locations, but it also offers distinct possibilities for those same activities in other seemingly less likely locations as human beings adapt and/or use their creative energies to expand the frontier of possibilities at virtually any location. Thus, it was recognized that civilizations were products of adversity, and that inhospitable climates could provide great stimulus to engage in adaptive or survival activities (a philosophy evident in J. Russell Smith's work entitled *Industrial and Commercial Geography*, published in 1913).

Opposition to the environmental determinist philosophy mounted as geographers began to dismiss the deterministic theory. At this time a "regional studies" movement emerged in mid-19th-century France, represented in the writings of such scholars as Paul Vidal de la Blache (1845–1918) and the historian Lucien Febvre (*A Geographical Introduction to History,* published in 1932), and later America (as represented in the writings of the geographer Carl Sauer 1889–1975). The regional studies movement was supported by those who rejected the extreme position held by the environmental determinists, as scholars recognized the power of human agents to overcome adversity in physical systems. They opposed the idea that the physical environment strictly determines the course and nature of human activity.

Attention then turned toward how each place on the earth's surface was unique or distinct in terms of its physical and social characteristics and how the combination of the physical and social processes and relationships served to direct change in the human landscape. Although all elements of this landscape were viewed as interrelated, the idea of the physical environment as the sole causal agent of human action was rejected in favor of a more realistic position, consistent with the belief that human agents can choose many possible courses of action. It was understood that the environment may direct and sometimes limit certain types of activities, but human beings have the ability to alter physical settings and manipulate the environment. Hence, it was well accepted that human agents could adopt new behaviors and/or technologies to cope with existing conditions and enhance their relationship with the physical or natural environment. If human agents were unable to cope with existing conditions or favorably redirect the outcome of the human-environment relationship, they could move from places with extreme and limiting conditions to places which offered a broader (or even more acceptable) range of alternatives and opportunities.

Rejecting the tenets of environmental determinism, Carl Sauer in *The Morphology of Landscape* (1925), and later others, discussed the existence of a "cultural landscape." The **cultural landscape** was viewed as *a by-product of the complex interplay between the human community, which embodies certain cultural preferences and possibilities, and a particular set of natural circumstances.* The focus turned toward the study of the differences between landscapes, which could not be attributed to natural forces but rather to human action and occupance. Hence, the physical environment was not viewed as the principal factor which influenced and guided human behavior, as human agents were largely accepted as adaptive entities. The environment did, nonetheless, serve as the backdrop or setting for human expression as the efforts of many generations sculpted the landscape in ways conditioned by culture and tradition. Subsequently, the philosophy of environmental determinism gave way to **possibilism**—*a viewpoint that the physical environment offers many opportunities, allowing human agents discretion to choose among many options and even expand on the nature and range of possibilities through adaptive behavior, the use of technology, and/or manipulation of the environment.*

Despite the many controversies and debates between possibilists and determinists in this period, a most important revelation on "limits to growth" emerged from the writings of T. Griffith Taylor (1880–1963) in his observations concerning settlement potential of the continent of Australia and its ability to support a given population base. Studies of this variety would later be labeled as inquiries into regional "carrying capacity."

Carrying capacity refers to *the maximum number of living agents that can be supported and sustained by a given set of land-based resources in a given geographic area under prevailing technologies and proper management.* This concept forms the very foundation of some of the most important theories in ecology, environmental science, and geography, in relation to the study of plant communities, sustainable agricultural systems, livestock production, recreational activity thresholds, and regional size limitations to human settlements.

The "limits to growth" argument is central to the human-environment relations theme in geography. As discussions concerning limits to growth began to appear throughout this period and beyond, scholars began to focus on examining carrying capacity in a regional and global context. Modern-day geographic inquiries into regional carrying capacity, such as the seminal work of Meadows et al.: *The Limits to Growth* (1972) and *Beyond the Limits* (1992), are examples of the influence and legacy of this period. While little attention was paid to global limits to growth in the early 1900s, it has become a preoccupation with many contemporary scholars.

3. Areal Differentiation

As the retreat from determinism quickened, the emphasis turned toward the study of geographic regions and the differences in those regions. This new focus gave birth to areal differentiation, a movement which began around the mid-1930s with the introduction of Richard Hartshorne's

The Nature of Geography (1939) and lasted roughly 40 years as a prevailing school of thought. The areal differentiation movement supported the idea that geographers should concern themselves with the construction of regions, and emphasize regional dissimilarities (to describe and interpret the variable character of the earth as it differed from place to place). Those differences were viewed as the fundamental objects of geographic inquiry.

Detailed writings on the world's regions were once again in vogue. Writers of the time produced fantastic accounts of human economic activities describing differences in the levels and intensities of hunting, fishing, livestock grazing, crop production, forestry, mining, manufacturing, and commerce over various areas of the earth's surface. The emphasis was on observation, data collection, and the identification of economic regions, exercises that were, for the most part, purely descriptive. As in the previous two periods, very little discussion surfaced in regard to limitations at the global level.

By the 1950s, geographers concentrated their efforts on two related, yet different, emphases: "regional geography" (also referred to as chorology)—the delimiting, mapping, classification, and description of regions; and **"landscape geography,"** in the tradition of Carl Sauer, where attention focused on how human beings modified and shaped their world.

4. Post-1950s Geography

By the late 1950s, economic geographers became dissatisfied with the ideographic approach and the mundane exercise of describing, differentiating, and delimiting regions. Geographers longed for more than just descriptive accounts of the world and its regions in their search for explanation, understanding, and meaning. Alternatively, they opted for the development of theory on the spatial organization of society and economy.

Geographers were stirred by the **quantitative revolution**—*a conceptual revolution and transformation of spirit and purpose, spearheaded by major technological breakthroughs in statistical, mathematical, and quantitative methods which allowed for advanced applications in the testing and development of formalized theories of location and spatial organization.* The focus of post-1950s geography has been on formalized theory-building and the utilization of statistical procedures (objective science) to validate and enrich location theory. It is a tradition that is alive and well even today as geographers continue to explore the locational aspects of settlements and activities in a rapidly changing world.

Post-1950s geography produced a flurry of location analysts who sought to understand and explain that which they observed, rigorously testing their hypotheses, and validating that which they theorized. The philosophy behind this approach became known as **logical positivism**—*a way in which to search for explanation through formalized theory building, mathematical reasoning, and scientific validation.* Logical positivists supported the belief that research statements were valid if and only if they were rendered as scientifically meaningful "empirically" (as established through verification), or if supported "analytically" (if judged to be true in a logical or mathematical sense).

Ongoing geographic research has established and contributed to **location theory**—*a general body of knowledge or set of connected statements, conditions, or empirically validated hypotheses that comprise a conceptualizing framework that is used to help explain the location tendencies of human economic agents and activities.* Attention was also drawn to the behavioral aspects of location-based phenomenon and the location decision-making processes. Classic modeling procedures characterize human economic agents as rational decision-making units who operate as economic maximizers or optimizers. Attempts to build on the seminal contributions of the German agriculturalist Johann Heinrich von Thunen (1783–1850) and his work to identify the variables which influenced the location of agricultural activities and the German geographer Walter Christaller, who developed a "central place theory" (during the 1930s) to explain the distribution of urban centers and the extent of their hinterlands, are prime examples of theory-building efforts supported by frameworks cast under certain limiting assumptions. Assumptions such as the portrayal of economic agents as optimizers operating under full information have recently been relaxed to allow for satisfying behavior and decision making under risk, uncertainty, and imperfect information.

In most instances, the evolution of a general location theory has been guided by the construction and evaluation of spatial models. A **spatial model** is *a conceptualization, representation, or abstraction of reality that captures the dominant cause-and-effect relationships responsible for generating a spatial pattern.* Modeling provides a way in which to simplify a complex reality by identifying the predominant forces and factors which were responsible for shaping that reality. Spatial modeling was viewed as a bold attempt to replicate or account for observable patterns in human and physical systems, as most modeling efforts were constrained by prevailing theory and technology. Nonetheless, a wide variety of models and modeling approaches surfaced during the quantitative revolution.

Intense opposition to the logical positivist school of thought also emerged. The use of quantitative methods and models were shunned by a large number of geographers, as is still the case today. The proliferation of the nonquantitative geography literature has led some scholars to contend that a "quantitative revolution" never really took place in geography. For instance, University of Arizona's Gordon Mulligan contends that geography experienced what could only be labeled as a **quantitative movement**—*a transformation of approach with emphasis on quantitative methods and models adopted by a limited circle of geographers doing research in but a handful of areas.* The fact that there are relatively few research publications in major geography journals over the past several decades which rely on quantitative techniques (in comparison to the vast amount of nonquantitative work) is evidence that a discipline-wide "revolution" never really took place. The idea that human behavior could be explained or replicated using models and simplifying assumptions is something which was largely dismissed by a large number of geographers in nonquantitative camps. Nonetheless, a series of alternative frameworks have been applied to explain human action and behavior.

Alternatives included **phenomenology**—*studies which focused on the experiential aspects of an individual's world, and the contention that life has meaning only through the eyes and experiences of the individual,* and **postmodernism**—*the rejection of behavioral theory and logical-positivism, a doctrine challenging the very existence of a common objective reality among human agents.* The use of phenomenology and postmodern geography is consistent with the belief that the tangible world is complex, enigmatic, and sometimes impossible to model. Having knowledge of setting, background, perspectives, experience, etc., provides the researcher with a broader field level of understanding about that which is observed, something that could not be achieved through neoclassical modeling.

Dissatisfaction with neoclassical economics in general has given rise to **radical geography**—*frameworks that encouraged critiques of quantitative and analytical models citing their failure to explain social injustice, regional economic disparity, and unequal distribution of power amongst the world's economies.* Radical factions in geography continue to argue that the theories of spatial organization have failed to provide adequate levels of understanding and that the dynamics of evolving regional economies must be more fully understood in terms of class, class conflict, disparities in wealth and power, unequal accessibility to resources, and patterns of exploitation (things that are largely ignored by quantitative analysts and location theorists). As a result, radical factions continue to develop their own set of theories based on a **dialectical approach**—*a way to achieve understanding of a system by drawing attention to the power relations within that system.* Radical factions have kept the focus on disadvantaged people and regions—those that lacked the control and access to resources—and the notion that the disadvantaged will likely remain in that position unless the mechanisms that preserve or intensify disparities are dissolved or reworked. While radical viewpoints have not been entirely accepted by the "mainstream," they have nonetheless greatly influenced the research agendas of geographers and other social scientists.

The appearance of **behavioral geography** and "environmental perceptions movement" in the post-1950s period has also influenced the nature of geographic inquiry. Behavioral geography can be defined as *a subfield of human geography that seeks to gain insights into human behavior by understanding how human agents perceived and, in turn, responded to their environment.* As action and behavior were known to be affected by how human beings processed information about their surroundings, the focus turned toward questions of cognition and perception. The use of **mental maps**—*individuals' internally stored images of the external world*—were thought of as not only reflecting an individual's knowledge about places and the organization of space, but also how a person feels about those places, things which can ultimately affect human interaction. The **environmental perceptions** movement was based on *a recognition that an individual's sense of place and recollection of place, no matter how accurate or distorted, provided valuable insight into decision-making processes, attitudes, and behavior.*

To summarize, post-1950s geography could be best characterized as eclectic in terms of approach and method. It can also be said that each of the major periods in the evolution of economic geography has had a dramatic and long-lasting impression on modern-day geographic research as geographers continue to borrow from each of these proud traditions.

GEOGRAPHY TODAY AND TOMORROW: AN OVERVIEW

Geographers view the world in spatial terms. They attempt to analyze the organization of physical and human phenomena in order to understand the patterns, processes, complexities, distributions, interdependencies, and workings of human and physical systems. As you may have gathered from the previous discussion on the formative periods of economic geography, the face of the discipline has changed markedly over the past century, and even more so over the past few decades. Despite the new direction and emphasis of contemporary geographic research, many misconceptions of what geography is and what geographers do still persist. Some still think of geography solely in terms of its more traditional roots—as the study of "where" things are. But as the discipline continues its steady evolution as a modern-day social science, geographers are hard at work seeking to understand the forces and factors which shape and change the world as we know it.

Geography is much more than name-place recognition. It is much more than just a descriptive science that allows one to pinpoint the location of a place or physical feature on a map. True, geographers are concerned with the locational attributes of phenomena; yet their concerns run much deeper than just answering the question of "where?". Contemporary geographers also seek to understand the "how" and, more importantly, the "why" of the matter (that is, to understand how things came to be and why they are the way they are). It is a relentless quest to dig deep to seek truth and understanding, and find reason and explanation for spatially variant phenomena.

Consider the incidence of famine in sub-Saharan Africa over the past several decades, a situation that can be attributed to more than just the physical characteristics of the region or its location. Population pressures and the subsequent burden on scarce agricultural land are only partly to blame. The effects of political unrest, war, drought, environmental degradation, and inadequate or disrupted distribution of food, have contributed to the unfolding of this drama. Without an understanding of the various economic and political instabilities and internal conflicts in the region, one may be tempted to place the blame solely on regional production shortages of staple food crops. Upon closer inspection, one may discover that the existence of civil war and unrest and economic turmoil (in fractured nations such as Somalia, Ethiopia, and Rwanda) are largely responsible for the disruption of food production and distribution. These ongoing problems have left hundreds of thousands to seek safe haven across international borders and millions of people impoverished and near starvation.

In contrast to the ideographic approach in geography, the search to form a deeper understanding of spatial phenomena constituted a **nomothetic approach**—*a law-seeking or explanation-oriented approach.* Such inquiries have allowed geographers to directly participate in policy analysis once an understanding of the underlying causal relationships have been adequately exposed. By offering an understanding of the processes and cause-and-effect relationships responsible for creating a given situation, and through examination of any "local peculiarities," geographers have been instrumental in providing unique insights on existing problems.

It is the very nature of the geographer to seek understanding of **process**—*the mechanisms that influence, sustain, reproduce, or transform a dynamic system and its structure.* This is done in order to gain insight into a given **pattern**—*a snapshot of an ever-evolving and dynamic system at a particular point in time.* Geographers recognize that attention must be paid to the spatial, temporal, and contextual features of a process, with an understanding that the forces which shape a given pattern are both internal and external to that pattern.

Finding variables which best explain a given pattern is no easy task. Analyses must proceed on two levels: (1) the identification of **site variables**—*localized features or characteristics that directly influence a given phenomenon or outcome;* and (2) the identification of **situation variables**—*contextual elements, conditions, or circumstances which indirectly or externally affect a given phenomenon or outcome.* Site and situation variables generally combine to generate local peculiarities, complicating the search for explanations of spatial patterns in relation to spatial processes.

Analysis of human and physical landscapes is something which requires a multidisciplinary emphasis. This realization has prompted geographers to borrow techniques, theories, and approaches from a wide variety of disciplines and literature. United under the rubric of "spatial analysis," modern-day geography is a field of study which focuses on explaining the location, orientation, and interrelationships of real world phenomena using whatever means necessary to get the job done.

By borrowing theory and techniques from outside disciplines (such as sociology, urban economics, political science, environmental studies, etc.) and through the development of new and exciting methods in which to display and analyze geographic data, geographers are now ready to take on problems of global proportions.

Armed with conventional quantitative methods, **spatial statistics**—*techniques geared toward analyses of spatial dependence, directional trends, surfaces, and point-pattern distributions,* and **geographic information systems (GIS)**—*computer-based systems for capturing, handling, storing, managing, or manipulating spatial information,* geographers are now better able to explore spatial phenomena. Recent advancements in computer technology (on both the hardware and software ends) have given geographers new tools for managing and displaying spatial data. GIS, in particular, has offered the geographer an effective way in which to map spatial patterns and highlight the interrelationships in spatial phenomena. As a result, geographers have improved their capacity to deduce process from pattern and explain that which they observe. The tools of information systems technology have also allowed geographers to more effectively evaluate outcomes of alternative courses of action and monitor change over time and space (Star and Estes, 1990, p.12).

ECONOMIC GEOGRAPHY RESEARCH AGENDA

Geography is an integrative discipline that brings together the physical and human dimensions of the world in the study of people, places, culture, economy, and the environment and the interconnectiveness of regions and spatial phenomena (*National Geography Standards: Geography for Life,* 1994, p. 18). Economic geography continues to focus on aspects of change in the human economic system. As part of this focus, a new research agenda in economic geography has emerged in the study of regional and global economies. It is an unprecedented commitment to seek understanding of regional and global limitations to growth and an attempt to explain regional variations and disparities as tied to past and present economic conditions.

As a result, economic geographers are now concentrating their energy on identifying the interdependent nature of change amongst regional economies. Regional and global change continues not only to influence the actions and reactions of regions, but it has produced a series of global problems which must be resolved. In line with this new research agenda, this book will discuss regional change and the challenges associated with uneven growth and change. More specifically, the objectives of this exercise are fourfold:

1. provide an overview of the mechanisms and forces which influence economic growth and change;
2. compare and contrast patterns of regional and national economic growth and change;
3. discuss the short- and long-term implications of the patterns and processes responsible for promoting uneven growth and change; and
4. explore various arguments concerning "limits to growth" and limitations imposed by uneven growth.

Expanding on Garrett Hardin's essay on the **tragedy of the commons**—*a statement concerning the potential incompatibility of individual and collective interests in the use and depletion of community resources,* one may argue that the isolated actions of economic agents, regions or nations, acting individually or in the interest of a select group, do not necessarily coincide with the collective interest of all nations of the world. Hence, this book will elaborate on the ironies inherent to a divisive global economic system, where regions are becoming increasingly interdependent through time, more globally integrated, and yet more distant in terms of relational economic space.

KEY WORDS AND PHRASES

absolute location   3
areal differentiation   11
behavioral geography   13
boundary   4
carrying capacity   11
cartography   1
commercial geography period   9
competing and intervening opportunities   8
complementarity   7
cultural landscape   4, 11
dialectical approach   13
distance-decay effect   7
economic geography   9
environmental determinism   10
environmental perceptions   13
friction of distance   7
functional region   5
geography   1

geographic information systems (GIS)   14
geographic scale   2
human-environment relations   5
human geography   2
ideographic approach   8
landscape geography (see cultural landscape)   12
location   3
location theory   12
logical positivism   12
mental maps   13
movement   7
nomothetic approach   14
pattern   14
phenomenology   13
physical geography   2
place   3
possibilism   11

post-1950s geography   12
postmodernism   13
principle of least effort   7
process   14
quantitative movement   13
quantitative revolution   12
radical geography   13
regional geography   9
regions   4
relative location   3
relational space   1
site variables   14
situation variables   14
spatial models   12
spatial statistics   14
spatial structure   8
tragedy of the commons   15
transferability   8
uniform region   4

REFERENCES

Abler, R., J.S. Adams, and P.R. Gould. *Spatial Organization: The Geographer's View of the World.* Englewood Cliffs, NJ: Prentice Hall (1971).

Anselin, L., and A. Getis. "Spatial Statistical Analysis and Geographic Information Systems." *Annals of Regional Science* 26 (1992), pp. 3–17.

Berry, B.L.J., E. Conkling, and D.M. Ray. *The Global Economy.* Englewood Cliffs, NJ: Prentice Hall (1993).

Billinge, M., D. Gregory, and R.L. Martin. *Recollections of a Revolution: Geography as a Spatial Science.* New York: St. Martin's Press (1984).

Chapman, G.P. *Human and Environmental Systems: A Geographer's Appraisal.* New York: Academic Press (1977).

Chisholm, G.G. *Handbook of Commercial Geography.* London: Longmans, Green (1889).

Chisholm, M. *Human Geography: Evolution or Revolution?* New York: Penguin (1975).

de Blij, H.J., and P.O. Muller. *Geography: Realms, Regions and Concepts* (7th edition). New York: J. Wiley & Sons (1994).

de la Blache, V. *Principles of Human Geography.* London: Constable (1926).

Downs, R.M., and D. Stea. *Maps in Minds: Reflections on Cognitive Mapping.* New York: Harper and Row (1977).

Ellis, W.S. "A Soviet Sea Lies Dying." *National Geographic.* Washington, D.C.: National Geographic Society (February, 1990), pp. 73–93.

Entrikin, J.N., and S.D. Brunn (eds.). *Reflections on Richard Hartshorne's: The Nature of Geography.* Washington, D.C.: Association of American Geographers (1989).

Febvre, L. *A Geographical Introduction to History.* London: Kegan Paul Trench Trübner.

Gaile, G.L., and C.J. Willmott (eds.). *Geography in America.* Columbus, Ohio: Merrill (1989).

Gatrell, A.C. *Distance and Space: A Geographical Perspective.* Oxford: Clarendon (1983).

Glazovsky, N.F. "The Aral Sea Basin." In Kasperson, J.X., R.E. Kasperson, and B.L. Turner II (eds.). *Regions at Risk.* New York: United Nations University Press (1995), pp. 92–139.

Gold, J.R. *An Introduction to Behavioral Geography.* New York: Oxford University Press (1980).

Gould, P., and R. White. *Mental Maps* (2nd edition). Boston: Allen and Unwin (1986).

Gregory, D. *Ideology, Science and Human Geography.* New York: St. Martin's Press (1978).

Haggett, P. *Locational Analysis in Human Geography.* London: Edward Arnold (1965).

Hanink, D.M. *The International Economy: A Geographical Perspective.* New York: J. Wiley & Sons (1994).

Hardin, G. "The Tragedy of the Commons." *Science* 162 (1968), pp. 1243–1248.

Hartshorn T.A., and J.W. Alexander. *Economic Geography.* Englewood Cliffs, NJ: Prentice Hall (1988).

Hartshorne, R. *The Nature of Geography: A Critical Survey of Current Thought in the Light of the Past.* Lancaster, PA: Association of American Geographers (1939).

Hartshorne, R. *Perspective on the Nature of Geography.* Chicago: Rand McNally (1959).

Harvey, D. *Explanation in Geography.* New York: Edward Arnold (1969).

Harvey, D. "The Geopolitics of Capitalism." In D. Gregory, and J. Urry (eds.) *Social Relations and Spatial Structures.* London: MacMillan (1985), pp. 128–163.

Hodder, B.W., and R. Lee. *Economic Geography.* London: Methuen (1974).

Holt-Jensen, A. *Geography: Its History and Concepts.* Totowa, NJ: Barnes and Noble (1980).

Huntington, E. *Civilization and Climate.* New Haven: Yale University Press (1915).

James, P.E., and G.J. Martin. *All Possible Worlds: A History of Geographical Ideas* (2nd edition). New York: Bobbs-Merrill (1981).

Johnston, R.J. *Multivariate Statistical Analysis in Geography.* New York: Longman (1978).

Johnston, R.J. "Applied Geography, Quantitative Analysis, and Ideology." *Applied Geography* 1 (1981), pp. 213–219.

Johnston, R.J. *Geography and Geographers: Anglo-American Human Geography Since 1945.* London: Edward Arnold (1983).

Johnston, R.J. *Philosophy and Human Geography: An Introduction to Contemporary Approaches* (2nd edition). London: Edward Arnold (1986).

Johnston, R.J., D. Gregory, and D.M. Smith (eds.). *The Dictionary of Human Geography* (2nd edition). London: Blackwell (1986).

Jones, J.P., W. Natter, and T. Schatzki (eds.). *Postmodern Contentions: Epochs, Politics, Space.* New York: Guilford Press (1993).

Kasperson, J.X., R.E. Kasperson, and B.L. Turner II (eds.). *Regions at Risk.* New York: United Nations University Press (1995).

Meadows, D.H., D.L. Meadows, J. Randers, and W. Behrens. *The Limits to Growth: A Report for the Club of Rome's Project on the Predicament of Mankind.* New York: Universe Books (1972).

Meadows, D.H., D.L. Meadows, and J. Randers. *Beyond the Limits: Confronting Global Collapse, Envisioning a Sustainable Future.* Post Mills, VT: Chelsea Green (1992).

National Geography Standards: Geography for Life. Washington, D.C.: National Geographic Research & Exploration (1994).

Norris, R.E. *World Regional Geography.* New York: West (1990).

Norton, W. *Human Geography.* New York: Oxford University Press (1992).

Peet, R. "The Social Origins of Environmental Determinism." *Annals of the Association of American Geographers* 75 (September 1985), pp. 309–333.

Pickles, J. *Phenomenology, Science and Geography.* New York: Cambridge University Press (1985).

Ratzel, F. *Anthropogeographie.* Stuttgart: J. Engelhorn (1882).

Relph, E.C. *Place and Placelessness.* London: Pion (1976).

Rubenstein, J.M. *An Introduction to Human Geography: The Cultural Landscape* (3rd edition). New York: MacMillan (1992).

Sauer, C.D. *The Morphology of Landscape* (1925). Reprinted in L. Leighly (ed.) *Land and Life: Selections from the Writings of Carl Ortwin Sauer.* Berkeley: University of California Press (1974), pp. 315–350.

Semple, E.C. *The Influences of Geographic Environment or The Basis of Ratzel's System of Anthropogeography.* New York: H. Holt & Co. (1911).

Smith, J.R., and M.O. Phillips. *Industrial and Commercial*

Geography (3rd edition). New York: H. Holt & Co. (1946).

Smith, N. *Uneven Development: Nature, Capital and the Production of Space.* Oxford: Basil Blackwell (1984).

Stamp, L.D. *Applied Geography.* New York: Penguin (1960).

Star, J., and J. Estes. *Geographic Information Systems: An Introduction.* Englewood Cliffs, NJ: Prentice Hall (1990).

Stutz, F.P., and A.R. de Souza. *The World Economy: Resources, Location, Trade, and Development.* Upper Saddle River, NJ: Prentice Hall (1998).

Taylor, T.G. *Australia: A Study of Warm Environments and Their Effect on British Settlement.* London: Methuen (1940).

Ullman, E. "The Role of Transportation and the Bases for Interaction." In W. Thomas, Jr. (ed.) *Man's Role in Changing the Face of the Earth.*

Chicago: University of Chicago Press (1956), pp. 862–880.

Wallace, I. *The Global Economic System.* London: Unwin Hyman (1990).

Wheeler, J.O., P.O. Muller, G.I. Thrall, and T.J. Fik. *Economic Geography* (3rd edition). New York: J. Wiley & Sons (1998).

Zipf, G.K. *Human Behavior and the Principle of Least Effort.* Reading, MA: Addison-Wesley (1949).

2

REGIONAL ECONOMY: CLASSIFICATION, GROWTH, AND DEVELOPMENT

OVERVIEW

Economic geographers have long held an interest in studying matters of regional economy. A **regional economy** may be defined as *a highly organized system which administers the flow of resources, information, technology and know-how, while simultaneously directing the type, the manner in which, and the locations where production occurs, and how goods and services are distributed over space.* While economists focus on markets and the relationships between competing ends and scarce means, economic geographers focus on the locational aspects of human economic activity; namely, production, exchange, and consumption. Regional variations in the levels of these activities can lead to variations in economic growth and change. As regional economies become more integrated and interdependent over time, economic geographers have shifted their interests toward studying the **world economy**—the global-wide system of production, exchange, and consumption.

Production *refers to any process that increases the economic value of a commodity beyond the value of its raw material inputs.* For example, the transformation of raw material inputs into steel or the conversion of steel into usable products like wire or sheet metal are production activities. Note that production involves an increase in **form utility**—*the value of a commodity that is attributed to its form or physical characteristics.* The greater the production input and subsequent change in form, the higher the end-product value. For instance, an assembled

and finished oak dining table is worth considerably more than a series of cut unfinished boards of equal weight. The form utility of a commodity at a given stage of production is directly associated with the amount of human and physical inputs necessary for the production of that commodity.

In general, we may identify six **factors of production**—*the fundamental components of any production process:*

1. *land* (including land-based materials);
2. *labor* (human input as defined in terms of the amount and/or quality of labor input);
3. *capital* (which includes "real capital" such as machinery, tools, storage, and other facilities, excluding money or "financial capital");
4. *entrepreneurial skills* (including know-how, management and operations skills, and craftsmanship);
5. *infrastructure*—temporary or permanent structures or installations for the purposes of assisting production or the transport or transfer of people, commodities, and information; and
6. *technology.*

Exchange refers to *the process(es) by which goods and services are transferred (or traded) from one location to another location where they have more worth to end-users* (be they producers or consumers). Consider the transport of iron ore to steel-milling operations located far from the site of extraction. This exchange is known to increase the **place utility** of a commodity—*the worth of a commodity at a given location.* As a commodity moves from a place of origin to a given destination, its value increases simply because it becomes more valuable to those who will ultimately use it. Note that the increase in value may be attributed not only to changes in location but also to changes in its form along the way. Consider the case of grain harvested on a farm in the Midwest, collected at a grain terminal near a large nearby town, shipped to a large milling operation in the Great Lakes region, and later purchased as flour by a consumer in an eastern Atlantic state. Surely the increase in the value of the flour is both production and exchange related; that is, the final product has gained value from changes in its form and location.

Generally, the value of a product increases as that product moves closer to consumers, as each transformation, refinement, exchange, or transfer will add to its value. In short, value is added by altering commodities' form and place utility. **Value added** generally describes *the difference between the cost of raw material inputs, labor, and other production and transport costs and the selling price of a product at intermediate or final production and/or transport stages.* Value added is a concept that is inherently geographical as raw materials, as well as intermediate and final products, must ultimately change locations before reaching the places at which they will be transformed or consumed. It is a sequence that involves the movement of commodities from extractors to producers to wholesalers to retailers and ultimately on to consumers. Note that the greater a region's ability to retain value added in production in this sequence, the greater the potential benefits of that production as defined in terms of its ability to stimulate growth and opportunities in the regional economy. Put another way, the retention of value added is a catalyst for regional economic growth.

Consumption refers to *the use (and possible final use) of goods, services, or resources to satisfy human needs and wants directly.* In general, it can be noted that large consuming regions also tend to be large producing regions. The propensity to consume goods and services in the world economy, however, is not something that is necessarily in direct correspondence with the spatial distribution of need or the world's population. The degree to which human agents are able to consume is highly associated with wealth, income, and access to production factors. Individual consumption decisions are typically carried out with little or no concern over the direct and indirect impacts of those decisions, especially to the environment.

Economic Activities and Sectors

The mix of economic activity in a region reveals much about that region's present and future potential for economic growth and change. To form a deeper understanding of a regional economy, it is important to highlight activity levels by **economic sector**—*divisions of a regional economy in which particular activities take place.* Typically, any activity in a regional economy may be associated with one of three economic sectors:

1. the primary sector;
2. the secondary sector; and
3. the tertiary sector.

Classification of human activity levels by economic sector provides an overall indication of region's **economic profile**—*a composite view of a region's activity levels by economic sectors.* This profile can tell us how a sector activity contributes to and shapes both the current and future economic landscape. Economic profiles also provide important information about a region's ability to retain value added; a feature which can be used to gauge a region's **economic status**—*how advanced or well-off a region is relative to other regions of the world.*

The **primary sector** *involves activities or industries which go directly to nature for resources, food, or other raw materials to satisfy a variety of human wants and needs or the demand for a production input.* Primary sector activities include: agriculture—the growing of crops and animal husbandry; mining—the extraction of minerals and natural resources from the earth; fishing—the harvesting of fish from fresh- and salt-water fisheries (including aquaculture or fish farming); forestry—the cutting and collecting of lumber from natural or planned forested areas; and various hunting and gathering activities such as tracking and trapping of animals, and the collection of fruits, vegetables, plants, insects, and other natural by-products.

The **secondary sector** *is characterized by activities or industries that increase the value of commodities by changing their physical form* (that is, those industries that increase the form utility of a commodity). The secondary sector, thus, includes construction and manufacturing. Manufacturing activities may be divided into two distinct categories: processing and fabrication.

1. **Processing** *refers to manufacturing activities that look to the primary sector for inputs.* For example, consider the milling of flour from grain or the production of raw steel from iron ore.
2. **Fabrication** *involves manufacturing activities that look to the secondary sector for inputs* (production activities that use already processed manufactured products as inputs). Consider the production of bread and baked goods from flour; the stamping of sheet metal into usable shapes and forms; and the assembly of automobiles from previously manufactured parts, all of which require processed inputs. It is important to note that secondary sector activities have traditionally been associated with

the greatest amount of value added in the value-added sequence. In addition, post-processing manufacturing (that is, fabrication) is known to account for the greatest share of value added in the secondary sector (Figure 2.1).

Secondary sector industries are said to constitute a region's **industrial base**—*the central or key industries which drive a regional economy.* The secondary sector is generally associated with high-income-generating, and typically, high-wage activities. If a large portion of secondary sector output is earmarked for export (that is, consumption outside of the producing region itself), the region's industrial base is said to be strong. In terms of wages and earnings, it is said that one manufacturing job is worth at least as much as two jobs in other sectors of the economy in terms of contributing to a region's total economic output and growth. Although the industrial base concept has changed dramatically over the last decade to include nonmanufacturing activities and exportables such as information and technology, there is little doubt among regional economic analysts as to the importance of manufacturing as a catalyst for economic growth.

The **tertiary sector** *depicts all remaining activities not associated with the primary or secondary sectors.* These activities include transportation, communications, and public utilities (TCPU); wholesale and retail trade (WART); finance, insurance, and real estate (FIRE); consumer and producer services (CPS) including personal, business/financial, professional and education services; and public administration and government (PAG).

FLOWS AND LINKAGES

Flows of goods, services, information, and resources in the world economy can be ultimately described as a series of intra- and intersectoral linkages. In general, **linkages** are *contacts, connections, and locational interdependencies which support and facilitate the movement of information, goods, or materials within or between economic sectors.* Linkages may be local, regional, interregional or global in scope. Moreover, linkages may be subdivided into three categories: forward, backward, and sideways linkages.

Forward linkages *involve the forward progression or movement of commodities through the value-added sequence.* Consider the flow of oranges from harvesters to processors and from processors to retail outlets. Oranges once harvested from groves in south-central Florida make their way to juice processors in the southeastern United States, where the juice is later sold to consumers in various regions in fresh or concentrated form (Figure 2.2).

Backward linkages, by contrast, *involve the backward progression or movement of commodities from the secondary sector to the primary sector or from post-*

Figure 2.1   Manufacturing: a secondary sector activity.

processors to processors. For example, consider the flow of farm machinery, tools, irrigation, and packing equipment from suppliers in the secondary sector to citrus growers in the primary sector. Such linkages are typically associated with the intersectoral movement of fabricated products.

Sideways linkages involve flows (typically information or production technology flows) between firms or economic agents involved in the same production activity. For example, consider the case of neighboring farmers who share information on their experience with a new commercially available fertilizer or seed, or an innovative harvesting technique or pest-control method. As such, a sideways linkage generally involves a flow that is intrasectoral.

The value-added sequence generally involves a complex web or network of flows and linkages. Consider the network of flows and linkages as displayed in Figure 2.3 for the production of steel. Note that the recycling of scrap metal represents a backward linkage in this network, as used products flow from consumers to recyclers. The flow of information amongst steel producers on the growing availability of scrap metal for recycling and the profitability of recycling is a good example of a sideways linkage (Figure 2.3).

The formation of a production network is a critical element in any regional development equation. A **network** may be defined as *a concentration of linkages, in a given geographic region, oriented toward one or more related or interdependent production activities.* As a production network takes shape over time, jobs are created as linkages are established within and between industries. It can be said that a region's competitive position and relative growth potential is directly proportional to the density and strength of linkages found within its production network. In general, the stronger and more

Figure 2.2   The movement of commodities from the primary to secondary sector and then on to market: oranges from the grove make there way to an orange-juice processing facility and on to store shelves in the form of juice and frozen concentrate or as fresh fruit.

numerous the linkages within a region, both intra- and intersectorally, the more likely that region is to experience and sustain growth. Flows and linkages not only create employment opportunities, but are largely responsible for determining a region's economic status.

GROWTH AND DEVELOPMENT

While economic geographers have focused much of their efforts on studying industrial location, the locational tendencies of industry, and changing patterns of production, they have also been concerned with issues and problems related to uneven economic growth and development. Note that the term "growth" implies an expansion or augmentation of something. **Economic growth** implies *the expansion or augmentation of a production network, industrial base, or regional economy and the subsequent increase of employment opportunities and earnings*. In general, economic growth may be viewed as a by-product of the proliferation of industries and linkages on the economic landscape. In addition, growth may be accompanied by the integration of a local or regional economy into a national or global economy.

On one hand, promoting regional economic growth is desirable as it provides enhanced economic and employment opportunities. On the other hand, economic growth may be undesirable if it is found to have a detrimental impact on the quality of life in a given region. For instance, consider an outcome where unbridled economic growth over time leads to industrial expansion and a deterioration of the physical environment where excessive congestion and pollution negatively affect output. The explosive growth of an economy may often lead to **runaway inflation**—*an out-of-control price spiraling that rapidly erodes consumers' purchasing power.* Consider the Latin American economy of Brazil and how the benefits of explosive growth during the 1980s and 1990s were less visible given the Brazilian economy's tendency to post double- and triple-digit inflation rates. Concerns over growth have led many regional analysts to question the desirability of promoting economic growth as a vehicle to positive change should the benefits of growth be negated by its costs. Yet, in an overwhelming number of cases, economic growth is more beneficial than it is costly to a regional economy, especially over the long haul.

"Jobs versus quality-of-life" or "growth versus envi-ronment" debates have emerged in the policy arenas of the world, where it is argued that the environment main-tains a finite capacity to sustain economic growth. Con-cerns over limits to growth have fostered a new realization amongst industry, government, and planning authorities on the importance of simultaneously promot-ing growth and environmental quality. Policy makers and analysts continue to discuss the dual roles and responsi-bilities that one must assume in outlining strategies for economic change as managers of the human economic landscape and as stewards of the environment. It is an awareness that excessive growth, beyond a region's ca-pacity to sustain that growth, may be viewed as detrimen-tal to the immediate quality of life within a region and

counterproductive in the long run as the potential for fu-ture growth is diminished. Consequently, the terms "growth" and "development" should not be confused with one another, nor should they be used synonymously.

Development *implies making progress toward desir-able goals and outcomes, the most important of which is improving the human condition.* **Economic development** *constitutes positive changes and progress in the human condition through economic means.* It is a process that brings about "positive" economic changes to a region. What constitutes positive change, however, is something that is highly dependent upon one's perspective. More-over, even positive economic change does not necessarily ensure that a region will be able to support and sustain future progress. Hence, there is much debate over what

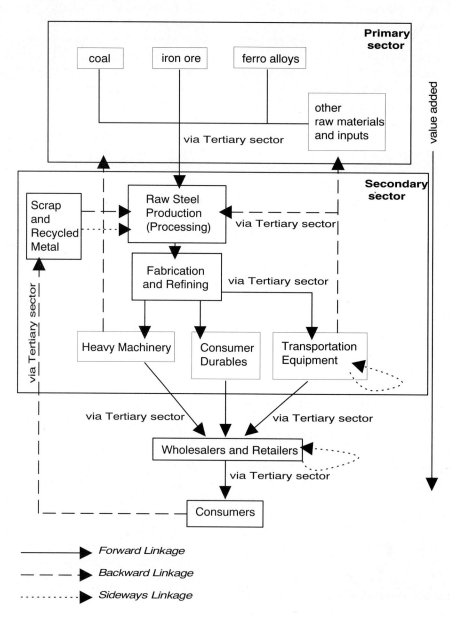

Figure 2.3   Economic flows and linkages in the production of steel.

constitutes development in a world economy composed of heterogenous cultures, regions, and perspectives.

In regard to regions of least economic status, the nature and purpose of development is best thought of as "the transformation of nations from a state of poverty to one of improved well-being, from a condition of adequate but self-limiting subsistence to one of continuing and self-generating growth." (E.S. Simpson, 1994, p. 6)

Economic growth and development takes place when a region experiences an increase in both the number and breadth of economic activities and opportunities. The speed at which growth and development occurs depends, however, upon many factors and conditions. Malecki (1997) views economic development as a process driven by linkages, investment, and entrepreneurial forces. Key to this equation is the formation of networks which encourage the flow of information and capital, enhance the efficiency of interaction amongst economic agents in the production setting, and promote the diffusion of enabling technologies. Information and technology flows are seen as vital to the formation and sustainability of **industrial agglomerations**—centers of innovation and production where interdependent agents and activities cluster on the economic landscape.

The new industrial landscapes of the 1980s and 1990s are a far cry from the old heavy industrial corridors established during the turn of the century. Nonetheless, the ideas behind their formation are the same. It is the creation and proliferation of linkages and the subsequent increase in geographically clustered entrepreneurial activity that foster growth and development of industry; a process that continues so long as a region is able to remain competitive. This, of course, is contingent upon a region's ability to attract, sustain, and retain globally competitive industry, as well as its ability to generate a positive business and entrepreneurial climate. Note that the stability and predictability of its government and the degree to which that region's culture will endorse or embrace innovation and change are factors which play an important role in the development process (Figure 2.4).

Michael Todaro (1985) contends that the objectives of development must encompass more than just progress on an economic front. Development strategies and efforts must not only allow for the provision of the basic necessities of life, but must encompass a broad spectrum of human-oriented development goals. These goals include the promotion of self-esteem and a sense of self-worth among members of society, improved educational attainment and cultural advancement, improved access to physical and financial capital, the preservation of human rights, and freedom from servitude, where human agents have an active voice in political, social, and economic matters.

This viewpoint is consistent with that of Mabogunje (1980) who argues that world development efforts must ensure that solutions be made for the collective good of all humanity, allowing both peoples and nations the ability to gain the self-respect that comes from the active and equal participation in economic matters and a growing sense of partnership. While economic growth can bring about significant improvements in the living and working conditions of some people in a regional economy, the benefits of growth do not necessarily "trickle down" to all people. Even if the average member of society is made better off from growth, growth can also lead to a growing rift between the richest and poorest members of that society. Subsequently, economic growth may not necessarily translate into economic development. This paradox is particularly true at the global scale, for as world markets, trade, and production networks expand, so too has it led to growing regional disparities.

The very notion of economic development has been associated with the viewpoint that nations or regions proceed through a series of development stages. Many believe that economic growth and industrialization is the passkey to development; a notion that has been

Figure 2.4   The changing face of industry: The emergence of high-tech industrial parks in places like Silicon Valley stand in contrast to the heavy industrial setting of places such as Gary, Indiana.

challenged given the ineffectiveness of these types of development policies to promote positive change in many regions. There is, nonetheless, a consensus that there is no universally acceptable development solution given the world's diverse set of development problems. Furthermore, and assuming that development does occur in stages, regional development policies must be recast in ways which recognize the needs of regions as they pass through various development stages.

Development efforts at the global scale must not only be coordinated, but they must be tailored to meet region-specific needs while sensitized to the interdependencies and dynamics of interregional economic change. The true worth of any regional development policy, regardless of scope, lies in the degree to which it allows the people of the affected region(s) to realize a better way of life. Despite the importance of social and political goals, most regional economic planning efforts are still centered around economic objectives; they tend to promote economic growth, even at the expense of broader developmental goals.

E.S. Simpson (1994) suggests that successful development hinges directly upon a region's ability to create and sustain a socioeconomic environment that is both attractive to the international investment community and able to exploit any existing advantages in production. However, it may be noted that in the highly competitive and dynamic global economic system, there are many regions which offer very similar production advantages, and as such, these competing regions must be viewed as substitutable destinations for international investment flows. This situation has lessened the potential leverage of any one given region in the world economy, reducing its ability to create and sustain a favorable business and entrepreneurial climate for a period of time that is sufficient to meet stated development objectives. The unstable and sometimes corrupt governments of underdeveloped nations make it difficult for them to be perceived as areas which offer suitable business climates. Many underdeveloped regions have, therefore, been branded as nonconducive for technology transfer, innovation development, and production from the perspectives of entrepreneurs and industry professionals; this is true despite the availability of cheap and abundant labor.

Morrill (1970) asserts that various **barriers to regional development** persist. These barriers include *conflicts of development policies with existing attitudes, traditions, and cultures (including religious practices and beliefs); the inability of industries in poor regions/nations to compete with industries in wealthy regions/nations; and the legacy of dependence created by existing patterns of commodity flows whose origins are deeply rooted in established colonial trade patterns.* With reference to trade patterns, it is widely accepted that poor regions remain at a disadvantage as they tend to export primary sector commodities (for example,

raw materials and minerals) while importing secondary sector goods (for example, finished manufactured products). Exporting goods of lower value and importing goods of higher value creates a no-win situation for underdeveloped regions as they are unable to benefit from the exchange. Moreover, **population imbalances** may further complicate the underdevelopment dilemma. Three basic types of population imbalances have been recognized:

1. when the ratio of population to agricultural resources (particularly land) is too high and demand for food cannot be met domestically; a situation often exacerbated by low levels of technology in the agricultural sector;

2. when population size overwhelms production capacity for food or other goods, and deficits cannot be satisfied by trade because the region lacks globally competitive industries and/or exploitable resources; and

3. when the rate of population increase is characterized as excessive beyond a region's long-term ability to support population growth. International trade, in and of itself, cannot be viewed as a cure-all for the development woes of the world's least-developed regions/nations.

Population imbalances such as those described above continue to plague many highly populated nations such as India, Pakistan, and Nigeria. Large populations and rapid population increases have largely negated the benefits of economic growth. To provide for their growing numbers, many underdeveloped nations must deal with the problems of finding new ways to both improve output in food production systems and create sufficient economic growth to generate employment opportunities to meet the growing demand for jobs. While technological advancements in agricultural settings have increased food availability worldwide, a critical shortage of employment opportunities in underdeveloped regions has created seas of unemployed and/or economically inactive members of society. The fact that the growth of opportunities has not kept pace with population growth has meant that underdeveloped nations must also deal with the backlash of related social and economic problems including crime, poverty, the spread of disease, and the exploitation of children. Moreover, there has been an increase in "black market" activities (referring to the undercurrent of illicit operations or illegal enterprises carried out by individuals or an organized group of individuals) and government corruption in places where legitimate economic opportunities are scarce. For example, consider the growing black markets in nations such as Colombia, Russia, and Peru. The reputation of regions which reluctantly support such activities further hampers those regions' abilities to develop.

Nurske (1953) suggested some time ago that increasing industrial efficiency and the use of advanced production technologies in developed regions have combined to give established production centers a competitive edge in the world economy. Industrial efficiency includes effective economizing on production inputs, the use of synthetic substitutes, and the recovery and recycling of materials. Subsequently, economic development is more likely to take place in advanced nations given their ability to employ industrial efficient methods and given that they have greater access to advanced production technologies. The growth and proliferation of increasingly efficient manufacturing and service sectors in the world's leading production zones has helped solidify the positions of the more-advanced economies of the world.

The formation of highly competitive industrial agglomerations in the United States, Europe, and especially Japan since World War II, along with the reproduction of self-reinforcing production and investment networks and limited technology spillovers, have contributed to persistent imbalances in the world economy. Areas containing these industrial agglomerations continue to flourish and attract (or retain) the predominant share of investment capital. By contrast, barriers of regional development and inefficient production systems have left many underdeveloped regions in a situation where they are unable to attract and accumulate sufficient amounts of investment capital. Such limitations further reduce the likelihood that underdeveloped regions will ever experience efficiency-related growth as their industrial bases remain small and noncompetitive.

Limited capital formation has continued to stifle underdeveloped economies, resulting in production inefficiencies. In addition, many underdeveloped economies suffer from low levels of worker productivity given that capital-starved industry supports less than optimal capital-to-labor ratios. As part of the vicious cycle of underdevelopment, limited capital formation both initially hinders the income-generating potential of high-value-added-for-export sectors and subsequently reduces the industrial performance of sectors stimulated by paltry investment flows. Facing large populations, population explosions, and lower capacities to save and invest, shortages of investment capital have restricted underdeveloped regions' investments per capita to lower-than-desirable levels.

To support the growth of domestic industry, development policies in these regions are cast in "protectionalist" terms, as policy makers attempt to shield domestic industries from outside competition. Protectionalist measures (such as import tariffs or quotas) are thought to lessen the adverse effects of import penetration. Such measures are designed to limit the loss of domestic sales revenues and profits, thus allowing the region to retain value added. The protectionalist mindset has been supported by the formation of regionalized production and free-trade zones. The manifestation of regionally based industrial clusters has led to supplementary efficiency gains for industry and regions within the production and free-trade zones. Regions and industry outside those production-and-trade loops, however, find it harder to compete.

As entrepreneurial forces in the world economy continue to exploit distinct production advantages in established or (already) spatially linked production zones, the complete integration of all regions in the world economy is something which is not likely to occur in the near future. Institutional, cultural, and environmental constraints, as well as variations in business climate, which continue to reinforce existing imbalances in power, wealth, and production capacity, also guarantee unequal access to capital. The distortions caused by imbalances have promoted directional biases in commodity and resource flows, uneven development, and a geography of surplus and deficit conditions which favor rich nations at the expense of poor nations. This development dilemma is something which is not remedied as the world's highly advanced economies continue to amass even greater concentrations of wealth and power over time.

Progress on the economic and social policy fronts has also been vigorously challenged by environmental concerns in most underdeveloped regions. Environmental damage, resource and land degradation, and decreasing productivity from increasing risk and susceptibility to extreme weather events and natural hazards have further complicated the development situation. Regions with unfavorable or deteriorating environmental conditions have become less desirable as production sites in the globalizing production network.

Asian-Pacific nations, as well as Africa, Latin America, and the Caribbean have been particularly hard hit by land degradation, forest and biodiversity loss, habitat fragmentation, atmospheric and freshwater pollution, degeneration of marine and coastal areas, and urban and industrial contamination and waste. As a result, these regions have become increasingly dependent on foreign assistance as they become increasingly predisposed to environmental crises. Although environmental policy responses have been swift or forthcoming in many of these regions, environmental crises continue to aggravate the poverty situation. Increasing pressures on land and food production systems from increasing populations, and losses of output in association with degradation and unsustainable management practices have led to unprecedented shortages of food.

While global food trade has allowed underdeveloped regions to meet production shortfalls, it has also created greater dependencies on outside suppliers for food imports. Access to safe drinking water remains a growing problem in most underdeveloped regions.

Areas most affected by environmental crises have acquired a reputation of having physically unstable production environments as they struggle to meet the most basic demand for food and water. Moreover, these regions face the problem of marginalization in the flow of investment capital in a world economy where investment dollars continue to gravitate toward areas which are perceived as "low risk" in terms of both business and physical settings.

Notwithstanding development barriers related to cultural differences, the goal of improving the human condition has been challenged on three major fronts. First, regional population growth (especially in nations of low economic development status) has placed an enormous pressure on global production systems in order to support a rapidly growing number of people in locations that are somewhat far removed from the world's major production regions. Second, continued polarization rather than equalization have led to a widening gap between the haves and the have-nots, further compounding problems such as poverty, inequality, and disparities in power, wealth, and well-being. This gap has resulted in uneven development and unequal access to world resources and capital, and a more pronounced inability for poor nations to compete in a divisive world economy. Third, the environmental limitations imposed on regional economies as a result of degradation, resource scarcity, and the cumulative effect of years of natural resources mismanagement have now begun to take a toll on the production systems of many of the world's least-developed regions. While the latter problem constitutes the very underpinning of the growth versus environment debate (and simply cannot be ignored), it is the first two problems that will form the primary focus of subsequent discussions found throughout this book. As such, regional variations in economic development status will be traced to inequalities, inefficiencies, and variations in production advantage. Before examining the forces that are constantly shaping and reshaping the global economic system and the implications of economic disparities, let us now turn toward classification as a means for making regional comparisons.

CLASSIFICATION SCHEMES

Early Classification Schemes

Attempts at classifying regional or national economies have produced a language of development, one that is sometimes confusing and inconsistent. As part of this language, classification schemes have emerged to describe the relative economic development status of nations and regions, each with their own set of strengths and weaknesses. While easy to use, two-tiered classification schemes have been labeled as inadequate in their design. Consider the dichotomy of commercial versus subsistence economies, whose attributes are listed in Table 2.1. **Commercial economies** tend to be open and outward looking in terms of production. These economies tend to be diverse in terms of economic scope, industry, and purpose as production responds to demand in both internal and external markets. **Subsistence economies** tend to be more closed and inward looking. With emphasis on survival activities, subsistence economies tend to be less varied in terms of industry as they produce almost exclusively in response to internal demand.

Note, however, that the description of regions as either commercial or subsistence is overly simplistic. Most regions or nations have attributes that would allow them to be classified as both commercial and subsistence at the same time. Furthermore, the degree to which a region is commercial or subsistence in its orientation is something which is highly variable both spatially and sectorally (Figure 2.5).

Perhaps a more effective classification scheme would involve the examination of economic development status by region. One such approach is the classification of "least-developed countries" (LDCs) versus "more-developed countries" (MDCs). **LDCs** are characterized by low production output, low per capita incomes, poor nutrition, inadequate health care, high rates of illiteracy,

TABLE 2.1	Attributes of Commercial and Subsistence Economies
Commercial Economies	**Subsistence Economies**
emphasis on commerce, trade, and finance	emphasis on subsistence or survival activities
secondary and tertiary sector orientation	primary sector orientation
open and highly integrated	closed or less integrated
information and technology driven	driven by agrarian endeavors
locations of major markets with the largest concentrations of wealth, power, and industry	locations of emerging markets with low concentrations of wealth, power, and industry
responsive to internal and external demand or markets	responsive to internal demand only
production of surplus for export	little or no surplus
high division of labor and a highly specialized labor force	low division of labor and a less specialized labor force
high degree of capital accumulation	low degree of capital accumulation
capital-intensive production, high capital-to-labor ratios	labor-intensive production, low capital-to-labor ratios

and other setbacks or disadvantages such as unfavorable resource endowments. **MDCs**, by contrast, are characterized by high and varied production outputs, high per capita incomes, etc. Consistent with the notion of a "developed world" and an "underdeveloped world," LDCs and MDCs are basically used to describe the rich and poor nations of the world. Note that the geography of the developed world versus the underdeveloped world has a peculiar NORTH and SOUTH orientation. This distinction is depicted in an updated version of the renowned **Brandt map**—*a map which sought to provide a geographic depiction of the location of the world's developed versus underdeveloped nations* (see Figure 2.6).

The Brandt map's **NORTH-SOUTH dividing line** is *a line which distinguishes the more developed nations of the NORTH from the underdeveloped nations of the SOUTH.* Though nearing two decades old, the Brandt map (and its dividing line) is still a fairly accurate depiction of the geographical locations of MDCs and LDCs. This two-tier classification, while stimulating from a geographic standpoint, is inadequate as it fails to distinguish between the development status or potential of nations of the SOUTH (Figure 2.6).

Another very popular classification scheme appeared shortly after World War II. It was a three-tiered scheme characterizing nations as belonging to one of three "worlds." The **First World** described *the advanced or developed economies of the "free world" (that is, noncommunist nations with the greatest concentrations of wealth, power, and industry).* The **Second World** described *the advanced or developed economies of the "nonfree world" (that is, nations that functioned under socialist or communist rule).* The **Third World** referred to *the less-advanced or underdeveloped economies of the world.*

The introduction and usage of the phrase "Third World" coincided with the formation of the United Nations (UN)—a coalition of nations aimed at keeping order in the international community. As a mediating force, the UN's primary role was to peacefully settle conflicts and geopolitical disputes. As a secondary function, the UN's General Assembly would focus on matters of economic development in Third World nations as part of poverty and crisis relief. In 1974, the UN General Assembly passed a resolution calling for a New International Economic Order (NIEO). This controversial initiative called for restructuring of the international economy and the eventual integration of all regional economies into the global economy. Recommendations were put forth on ways in which to improve economic conditions in the so-called Third World at a time when many of those nations were not so rigidly aligned with the world's economic superpowers. The promotion of a New International Economic Order is something that continues to be championed by globalists throughout the world, despite the backlash from those supporting more nationalistic ideals (Figure 2.7).

Ironically, the NORTH-SOUTH dividing line of the Brandt map served as a guide for the NIEO initiative, as it has allowed the world economy to be separated geographically into the haves and the have-nots. By its very design, however, the Brandt map renders the concept of a Second World meaningless from a geographic perspective as both First and Second World nations are part of the NORTH. To complicate matters, many social scientists have begun to describe the existence of a so-called Fourth World. This addition has helped distinguish between the Third World (underdeveloped nations that have significant future development potential as tied to existing resources and/or a production advantage in the world economy) and the **Fourth World**—*underdeveloped nations which lack the means to elevate their economic development status due to unfavorable resource endowments or the lack of a competitive production advantage.*

Nations of the Fourth World are frequently characterized as economically and sociopolitically unstable. Furthermore, Fourth World nations are also thought of as supporting insufficient industry, an insignificant export

Figure 2.5   Commercial versus subsistence farming activities in the America's: A combine crew hard at work harvesting wheat in the Palouse region of eastern Washington, while a peasant farmer in central America tends to a field.

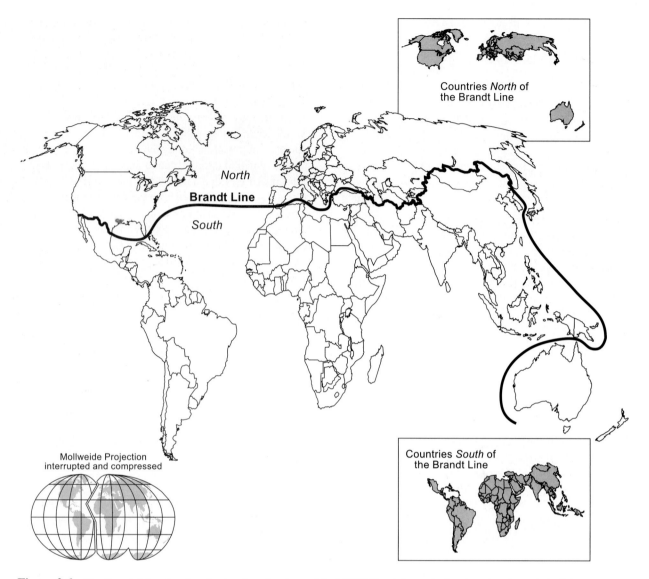

Figure 2.6   The Brandt Map showing the developed nations of the NORTH and the underdeveloped nations of the SOUTH.

base, and an inadequately trained labor force, with little or no hope of attracting investment capital. In short, the Fourth World is comprised of underdeveloped nations with insufficient development potential. Fourth world nations, therefore, have little chance of experiencing any real benefits of growth given that they face far too many obstacles which preclude them from successful integration into the world economy.

Despite the wide usage of the Western-based "worlds" classification scheme, this terminology is viewed as largely outmoded. References to the Second World have become less meaningful with the breakup of the Soviet Empire and the end of the Cold War, not to mention the trend toward **mixed economies**—*economies which support both private and state-owned production activities* (for example, China), and the increase in free-market activities around the globe.

An Eastern-based "worlds" classification system has been credited to the late Chairman Mao Tse Tung (of China). Mao described the world economy as a hierarchy of power structures. He viewed the First World as an entity composed of controlling superpowers (for example, the United States and the former Soviet Union). In this scheme, the Second World contained all nations that were politically aligned with the superpowers, nations that achieved integration by way of their political and economic orientation. According to Mao, the Third World was composed of all remaining nations of the world; typically those that were poor and powerless. This hierarchical classification, although seldom used by Western scholars and analysts, is still highly applicable today as a way in which to describe nations and their leverage in the global economy. As intended, its strength lies in its ability to draw attention to the stark contrasts that exist between nations residing in the upper versus lower positions of the hierarchy. Its ineffectiveness can be traced to its limited geographical content.

The Core and Periphery Distinction

The NORTH-SOUTH (Brandt map) division is consistent with an alternative classification based upon the "core-periphery" model. The distinction between "core" and "peripheral" regions was an idea concurrently introduced by Raul Prebisch (an Argentinean economist) and Andreas Predohl (a German economist). It was a way in which to characterize regions in terms of production, trade, and economic development potential. As discussed by Friedmann (1966, 1972), the core-periphery model brought attention to unequal divisions of power that resulted from the uneven distribution of economic activity over space, specifically, the concentration of economic activity in the core and the lack of economic activity in the periphery. The **core-periphery model** *describes variations in production and trade possibilities and development potential as outcomes of the relationship between a dominant core and a dependent periphery.*

A **core region** can be thought of as *a region which contains a large agglomeration of "vertically linked" industries or one which supports a spatially concentrated production network.* **Vertical linkages** *describe the movement of goods between producers (and ultimately to consumers) in the complex chain of production linkages in the extraction and processing of raw materials to the fabrication of finished products as goods make their way through the value-added sequence.* Vertically linked industries are generally established in a given region to satisfy both local demand and as producers strategically position themselves to take advantage of demand at several dispersed geographic locations. This allows them to operate at larger scales and internally support services and activities that would normally be contracted out.

Hence, the emergence of a diverse and interdependent production network allows the region to secure and retain value added across many sectors and industries. Industrial agglomerations not only promote minimization of transport costs amongst firms and industry in the area, but also allow for **increasing returns to scale**—*output and efficiency gains due to the increasing size of firm- or industry-level operations* and **increasing returns to network size**—*output and efficiency gains associated with a larger and highly integrated production network.* This enables a production region to amass a diverse industrial base with a large internal production capacity. Typically, industry is concentrated in or near urban areas. Urban areas are in themselves agglomerations of economic activities, and are generally thought of as core areas. Urban areas provide a distinct locational advantage to producers given that they enhance access to factors of

Figure 2.7   Debate and deliberation: The United Nations General Assembly at a recent Economic Summit.

production and at the same time provide a highly concentrated end market for goods and services produced.

By contrast, a **peripheral region** is *a region which does not contain a large or dense vertically integrated production system.* In other words, it does not possess a large production capacity, nor does it support a large number of spatially concentrated activities. Production activities are generally limited to the extraction and processing of raw materials for export, or subsistence activities to meet local demand, or some form of labor-intensive manufacturing (typically low-value goods). All in all, peripheral regions tend to be less diversified in their industrial base. On average, peripheral regions also tend to have less capacity to produce high-end or high-value manufactured goods.

The core-periphery distinction suggests that underlying dependency relations exist in the world economy. In effect, the core offers the periphery a wide range of finished products, technical expertise, know-how, and information. The core is also a major source of funding for regional development projects. The **periphery**, by contrast, offers a low-cost production environment (cheap and plentiful labor) and, in turn, is able to produce a wide range of low-cost consumer goods to meet demand in the core. The periphery also offers the core resources and materials in raw and processed forms, as well as external markets for the higher-end goods produced in the core.

Overall, the core-periphery model is highly effective in its ability to depict both the hierarchical aspects of economic development status and the underlying dependency relations which exist in the global economic system. The core-periphery framework is flexible in that one may describe core regions which exist within peripheral regions and peripheries which exist within the core. For example, parts of the rural southeastern United States may be viewed as peripheral in comparison to the urbanized corridor of the Northeast, while underdeveloped nations are most certainly considered to be peripheral to the United States. Note also that the core-periphery distinction is inherently geographical as it is determined by the spatial concentration (or lack thereof) of vertically linked industries.

In reference to the Brandt map, one could easily describe the economies of the NORTH as core economies and the economies of the SOUTH as peripheral economies. In addition, the NORTH-SOUTH/CORE-PERIPHERY distinction could be expanded to identify a middle tier of economies which fall under the heading of semiperipheral. **Semiperipheral economies** may be defined as *economies having a development status that falls below that of nations of the core yet above that of the least-developed nations of the periphery, and as such, may be described as "core regions of the periphery," or "peripheral regions of the core."* Thus, semiperipheral nations may be viewed as the core economies of the SOUTH or the peripheral economies of the NORTH as illustrated in the tri-colored choropleth map in Figure 2.8.

Core economies of the world include the United States, Canada, Japan, the economies of Oceania (Australia and New Zealand), the nations of Western and Central Europe (Germany, Italy, France, the United Kingdom, Norway, Sweden, Belgium, the Netherlands, Spain, Denmark, Switzerland, and Austria).

Semiperipheral economies would include the **big-five economies** of the SOUTH: Mexico, Brazil, India, South Korea, and China. Each of these economies boasts a 1995 GDP that exceeds $250 billion. Also included as semiperipheral are the economies of Russia, Ukraine, and other resource-rich republics of the former Soviet Union. The list of semiperipheral economies would most certainly include oil-exporting nations of the Middle East (Saudi Arabia, Kuwait, United Arab Emirates); the bulk of Southeast Asian and Pacific Island nations such as Taiwan, Singapore, Malaysia, Indonesia, Thailand, and the Philippines; the more industrialized eastern European economies (namely, Poland, Romania, Hungary, Czech Republic, etc.); the larger South American economies (for example, Venezuela, Argentina, Uruguay, and Chile); and a host of other significant members such as Turkey, Israel, Iran, Pakistan, Egypt, South Africa, Iceland, Ireland, and Portugal.

The remainder of nations in the SOUTH would, by default, be characterized as forming the periphery of the world economy.

The NORTH-SOUTH dividing line provides a useful boundary for distinguishing the "rich" core from the

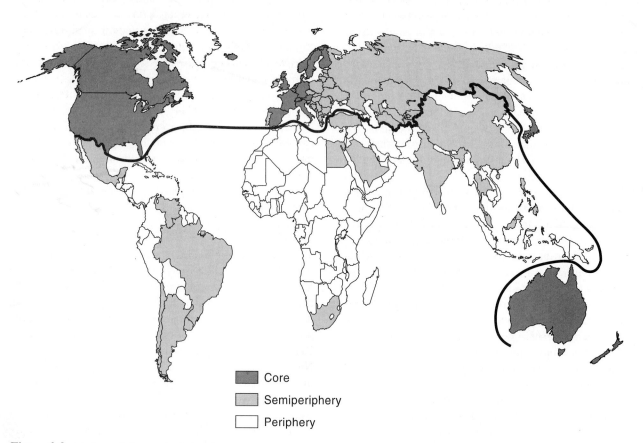

Figure 2.8 A choropleth map depicting the core, semiperipheral, and peripheral nations of the world economy.

"poor" periphery. The introduction of semiperipheral economies is necessary when one considers the geographic locations of the big-five economies in the SOUTH and their immense production capacities. Although the largest economies in the SOUTH are technically defined as the core economies of the periphery, from a development standpoint, semiperipheral economies are still ranked behind the core economies of the NORTH in terms of both economic status and power.

MEASURES OF ECONOMIC DEVELOPMENT STATUS

There are many ways in which to gauge a region or nation's economic development status. Several key indicators are typically used to examine the performance and the general well-being of an economy. In particular, measures of "production output" tend to be used with great regularity amongst scholars and regional analysts. The production output of an economy in a given time period (usually over a year) may be measured by the income or the value of products produced in that time period. Two such measures which describe the size and strength of an economy are the Gross Domestic Product and the Gross National Product.

Gross Domestic Product (GDP) refers to *the monetary value of all final goods and services produced by individuals, businesses, and government within a nation's domestic borders in a given year.*

Gross National Product (GNP) refers to *the monetary value of all final goods and services produced by a nation both domestically and abroad in a given year.*

Recall that value added was defined as the difference between the value of a commodity and the cost of materials, inputs, and services required to produce that commodity (including wages, interest, rents, and profits). In value-added terms, GDP is the sum total of value added in all sectors of an economy.

While measures such as GDP and GNP are useful in making comparisons in absolute terms, they tell us little about an economy's true relative production clout as defined by the average contribution to income by individuals which make up that economy. The problem is that larger economies (as depicted in terms of geographic size or population size) tend to have larger GDPs and GNPs. This requires that production output measures be "standardized"; something which is easily accomplished by dividing such measures by population size. Per capita production measures reflect the monetary value of all final goods and serviced produced by the average person in an economy.

The two most widely used per capita production measures are **GDP per capita**—GDP divided by population size, and **GNP per capita**—GNP divided by population size. In general, the higher the per capita output, the higher the development status of the economy in question.

Note that the United States and other highly developed countries such as Germany, Japan, France, and the United Kingdom, have GDP per capita measures that exceed $12,500 of output per person, whereas most underdeveloped nations have GDP per capita measures that are typically less than $1000 of output per person (as measured in 1995 U.S. dollars).

The occupational structure of a region's labor force is another important economic indicator. Labor force participation statistics which define a region's **employment profile**—*the activity levels of labor or percentage of labor force involvement by economic sector*—offer much information about a region's development status. In general, the higher the percentage of a region's labor employed in services, the higher that region's economic development status. Following similar lines of reasoning, the lower the percentage of a region's labor involved in primary-sector activities, the higher the economic development status of the region. These relationships will be explored further in an upcoming section.

Trade statistics such as the structure of exports and imports or the monetary value of exports also tell us a good deal about a region's economic development status. Both the type of exports and the value of exports (and exports per capita statistics) provide useful measures of a nation's production capacities and its ability to compete for shares of external geographic markets. In general, the greater the value of exports and the higher the value of exports per capita, the higher the region's economic development status as defined by its ability to retain value added in production.

Miscellaneous measures such as infant mortality rates, literacy rates and educational attainment levels, caloric intake per capita, savings/investment per capita, external debt, physicians per capita, life expectancies, and population growth rates are also useful in explaining variations in economic development.

UNEVEN DEVELOPMENT

There is evidence to suggest that the nations which reside in the SOUTH (in reference to the Brandt map) vary tremendously in terms of their economic development status and potential. Consider the comparison of selected indicators for several "underdeveloped" nations of similar population size; namely, Bangladesh, Nigeria, Brazil, and Pakistan. The data presented in Table 2.2 suggest that Nigeria maintained a rather precarious economic position over the period in question, given negative average annual growth rates in GNP per capita and exports, combined with double-digit inflation and a low GNP per capita.

Nations such as Bangladesh, Brazil, Pakistan, and Nigeria each face a very unique series of development challenges. One could argue that the development potential of a Bangladesh or Pakistan is higher than that of a nation such as Nigeria or Brazil given the statistical evidence. The data imply very different levels of economic growth and trade positions, as well as different economic development problems. Although the economy of Brazil appears as dominant in terms of GNP per capita (a figure which is roughly three times that of the others listed), with relatively high life expectancies, economic and export growth has been sluggish in comparison to its West Asian counterparts over the period examined. In addition, Brazil had to contend with a formidable external debt and runaway inflation problem; things that most likely contributed to its low overall economic growth rate.

Consider the case of China, the largest semiperipheral economy in the SOUTH, an economy that continues to show enormous growth and development potential. By the mid-1990s China accounted for about one-tenth of the world's production output. Although China's external debt was approximately $80 billion in 1993, it was a debt equal to a mere 3% of China's GDP. The sub-Saharan African nation of Sudan, by contrast, held a $17 billion external debt in 1993—a figure equal to roughly 80% of its GDP. This situation is not uncommon to underdeveloped regions. Large external debt continues to plague a large majority of economies in Africa and Latin America. Many regional analysts question the ability of these economies to overcome their long-term financial woes as short-term economic growth and development has been largely financed by debt.

Certainly, the implications of large external debt extend far beyond the confines of the SOUTH. Note that the national debt of all nations in the SOUTH stood at approximately $1 trillion in 1997. This figure is equal to roughly 3.5% of the **Gross World Product (GWP)**—*the monetary value of all final goods and services produced by the world economy*. The national debt of the U.S. economy in 1997 was approximately $5 trillion—five times the debt of all developing nations together, a figure equal to roughly 16% of GWP.

Uneven development in the SOUTH does not mean that some nations will never elevate their economic development status. The preconditions for development and the forces which dictate where and when development may occur continually change over time. As the global economy evolves, peripheral economies will also mature. Improving the economic conditions of the human agents which occupy those regions depend, in part, on whether those regions can establish competitive industries and trade linkages along with investment in infrastructure to support these industries and associated commodity flows. The nations of the underdeveloped world are likely to render the notion of a Fourth World obsolete if global development efforts prove successful. Those optimistic about the long-term impacts of regional development initiatives would support abandoning the use of the term "underdeveloped" altogether in favor of declaring all semiperipheral and peripheral economies as "developing." Unfortunately, this term may not be currently representative of all nations of the SOUTH due to the simple fact that many nations are not making progress. Some analysts have attributed the lack of progress to the interdependent nature of regional development.

Robert Bennett (1991) has elaborated on the growing interdependence of national economies, and the series of development problems associated with interdependence. Individual nations are becoming less able to control their own destinies as the events that affect their development and prosperity are largely dictated by global economic change and external forces. As nations vary greatly in terms of their relative size, attributes, resource endowments, cultural identity, strategic and competitive

TABLE 2.2	A Comparison of Economic Development Indicators for Selected Underdeveloped Nations			
Indicator	**Bangladesh**	**Nigeria**	**Brazil**	**Pakistan**
Population 1993 (in millions)	115.2	105.3	156.6	122.8
GNP per Capita (1993)	$220	$300	$2930	$430
Annual Growth Rate of GNP (1980–93)	2.1%	–0.1%	0.3%	3.1%
Life Expectancy at Birth (in years)	56	51	67	62
Average Annual Rate of Inflation (1980–93)	8.6%	20.6%	423.4%	7.4%
Average Annual Growth of Exports (1980–93)	9.8%	–0.6%	5.2%	10.1%
Total External Debt per Capita (1993)	$120.4	$308.9	$848.2	$212.1

Source: *World Development Report 1995,* pp. 162–163, pp. 186–187, pp. 200–201.
Note: Monetary data shown in 1993 U.S. dollars.

positions, social and political objectives, and ability to be influenced by or accept economic growth and industrialization, it is logical to assume that such differences will lead to unique perspectives on national development and, in turn, varying capacities to adapt and endure change in the global economy.

If peripheral economies are to be successful in adapting to an increasingly interdependent global economic system where competition continues to intensify over time, they must have increased access to production factors and an inherent flexibility in their production regimes. In short, industry in peripheral economies must be receptive to changing needs of the market and adaptive in terms of its production technologies. To accomplish these tasks, it is essential that underdeveloped regions procure not only a competitive labor force (in terms of skills and productivity), but secure a financial base to support infrastructure and industrial growth, an entrepreneurial and management capacity to implement and support new production technologies, and a level of government involvement that is necessary to improve a region's general business climate.

The formation of social, cultural, and business information networks are also essential to the development process. Transportation and communications infrastructure are of paramount importance here as peripheral economies seek to attract and support a critical mass of producers and suppliers. This must be accompanied by investment in human resources (by educating and up-skilling the labor force) to improve standards and worker productivity. As the success of many regional development efforts are contingent upon the success of local development efforts, it is imperative that local development processes and policies be coordinated in such a way as to meet broader regional development objectives. The success of economic development efforts hinge greatly upon the cooperation and coordination of labor, investors, industry, and government.

CLASSIFICATION BY INCOME

The World Bank, a premier investment and lending institution, divides up the world into groups on the basis of income using output per capita measures. Economic development status is defined in terms of GNP per capita (the gross national product of a nation divided by the total population). Table 2.3 highlights those divisions as shown in the *World Development Report 1995* (see Appendix A for a comprehensive listing of low- and middle-income economies by region and income). A comparison of selected economic development indicators for nations which comprise the various income groups are also listed. The data provide evidence of strong regularities in the relationships between GNP per capita and mean GNP per capita and other key development indicators such as life expectancy and literacy. These data support the argument that the higher the level of income or development status, the higher the quality of life.

There is a direct relationship between investment and economic growth. Consider the data in Tables 2.4 and 2.5, which highlight the relationship between Gross Domestic Product (GDP) and Gross Domestic Investment (GDI). In general, investment refers to spending devoted to maintaining or increasing the stock of capital; that is, investment in factories, machines, office equipment, and other products that are used in the production process plus additions to inventories. More formally, **Gross Domestic Investment (GDI)** may be defined as *outlays on additions to the fixed assets of an economy plus net changes in the level of inventories* (*World Development Report 1995*, p. 234). The adage that "investment equals growth" applies well here as there is ample statistical

TABLE 2.3	1995 World Bank Income Groups and Development Indicators

Income Group	GNP per Capita (in $)	Mean GNP per Capita (in $)	Life Expectancy (in years)	Illiteracy (% population)
World (all nations)	**200–27,000**	**5500**	**66**	**n/a**
Low (excluding China and India)	<700	300	56	49%
Low	<700	400	62	41%
Low-Middle	700–2500	1600	67	19%
High-Middle	2500–10,000	4400	69	14%
High	>10,000	23,100	77	n/a

Source: *1995 World Development Report*, pp. 162–163.

Note: GNP per capita and mean GNP per capita shown in 1993 U.S. dollars rounded to the nearest $100.

n/a = not available.

TABLE 2.4	Average Annual Growth Rates of GDP and GDI by Income Group

	Growth Rates (1980–1993)	
Income Group	**GDP**	**GDI**
World (all nations)	**2.9**	**3.2**
Low (excluding China and India)	2.8	–0.2
Low	5.8	6.1
Low-Middle	1.6	0.8
High-Middle	2.7	2.1
High	2.9	3.2

Source: *1995 World Development Report*, pp.162–163.

TABLE 2.5	Average Annual Growth Rates of GDP and GDI for 42 Selected Low- and Middle-Income Nations

Nation	Average Annual Growth Rate of GDP 1980–1993	Average Annual Growth Rate of GDI 1980–1993
WORLD	**2.9**	**3.2**
China	9.6	11.1
S. Korea	9.1	11.8
Thailand	8.2	11.4
Malaysia	6.2	6.3
Pakistan	6.0	5.0
Indonesia	5.8	7.1
India	5.2	5.7
Chile	5.1	9.6
Turkey	4.6	5.6
Egypt	4.3	1.2
Bangladesh	4.2	1.6
Sri Lanka	4.0	2.4
Kenya	3.8	–0.7
Colombia	3.7	2.1
Tunisia	3.7	1.2
Costa Rica	3.6	5.5
Ghana	3.5	9.8
Portugal	3.0	4.1
Paraguay	2.8	0.8
Congo	2.7	–11.1
Nigeria	2.7	–5.5
Zimbabwe	2.7	3.0
Ecuador	2.4	–1.7
Jamaica	2.3	0.7
Algeria	2.1	–3.6
Brazil	2.1	–0.3
Venezuela	2.1	–1.1
Mexico	1.6	0.1
Philippines	1.4	–0.1
Namibia	1.3	–6.5
Panama	1.3	–1.5
Uruguay	1.3	–3.4
Bolivia	1.1	–4.3
Bulgaria	0.9	–2.3
South Africa	0.9	–4.7
Argentina	0.8	–1.3
Poland	0.7	–1.1
Cameroon	0.0	–4.0
Zambia	0.0	2.5
Hungary	–0.1	–1.6
Peru	–0.5	–2.3
Nicaragua	–1.8	–5.5

Source: *World Development Report 1995*, pp. 164–165; pp. 176–177.

evidence of a fairly strong positive correlation between the investment and regional economic growth.

Geographic patterns in the relationship between the average annual growth rates of GDI and GDP are also evident. Asian nations lead the pack in terms of domestic investment and economic growth. Investment and capital formation, along with an inexpensive, abundant, and productive labor force has enabled many Asian nations to rapidly improve their economic development status. Marked geographic differences are observed in both the accumulation of production capital and wealth (see breakdown of nations by region and income group in Appendix A). Note that the World Bank classification scheme suggests geographic regularities in the wealth and well-being of nations. Not surprisingly, this pattern is similar to that of the Brandt map and the core-periphery models discussed earlier.

High-income nations are typically found in the NORTH, with low-income nations in the SOUTH. Middle-income designations are mostly associated with nations of the SOUTH and/or those of semiperipheral status (with only a few minor exceptions). Note, for instance, that the GNP per capita of China and India are consistent with nations classified as low-income, yet the data suggest that China and India are unlike their low-income counterparts, largely attributed to the sheer size of their economies. When China and India are excluded from the low-income category, low-income nations on average tend to exhibit statistics that are less impressive. For example, compare the average annual growth rate statistics for the low-income group versus the low-income group excluding China and India as shown in Table 2.4. The low-income group (excluding China and India) exhibits growth rates that tend to be consistently lower than the world average.

Note that the investment and growth measures discussed above are defined purely in economic terms. These measures fail to take into account development-related investments in education, health, and national security, and the subsequent long-term social benefits that

TABLE 2.6	Estimated World Labor Force Participation by Sector (1998)

Sector	Estimated Percentage of World Labor Force
Mining, fishing, and forestry	< 1%
Agriculture	47%
Manufacturing and handicrafts (and cottage industries)	20%
Services and other	32%

Source: Forecast estimates obtained using *World Resources 1994–95* data on pp. 286–287 and p. 300, and rounded to sum to 100%.

are nonquantifiable in most cases. Nonetheless, short-term changes in economic output are useful measures of the rate of economic development.

LABOR FORCE PARTICIPATION STATISTICS

The world's total population in 1998 is estimated at approximately 6 billion people. The world's total labor force represents roughly half of the world's population or approximately 3 billion people. Note that a **labor force** *is comprised of people actively engaged or employed in production of goods and services in a given region.* The sectoral distributions of the world's labor force varies dramatically from region to region. Table 2.6 shows labor force participation statistics by economic sectors. Note that the largest share of the world's labor force is engaged in primary sector activities (accounting for almost half of the total labor force). Note that approximately 1.4 billion people (that is, 47% of the world's labor force) are directly involved in **agriculture**—*the growing and harvesting of crops and the tending and rearing of animals for food and other products.*

Despite characterization of the modern world economy as postindustrial, this in reference to a postindustrial-revolution period where traditional manufacturing (heavy industry) has given way to the growth of high-tech sectors, government, and information, producer and consumer services, the majority of the world's labor force are involved in agrarian endeavors. This situation is unlikely to change as population pressures continue to place enormous demands on the world's food production systems. To characterize the world economy as postindustrial seems illogical given that only one-third of the world's labor force is involved in tertiary sector activities. Although the postindustrial label is highly applicable in the case of the world's core economies, it is, nonetheless, a label which misrepresents the large majority of peripheral and semiperipheral nations where agriculture is the dominant economic activity.

Labor force participation by sector and region shows a very definitive geographic pattern, one that is highly correlated with development status and income. The percentage breakdowns of employment by sector and region are shown in Table 2.7.

In general, several geographic regularities are revealed in the analysis of regional employment profiles and economic development status. First, as the percentage of the labor force in agriculture decreases, economic development status increases (an inverse relationship between primary sector activities and status). Second, as the percentage of the labor force in manufacturing increases, economic development status generally increases. In other words, there is a positive relationship between labor force participation in secondary sector activities and economic status. Last, as the percentage of the labor force in services increases, economic development status also increases. Hence, there is statistical evidence of a direct positive relationship between the percentage of tertiary sector activities and economic development status or income.

In the case of the U.S. economy, only 3% of its labor force is directly engaged in primary sector activities, with roughly 2.5% in agriculture. Efficiency in U.S. agricultural production, the presence of a globally competitive secondary sector, and an emphasis on tertiary sector activities stand in stark contrast to patterns observed in underdeveloped economies where production inefficiencies abound and subsistence orientation is a way of life.

Consider the nations of **sub-Saharan Africa**—*nations south of the great Saharan Desert (excluding South Africa)*—where as high as 75% of the labor force is involved in primary sector activities. Note, once again, that the Brandt map's NORTH-SOUTH dividing line provides an adequate division between the highly agrarian economies of the SOUTH (where 50% or more of the labor force is directly involved in agriculture) and the postindustrial economies of the NORTH. Despite the high percentages of labor force in primary sector activities in both India and China, these economies are considered as having formidable manufacturing bases given the sheer size of the labor force (that is, the number of people) involved in the secondary sector. Together, China and India accounted for almost 15% of the world's total production output during the mid-1990s.

Figure 2.9 illustrates the general relationships between employment levels by economic sector and economic development status. Note the precipitous drop in primary sector employment, the subtle rise (and later leveling off or drop) in secondary sector activities, and the dramatic rise in tertiary sector employment as a percentage of total employment as economic development status increases. The observed patterns are attributed to five major factors:

1. production orientation;
2. efficiency in production due to scale, technology, and access to factors of production;

TABLE 2.7	Employment Profiles for Selected Nations and Geographic Regions (1989–1991)

Region	Percentage of Labor Force (%)			Income group		
	Primary	Secondary	Tertiary	Low	Middle	High
North America	4	25	71			*
United States	3	26	72			*
Australia	5	24	70			*
North and Central America	12	29	58		*	*
Mexico	23	29	48		*	
Europe	14	36	50			*
UK	2	29	68			*
Denmark	6	27	67			*
Poland	27	37	36		*	
Commonwealth (former Soviet Union)	20	39	47		*	
South America	39	26	45		*	
Uruguay	4	30	66		*	
Venezuela	13	25	62		*	
Chile	19	26	55		*	
Brazil	31	27	42		*	
Asia	60	18	22	*	*	*
Japan	7	35	59			*
S. Korea	17	36	48		*	
Malaysia	26	28	46		*	
Philippines	45	16	39		*	
Pakistan	47	20	33		*	
Indonesia	56	14	30		*	
India	70	13	17	*		
China	74	14	12	*		
Bangladesh	75	6	19	*		
sub-Saharan Africa (excluding S. Africa)	69	12	19	*	*	
Nigeria	68	12	20	*		
Sudan	71	8	21	*		
Zambia	73	10	17	*		
Zimbabwe	73	11	17	*		
Kenya	81	7	12	*		

Source: Estimates (regional averages) obtained from data in *World Resources 1994–95*, pp. 286–287.

Note: Some percentages do not sum to 100% due to rounding errors.

3. trade and the degree of global integration;
4. linkage enhancement, network formation, and associated spin-offs; and
5. the growth of nonagricultural sectors due to urbanization and subsequent shifts in the labor force.

Note that high-income economies have the ability to implement cost-saving production techniques and take advantage of productivity gains associated with a highly skilled labor force, high capital-to-labor ratios, and the use of state-of-the-art technology. This allows industry in the core to operate efficiently and remain competitive. This is especially true for commercial agricultural sectors which continually boast the production of surplus. Surplus production in commercial systems frees up labor from primary sector activities, permitting labor to seek opportunities in other sectors. Similar production advantages are obtained in high-wage manufacturing sectors

which produce high-value commodities for export. The ability of the high-income nations to compete for a share of the global market for manufactured products not only sustains the secondary sector but stimulates demand for secondary sector inputs and producer services. Hence, the expansion of efficient manufacturing sectors increases the efficiency of linked secondary and tertiary sectors.

Efficiency in production has had profound impacts on the economic landscape of advanced nations. Industrial agglomerations have become firmly rooted in the core, where emphasis is on commercial production. By contrast, the majority of largely populated underdeveloped regions are subsistence oriented. A large percentage of their labor force is involved in the production of agricultural commodities for direct consumption. The subsistence orientation and inadequacy of industries in these areas combine to increase the amount of labor required to produce the basic foodstuffs and other consumer products necessary to support their growing populations. The vast majority of underdeveloped regions tend to be deficit food-producing regions, making them reliant on production surplus from the more-efficient agricultural systems of the core.

International trade and integration in the global production network play important roles in expanding the tertiary sector of most regional economies. Imports and exports increase the number of employment opportunities in wholesale, retail, and transportation-related sectors, not to mention other sectors such as consumer and producer services. Subsequent increases in the demand for goods and services produces a ripple effect throughout the economy and an expansion of the entire regional employment base. New employment opportunities arise and more income is generated, distributed, and redistributed throughout the economy, where growth fuels growth.

As production and services activities are sustained in and around urban areas and as large urban population increases transform the economic landscape, the benefits and spin-offs of urban economic growth and development translate into the growth of all nonagricultural sectors. This includes the increase in the demand for public services, public administration, and government, as well as stimulation of the construction, transportation, communication, and utilities sectors. Industrial agglomerations become commonplace and production-cost savings further attract regionally and globally competitive industries.

SECTORAL DISTRIBUTION OF GDP

Economic activity in each sector of a regional economy contributes to overall output. The contribution of each sector differs dramatically over space and by economy. The sectoral distribution of the GDP by income group is shown in Table 2.8. Note that these statistics show a pattern slightly different from labor-force participation statistics discussed earlier. This is especially true for middle- and low-income economies. Although lower-income economies have a relatively high percentage of their labor force in primary sector activities, when compared to high-income economies, the greatest share of their GDP is associated with the secondary and tertiary sector activities. The predominance of the secondary sector as a contributor to GDP is a reflection of (a) the relatively low value of primary sector outputs (and the relatively high value of products from the secondary and tertiary sectors); (b) the existence of inefficiencies in agricultural sectors in low- and middle-income economies; and (c) the importance of manufacturing and services in the regional value-added process.

Employment profiles and sectoral contribution to GDP are useful indicators of economic development status. These profiles are highly indicative of a region's eco-

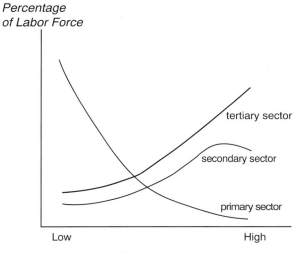

Percentage of Labor Force

Economic Development Status (Income)

Figure 2.9 Labor force participation and economic development status.

TABLE 2.8	Distribution of GDP (1991) for Selected Nations		

	Percentage of GDP by Sector		
Nation	**Primary**	**Secondary**	**Tertiary**
United States	2	30	69
Australia	3	31	66
Mexico	8	31	61
United Kingdom	2	36	62
Denmark	5	28	67
Poland	8	54	37
Uruguay	11	32	57
Venezuela	5	50	44
Brazil	10	39	51
Japan	3	42	56
S. Korea	9	45	46
Philippines	22	35	43
Pakistan	26	26	48
Indonesia	21	39	39
India	31	27	42
China	28	39	33
Nigeria	37	38	25
Zambia	18	45	37
Zimbabwe	20	32	48

Source: Estimates from data in *World Resources 1994–95*, pp. 256–257.
Note: Some percentages do not sum to 100% due to rounding errors.

nomic development status and its ability to sustain an industrial base. In fact, many regional economic growth and development problems can be traced to the lack of a globally competitive secondary sector.

REGIONAL ECONOMIC GROWTH: BLESSING AND BURDEN

As with other economic development indicators, there are noticeable variations in the average annual economic growth rates of nations by region and income group (see Table 2.9). Low-income nations (excluding China and India) have experienced roughly the same economic growth rates as upper-middle- or high-income nations. By contrast, East and South Asian economies have done remarkably well in terms of sustaining growth over the past few decades, while economies in sub-Saharan Africa, Latin America and the Caribbean, Europe and Central Asia, and North America have exhibited a steady decline in their growth rates. With the exception of Asian economies, the world's regions have experienced an overall decline in economic growth over the past several decades. In fact, the majority of the world's regions have posted annual economic growth rates that have been less than the world average (of 2.9%) from 1980 to 1993. High economic growth rates in the Asian corridor throughout this same period have been attributed to several factors: (1) the dramatic increases in domestic and foreign investment in the area; (2) internal economic reforms and adopted trade policies; (3) the rapid expansion of consumer markets in Asia; (4) the rise of the Pacific Rim in the global production network; (5) Asia's production advantage in labor-intensive sectors; and (6) the rapid growth of global trade and the increase in demand for inexpensive products coming out of Asia.

While the benefits of domestic investment and sustained economic growth of East and South Asian economies have been many, growth has also come at a high cost. Economies that once flourished, now stand on the brink of disaster. The late 1990s has brought about significant reductions in annual economic growth rates throughout Asia (to levels at or below the world average). Over-expanded industries, excessive production capacity, and corporate debt have left many Asian economies on the verge of collapse. The growth in consumer demand has not kept pace with the growth of production capacity. Banking and finance sectors, which provided much of the initial investment capital to fuel the industrialization frenzy, now face the challenge of dealing with over-extended credit lines and defaulting loans as producers fail to generate sufficient revenue and profits. The decline in export income, along with the added burden of foreign debt and reduced in-flow of foreign investment, have left many Asian nations to struggle with failing currencies and plunging stock markets. Consumers have also been hit with marked reductions in purchasing power as unexpectedly high rates of inflation have emerged—a consequence of two decades of unbridled growth. Over-industrialization and sluggish growth in exports have also meant less new job opportunities, rising unemployment levels, social unrest, and reduced future development

TABLE 2.9	Average Annual Growth Rates of GDP	
	Average Annual (GDP) Growth Rates (%)	
Region or Income Group	**1970–1980**	**1980–1993**
Low-income economies	4.3	5.7
Low-income economies (excluding China and India)	4.4	2.9
Lower-middle-income economies	5.1	1.6
Upper-middle-income economies	5.9	2.7
Severely indebted economies	5.8	1.5
High-income economies	3.2	2.9
sub-Saharan Africa	3.8	1.6
Middle East and N. Africa	n/a	2.2
Latin America and Caribbean	5.4	1.9
Europe and Central Asia	5.4	0.4
North America	3.7	2.7
South Asia	3.5	5.2
East Asia and Pacific	6.9	7.8
World	3.6	2.9

Source: *World Development Report 1995*, pp. 164–165.

potential. The recent Asian financial crisis has sent a wake-up call to economies around the world. It is a testimony that economic growth is not a panacea for development problems. In the case of Asian economies, economic growth has led to economic contraction. Hardest hit have been the economies of Japan, Indonesia, Malaysia, Thailand, and the Philippines.

While the devaluation of Asian currencies over the past several years has led to a short-term export revival, enhanced export income offers only temporary reprieve to most debt-riddled companies. Facing insolvency, many Asian companies have resorted to aggressively buying American dollars in currency markets to repay loans before their domestic currencies bottom out. Adding further complications is the fact that interest rates in Asian economies are likely to rise in order to combat the effects of falling currencies. As a result, domestic investment and growth are likely to slow even further. Note too that the financial crisis in Asia is likely to have a serious impact on the U.S. economy and elsewhere. Rising demand for the U.S. dollar, and a stronger dollar in currency markets, means that American goods will be more costly to Asian consumers. This will hit U.S. industry particularly hard as roughly 30% of all U.S. exports go to Asia. Recent slowdowns in demand for high-end U.S. goods such as aircraft, transportation equipment, heavy machinery, and high-tech products in Asia will have a long-lasting negative impact on U.S. industry.

To place the bittersweet success story of the Asian economies in its proper perspective, and to gain understanding of the growth and development dilemma, we must proceed to highlight the importance of manufacturing in the industrialization process. Discussion in Chapter 3 will center on manufacturing and trade as engines of economic growth and development.

KEY WORDS AND PHRASES

agriculture   35
backward linkages   20
barriers to regional development   24
big-five economies (of the SOUTH)   30
Brandt map   27
commercial economy   26
consumption   19
core region (or core economy)   29
core-periphery model   29
development   22
economic development   22
economic growth   21
economic profile   19
economic sectors   19
economic status   19
employment profile   31
exchange   18
fabrication   19
factors of production   18
First World   27
form utility   18
forward linkages   20
Fourth World   27
Gross Domestic Investment (GDI)   33

Gross Domestic Product (GDP)   31
Gross Domestic Product per capita (GDP per capita)   31
Gross National Product (GNP)   31
Gross National Product per capita (GNP per capita)   31
Gross World Product (GWP)   32
increasing returns to network size   29
increasing returns to scale   29
industrial agglomeration   23
industrial base   20
labor force   35
LDCs (least-developed countries)   26
linkages   20
MDCs (more-developed countries)   27
mixed economy   28
network   20
NORTH (in reference to Brandt map)   27
NORTH-SOUTH dividing line (in reference to Brandt map)   27
peripheral region (or peripheral economy)   29

periphery   29
place utility   19
population imbalances   24
primary sector   19
processing   19
production   18
regional economy   18
runaway inflation   21
secondary sector   19
Second World   27
semiperipheral economy   30
sideways linkages   20
SOUTH (in reference to Brandt map)   27
sub-Saharan Africa   35
subsistence economy   26
tertiary sector   20
Third World   27
value added   19
vertical linkages   29
world economy   18

REFERENCES

Alexander, J.W., and L.J. Gibson. *Economic Geography* (2nd edition). Englewood Cliffs, NJ: Prentice Hall (1979).

Beckerman, W. *In Defence of Economic Growth.* London: Cape (1974).

Bennett, R. "National Perspectives on Global Economic Change." In R. Bennett and R. Estall (eds.), *Global Change and Challenge: Geography for the 1990s.* New York: Routledge (1991), pp. 103–117.

Brandt, W. *North-South: Report of the Independent Commission on International Development.* London: Pan (1980).

Brown, L. *The State of the World 1996, A World Watch Institute Report on Progress toward a Sustainable Society.* New York: W.W. Norton (1996).

Brubaker, S. *To Live on Earth: Man and His Environment Perspective.* Baltimore, MD: John Hopkins (1972).

Bruton, H.J. *Principles of Economic Development.* Englewood Cliffs, NJ: Prentice Hall (1965).

Coates, B., R.J. Johnston, and P. Knox. *Geography and Inequality.* Oxford: Oxford University Press (1995).

Cohen, S.S., and J. Zysman. *Manufacturing Matters: The Myth of the Post-Industrial Economy.* New York: Basic Books (1987).

deSouza, A.R., and J.B. Foust. *World Space Economy.* Columbus, OH: Merrill (1979).

deSouza, A.R., and F.P. Stutz. *The World Economy: Resources, Location, Trade, and Development* (3rd edition). New York: Prentice Hall (1998).

Dornbusch, R., and S. Fischer. *Macroeconomics* (5th edition). New York: McGraw-Hill (1990).

Forrester, J. *World Dynamics.* Cambridge, MA: Wright-Allen Press (1971).

Friedmann, J. *Regional Development Policy: A Case Study of Venezuela.* Cambridge, MA: MIT Press (1966).

Friedmann, J. *Urbanization, Planning, and National Development.* Beverly Hills, CA: Sage (1972).

Haggett, P., and R.J. Chorley. *Network Analysis in Geography.* London: Edward Arnold (1969).

Hansen, N.M. *Growth Centres in Regional Economic Development.* New York: The Free Press (1972).

Hartshorn, T.A., and J.W. Alexander. *Economic Geography* (3rd edition). Englewood Cliffs, NJ: Prentice Hall (1988).

Hewings, G.J.D. *Regional Industrial Analysis and Development.* New York: St. Martin's Press (1977).

Hirschman, A.O. *The Strategy of Economic Development.* New Haven CT: Yale University Press (1958).

Isard, W. *Location and Space Economy.* Cambridge, MA: MIT Press (1956).

Jacobs, J. *Cities and the Wealth of Nations.* New York: Vintage Books (1984).

Johnston, R.J., D. Gregory, and D.M. Smith (eds.). *The Dictionary of Human Geography* (2nd edition). London: Blackwell (1986).

Kelly, A.C., J.G. Williamson, and R.J. Cheetham. *Dualistic Economic Development Theory and History.* Chicago: Chicago University Press (1972).

Kemp, M. *The Pure Theory of International Trade and Investment.* Englewood Cliffs, NJ: Prentice Hall (1969).

Knox, P., and J. Agnew. *The Geography of the World Economy* (2nd edition). London: Edward Arnold (1994).

Leibenstein, H. *Economic Backwardness and Economic Growth.* New York: J. Wiley (1957).

Lippit, U.D. *The Economic Development of China.* New York: M.E. Sharpe (1987).

Mabogunje, A.L. *The Development Process.* London: Hutchinson (1980).

Malecki, E.J. *Technology and Economic Development: The Dynamics of Local, Regional and National Change* (2nd edition). New York: J. Wiley & Sons (1997).

Marshall, A. *Principles of Economics.* London: Macmillan (1920).

Meier, G.M., and R.E. Baldwin. *Economic Development.* New York: J. Wiley (1957).

Morgan, T. *Economic Development: Concepts and Strategy.* New York: Harper & Row (1975).

Morrill, R.L. *The Spatial Organization of Society.* Belmont, CA: Wadsworth (1970).

Myrdal, G. *Economic Theory and Underdeveloped Regions.* London: Duckworth (1957).

Nurske, R. *Problems of Capital Formation in Underdeveloped Countries.* Oxford: Blackwell (1953).

Perloff, H., E. Dunn, E. Lampard, and R. Muth. *Regions, Resources, and Economic Growth.* Baltimore, MD: John Hopkins (1960).

Richardson, H.W. *Regional and Urban Economics.* New York: Penguin (1978).

Robinson, J. *Aspects of Development and Underdevelopment.* Cambridge, MA: Cambridge University Press (1979).

Rostow, W.W. *The Stages of Economic Growth: A Non-Communist Manifesto.* Cambridge, MA: Cambridge University Press.

Seitz, J.L. *Global Issues: An Introduction.* Cambridge, MA: Blackwell (1995).

Simpson, E.S. *The Developing World* (2nd edition). London: Longman (1994).

Smith, D.M. *Industrial Location: An Economic Geographical Analysis* (2nd edition). New York: J. Wiley & Sons (1981).

Solow, R.M. *Growth Theory: An Exposition.* New York: Oxford University Press (1970).

Todaro, M. *Economic Development in the Third World* (3rd edition). London: Longman (1985).

United Nations Environmental Program (UNEP). *Global Environmental Outlook.* New York: Oxford University Press (1997).

Watts, H.D. *Industrial Geography.* New York: J. Wiley & Sons (1987).

Wheeler, J.O., P.O. Muller, G.I. Thrall, and T.J. Fik. *Economic Geography* (3rd edition). New York: J. Wiley & Sons (1998).

World Development Report 1995: Workers in an Integrating World (published for the World Bank) New York: Oxford University Press (1995).

World Resources 1992–93 (a report by the World Resources Institute) New York: Oxford University Press (1992).

World Resources 1994–95 (a report by the World Resources Institute) New York: Oxford University Press (1994).

Classification of Economies by Income and Region, 1995

Income Group	sub-Saharan Africa (East, South)	(West)	South and East Asia	Europe and C. Asia	Middle East and N. Africa	Americas
Low-Income	Burundi	Benin	Afghanistan	Albania	Egypt, Arab	Guyana
	Comoros	Burkina	Bangladesh	Armenia	Rep.	Haiti
	Eritrea	Faso	Bhutan	Bosnia and	Yemen Rep.	Honduras
	Ethiopia	Central Afr.	Cambodia	Herzegovina		Nicaragua
	Kenya	Rep.	China	Georgia		
	Lesotho	Chad	India	Tajikistan		
	Madagascar	Côte d'Ivoire	Laos			
	Rwanda	Equatorial	Mongolia			
	Somalia	Guinea	Myanmar			
	Sudan	Guinea-	Nepal			
	Tanzania	Bissau	Pakistan			
	Uganda	Liberia	Sri Lanka			
	Zaire	Mali	Vietnam			
	Zambia	Mauritania				
	Zimbabwe	Niger				
		São Tomé				
		and Principe				
		Sierra Leone				
		Togo				
Lower-Middle Income	Angola	Cameroon	Fiji	Azerbaijan	Algeria	Belize
	Botswana	Cape Verde	Indonesia	Bulgaria	Iran	Bolivia
	Djibouti	Congo	Kiribati	Croatia	Iraq	Colombia
	Namibia	Senegal	Korea,	Czech Rep.	Jordan	Costa Rica
	Swaziland		Dem. Rep.	Kazakhstan	Lebanon	Cuba
			Maldives	Kirgiz Rep.	Morocco	Dominica
			Marshall	Latvia	Syrian	Dominican
			Islands	Lithuania	Arab, Rep.	Republic
			Micronesia	Macedonia	Tunisia	Ecuador
			N. Mariana Is.	Moldova	West Bank	El Salvador
			Papua New	Poland	and Gaza	Grenada
			Guinea	Romania		Guatemala
			Philippines	Russia Fed.		Jamaica
			Solomon Is.	Slovak Rep.		Panama
			Thailand	Turkmeni-		Paraguay
			Tonga	stan		Peru
			Vanuatu	Ukraine		St. Vincent
			West. Samoa	Uzbekistan		and the
				Yugoslavia		Grenadines
				Fed. Rep.		Suriname

Income Group	sub-Saharan Africa (East, South)	(West)	South and East Asia	Europe and C. Asia	Middle East and N. Africa	Americas
Upper-Middle Income	Mauritius Mayotte Reunion Seychelles South Africa	Gabon	American Samoa Guam Korea, Rep. Macao Malaysia New Caledonia	Belarus Estonia Gibraltar Greece Hungary Isle of Man Malta Portugal Slovenia	Bahrain Libya Oman Saudi Arabia	Antigua and Barbuda Argentina Aruba Barbados Brazil Chile French Guiana Guadaloupe Martinique Mexico Netherlands Antilles Puerto Rico St. Kitts and Nevis St. Lucia Trinidad and Tobago Uruguay Venezuela

Source: *World Development Report 1995* (World Bank), p. 248.

CHAPTER

EXPORTS AND ECONOMIC GROWTH

3

EXTERNAL GROWTH THEORY

The **globalization of production**—*the diffusion of regional production networks across international boundaries* and subsequent increase in the international exchange of money and commodities have been brought about by new technologies, improvements in transportation and telecommunications, and the rise of international investment. Technological advancements have greatly reduced barriers to moving money, raw materials, finished products, and financial assets around the globe. These changes facilitated an unprecedented growth in exports.

Although regional economic growth is highly dependent upon many forces and factors, it is difficult to challenge the contention that economic growth is greatly influenced by the income-generating potential of a region's **export base**. A regional export base is *comprised of industries where a large proportion of production is earmarked for consumers outside the region (that is, production to satisfy demand in external markets)*. Economic growth is directly related to growth of a region's export base and interregional **trade**—*the exchange of commodities between regions*.

The realization of trade potential and the mobilization of resources and production factors to take advantage of trade conditions when and where they arise are vital elements in the economic growth scenario. Given the highly variable and dynamic nature of regional economies, it is difficult to ensure longevity of a region's export base as trade is dependent upon both internal and external market conditions. Industries and regions most able to compete for a share of global markets are most likely to secure the greatest benefits of trade in the development of a regional export base. As such, regional variations in export potential and changing external market conditions are of great interest to the economic geographer.

Theory suggests that economic growth is intimately related to the rate of **capital accumulation**—*the process by which production capital is assembled and integrated in the region's production network*. As regional economies mature and accumulate capital, they become more efficient and more likely to produce surplus. Capital accumulation also increases productivity of the labor force. Increasing productivity improves a region's ability to compete with other regions. In short, economic growth occurs as regions accumulate capital and become more efficient and productive in their attempt to increase exports and produce surplus to meet demand in deficit-producing regions.

Questions arise as to the sustainability of an export base. **Sustained economic growth** *is a condition supported by the continuation of external conditions which favor exports of surplus from efficient and capital-accumulating production regions*. Export-led growth brings not only prosperity to workers and industry, but increased investment and employment opportunities for the region. The greater the number or density of production-for-export industries in a region's production network, the higher the region's propensity to sustain economic growth and the more able that region is to further integrate itself in the global economy.

These principles form the basis behind **external growth theory**—*a philosophy which maintains that interregional trade is the primary engine of economic growth.*

The interregional trade phenomenon is of particular importance to developed countries of the core and semi-periphery. These nations accounted for approximately 90% of the world's $4 trillion export base in 1996. Before proceeding to highlight some empirical evidence that supports external growth theory and the importance of establishing an export base, let us first examine how regional economic growth is affected by various types of transactions.

TRANSACTIONS IN A REGIONAL ECONOMY: AN ECONOMIC BASE ANALYSIS

There are four fundamental types of transactions that affect regional economic and employment growth. These transactions are illustrated in Figure 3.1.

1. Region-Forming Transactions   EXPORT

Region-forming transactions are *flows associated with export-oriented activities.* They involve exports of goods and services in exchange for money. Simply put, commodities move out of a region, and dollars come into the region. By satisfying demand that is external to a region, region-forming transactions supply the fuel which drives regional economic growth—hard currency. Export activities are region forming in the sense that exports of goods and services provide money income to support **industrialization**—*the expansion of an existing industrial base and/or the formation and growth of new industry in a given region.*

2. Region-Serving Transactions NON-EXPORT

Region-serving transactions are *internalized flows associated with non-export-oriented activities.* This type of transaction allows money to recirculate through a regional economy and the economy to grow as a result of that recirculation. Region-serving transactions are associated with internal production to meet internal demand. The recirculation of money (and wages) from demand-related exchange helps promote economic growth and expansion of the employment base. The larger the internalized production network, the greater the recirculation potential of dollars in the regional economy. A large production network also translates into greater diversity in production. The greater the diversity of end products produced within the region, the more likely that dollars are to "stay home" (that is, remain within the region as opposed to flow out of the region) as dollars are ex-

changed for goods and services. The larger and the more diverse a region's production network, the greater that region's ability to retain value added. Note that non-export-oriented activities are region serving in the sense that they support the growth of industries which produce to satisfy internal demand. In other words, region-serving transactions are associated with production activities that serve the needs of consumers and producers within the regional economy.

3. Income Transfers $ internal - ∅ expt related

Income transfers, also known as *transfer payments and direct monetary transfers,* are *inflows of money income, savings, investment capital, and/or other public or private transfers that are not associated with export-related activities.*

Income transfers at the regional level include such things as social security payments, retirement benefits, unemployment insurance, welfare benefits, medicare and medicaid payments, as well as interest, dividends, and rents paid on investments. Examples of transfers at the national level include income and profits which filter back to a company or nation from investments made abroad. Income transfers also include economic development loans, foreign aid, or monetary assistance from sources outside a region or nation. Money transfers may be from public or private sources or lending institutions. Direct transfers of financial capital across international boundaries for investment purposes generally fall under this heading, although more commonly referred to as foreign direct investment (FDI). Note that development loans ultimately require repayment and a loss of income due to interest paid on those loans. Hence, development loans may be viewed as a short-term income transfer.

Income transfers represent the inflow of money to a regional economy without a complementary outflow (with the exception of loans). Thus, income transfers generally have a positive impact on a regional economy. Income transfers also provide a great stimulus to a region's employment base as dollars are received and recirculated. Note that the recirculation of transfer income is typically associated with the expansion of the tertiary sector, particularly, retail sales and services. Consider the impact such transfer payments have had on the sales- and service-oriented economy of Florida. Income transfers have played a pivotal role in the economic and population growth Florida experienced in the 1970s and 1980s. In part, the economic boom was credited to income transfers associated with a large in-migrating population (composed of elderly retirees, seasonal migrants, tourists, and opportunity-seeking members of the labor force). The movement and relocation of individuals to Florida was propelled by a thriving tourist trade and the amenity-driven Snowbelt-to-Sunbelt migration trend in which large segments of the U.S. population moved out of the

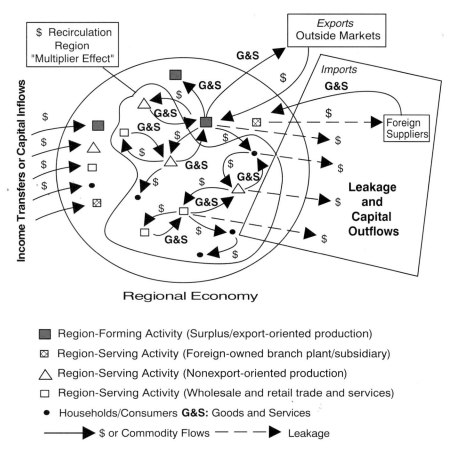

Figure 3.1 The regional economic base model: flows and transactions.

cold northern states to warmer places in the south. The influx of money from retirees and tourists fueled the remarkable growth of Florida's cities and coastal areas. Note that by the mid-1980s, roughly 50% of Florida's nonagricultural income was derived from income transfers.

4. Leakage Outflow of publpri $

Leakage is defined as *the outflow of money, investment capital, and/or other public or private monetary transfers from a region; (lost capital due to perceptions of abnormal risk at home or lost income from purchases made outside the region).* In short, leakage refers to money leaking out of a regional economy. Money income can be lost from investments made abroad or from consumers making purchases of goods and services from firms located outside the regional economy.

Leakage may bring about and accelerate the general decline of domestic industry. As internal demand and sales revenue fall and as a regional economy restructures in response to decreasing demand and revenue, a regional economy is likely to experience the erosion of its industrial base and a loss of jobs. The physical uprooting, exodus, and/or relocation of industry and the subsequent loss

of production and income can have devastating long-term impacts. Capital and money outflows limit the future expansion potential of a regional economy as lost earnings and income are not able to recirculate and generate new employment opportunities.

The capital flight phenomenon represents a significant form of leakage to most peripheral economies. **Capital flight** is the flight of capital (generally from an underdeveloped region to one or more developed regions). It is *the loss of domestic savings and/or investment capital to outside regions or nations from expectations of higher and more stable rates of return abroad*—a form of leakage that continues to plague much of the peripheral nations of the world. The *flight of human capital,* sometimes referred to as **brain drain**, is *a nonmonetary form of leakage associated with the exodus of highly skilled portions of the labor force.* As labor redistributes itself, moving toward regions with greater perceived earnings potential, the future income-generating potential of the region losing labor is greatly reduced. Brain drain is a subtle and destructive form of leakage from a human development standpoint.

Leakage may also come about from trade deficits. A **trade deficit** is *a situation which occurs when the overall value of imported goods and services exceeds the overall* *value of exported goods and services in a given region in a given time period.* In short, trade deficits represent a net loss of money income from the uneven exchange of goods and services. Many trade deficits are a direct result of trade imbalances.

A **trade imbalance** may be defined as *a trade deficit of a bilateral nature.* Note that the term bilateral, as it is used in this case, is in reference to *two trading regions or partners.* Trade imbalances are commonly brought about by irregular trade policies set in motion by a single trading partner to protect domestic industry. Trade imbalances may also be brought about by exchange rates which continually favor the sales of imports over domestically produced goods and services. Consider, for example, how the devaluation of the Mexican peso in 1995 created a situation that favored the flow of U.S. dollars to Mexico and the flow of Mexican goods to the United States. Peso devaluation (and a relatively strong U.S. dollar) made American goods and services more expensive to Mexican consumers at a time when the U.S. demand for imports from Mexico were on the rise. As a result, the 1995 U.S. bilateral trade deficit with Mexico reached a record level of over $17 billion. Though radical fluctuations in exchange rates are typically a short-term occurrence, the effects may be far reaching. This is particularly true if a significant amount of leakage occurs.

LEAKAGE AND THE DEINDUSTRIALIZATION OF THE U.S. ECONOMY: TRADE AND POLICY IMPLICATIONS

Ongoing trade deficits and regional trade imbalances are perhaps the most inconspicuous and devastating forms of leakage. An overall trade deficit (occurring when the total value of imports exceeds the total value of exports over a period of time) can be brought about by one or more bilateral trade deficits. Consider the case of the U.S. economy, posting large trade deficits from 1988 to 1998 as shown in Table 3.1. Trade deficits of this magnitude promote not only the loss of jobs and money income, but they also contribute to the noncreation of jobs in secondary and tertiary sectors of the economy. The nagging U.S. trade deficit has been attributed to consistently large trade imbalances with powerful trading partners like Japan.

Consider that the U.S. economy, which produced roughly 22% of world output, posted a $108 billion trade deficit in 1994. Note that 60% of that deficit was due to a trade imbalance with Japan. By contrast, Japan (an economy which accounted for 8.2% of world output) boasted a $121 billion trade surplus in that same year. Roughly 97% of Japanese exports are high-valued manufactured products: machinery and parts, motor vehicles, and consumer electronics, goods which account for the majority of Japan's exports. Of the $396 billion in exports from

TABLE 3.1	U.S. Trade Deficit Statistics (all goods and services)

Year	U.S. Trade Deficit (in billions $)
1988	115
1989	90
1990	78
1991	28
1992	40
1993	76
1994	108
1995	110
1996	115
1997	125
1998	150
1999	160 projected

Source: *U.S. Global Trade Outlook 1995–2000* (Department of Commerce), U.S. Government Printing Office, Washington D.C., pp. 200–201 (1995), CIA Factbook (on-line).

Note: Figures expressed in 1990 U.S. dollars.

Japan in 1994, approximately 30% (or $120 billion worth) were earmarked for the U.S. market. This stands in stark contrast to the $53.5 billion worth of goods (from the $513 billion in U.S. exports) sold by U.S. firms in the Japanese market.

Note that the U.S. trade deficit for 1997 was estimated at a whopping $115 billion. This deficit is largely accounted for by excessively large bilateral trade deficits with China and Japan. In the late 1990s, China exported (on average) about $40 billion more in goods to the United States than it imported from the United States on a yearly basis. Note also that the U.S. trade deficits of the 1990s are largely accounted for by the uneven exchange of manufactured goods.

Table 3.2 contains U.S. manufacturing trade deficit statistics. The manufacturing trade deficit is the largest contributor to the overall U.S. trade deficit. Astonishingly, the $126 billion deficit in manufactured goods trade was larger than the overall trade deficit in 1994. Small trade surpluses in other sectors helped to reduce the amount of leakage associated with the loss of value added in this high-value-added sector. Although trade imbalances have been observed between the United States and several of its top-ten trading partners during the 1990s, including Canada, Mexico, Germany, and Taiwan, these imbalances accounted for only a small percentage of the overall U.S. trade deficit. The economic losses are small when compared to the sizeable and increasing bilateral trade deficits with Japan and China.

According to the U.S. Department of Commerce's Office of Trade and Economic Analysis, there are two problem manufacturing-for-export sectors in the U.S.

TABLE 3.2	U.S. Manufactured Goods Trade Deficit Statistics for Selected Years (1970-1997) with Forecasts

Year	Exports	Imports (in billions of U.S. $)	Balance
1970	31.7	27.3	+ 4.4
1975	76.9	54.0	+ 22.9
1980	160.7	133.0	+ 27.7
1985	168.0	257.5	–89.5
1986	179.8	296.7	–116.8
1987	199.9	324.4	–124.6
1988	255.6	361.4	–105.7
1989	287.0	379.4	–92.4
1990	315.4	388.8	–73.5
1991	345.1	392.4	–47.3
1992	368.5	434.3	–65.9
1993	388.7	479.9	–91.2
1994	431.0	557.9	–126.8
1995	450.0	600.0	–150.0
—	—	—	—
1997	510.0	670.0	–170.0*

Source: *U.S. Global Trade Outlook 2000,* U.S. Department of Commerce (1995).
*Forecast based on 20-year time-series trend.

economy. The trade-related leakage associated with the motor vehicles and parts (where imports exceeded exports by over $46 billion in 1994, up from $43 billion in 1993) and textiles and apparel (where imports exceed exports by over $42 billion in 1994, up from $39 billion in 1993) are excessively large. Over two-thirds of the motor vehicle and parts deficit is with Japan, whereas Asian nations such as China and India account for a growing proportion of the textiles and apparel trade deficit. Deficits in these two sectors alone account for 80% of the overall U.S. manufacturing trade deficit. If trends continue, it is expected that these sector-specific deficits will each exceed the $50 billion mark by the turn of the century.

As the United States continues to open its borders to foreign products in the spirit of promoting free trade and global competition, larger trade deficits could be expected. This is especially true as the Japanese and the Chinese continue to support practices and policies that perpetuate their own trade advantage at the expense of the U.S. economy.

The Japanese and Chinese continue to shelter their industries from competition. Both nations seek to sustain trade imbalances through the imposition of trade barriers, import restrictions and import taxation, and their inclination to support and patronize firms from their respective homelands. The Japanese are particularly firm in the belief that benefits of value added and trade should be kept within their own globally expanding production network. With limited access to the Japan's growing market

and the nationalistic tendencies of Japanese consumers and producers, American firms have found it most difficult to sell their products in Japan.

Ironically, the Japanese consumers are most likely to suffer as less competition keeps price levels deceptively high (adding to increases in the cost of living) and limiting the quality and variety of products available to them. Note that foreign competition, particularly from Japanese producers, did wonders to improve the quality of U.S. automobiles once American consumers began to purchase foreign models in the early 1980s.

The switch to fuel-efficient imports was largely in response to (a) the poor quality of automobiles produced in the 1970s by America's Big-Three automakers—General Motors, Ford, and Chrysler; and (b) consumers' growing preference for fuel economy when high rates of inflation, high unemployment levels, and the perceptions of an energy crisis loomed overhead. Unlike the early years of the industry, Big-Three automakers were forced to make drastic changes in the way they designed and produced automobiles as they faced the growing deluge of inexpensive high-quality imports. Today, American automobiles and trucks are the finest on the road (and price competitive); a situation undoubtedly brought about by the global auto trade and the intense competition from Japanese, German, and Korean auto manufacturers.

As part of an overall trade and development strategy, the Japanese have also been very active in "dumping" products to external markets. **Dumping** is *the selling of goods in a given geographic market at prices which fall below comparable goods sold in the exporter's home market or at prices that are actually lower than the cost of production and delivery.* Japan was regularly accused of dumping steel, autos, and electronic devices in the U.S. market throughout the 1980s. Their goal was to gain noticeable entry and capture a significant share of the world's premier market for consumer goods.

Nations that engage in dumping, do so with the intent of (1) capturing a greater or larger-than-competitive share of a given market; (2) artificially sustaining non-competitive industries in an attempt to stabilize industrial output and increase the growth of their domestic economy (an action that is typically accompanied by government subsidies of targeted industries); and (3) promoting bilateral trade imbalances with selected trading partners to maximize the benefits and growth potential of dollars brought in by region-forming activities while simultaneously encouraging a breakdown of the production network in the region where dumping is taking place.

The U.S. trade predicament is something that is unlikely to change without a firm commitment on the part of government leaders to adopt more aggressive trade policies aimed at leveling the playing field. Many analysts cite the unfair trade practices of Japan and China as a major stumbling block for the U.S. economy. It is a recognition that nations must promote "free and fair trade"—*trade on equal terms*—to minimize leakage and imbalance-related job losses in affected industries. This may prove difficult with Japan's record of advocating one-sided trade practices and policies, which encourages Japanese consumers and producers to buy Japanese products and America's ongoing fascination with Japanese automobiles. Minimizing leakage has also been challenged by America's insatiable appetite for inexpensive clothing, footwear, and consumer electronics from Asia.

Furthermore, ongoing trade deficits and imbalances have not only restricted the flow of U.S. exports but have also contributed greatly to the steady erosion and stagnation of industries supported by export-oriented income. In particular, losses in high-value-added sectors have delivered a striking blow to the U.S. economy. Major manufacturing network linkages have been dissolved in steel and auto production and textiles, domestic production capacity has been diminished, and region-serving sectors have been constrained in terms of growth. Note that a large number of American jobs have been permanently lost in industries directly *and* indirectly affected by leakage.

At the expense of American jobs, Japan has enjoyed a surplus in the exchange of automobiles and auto parts throughout the 1980s and early 1990s and a steady surge in auto-related revenues and profits. Automobile-related trade between the United States and Japan accounted for approximately two-thirds of the bilateral trade deficit from 1980 to 1995. Consider the fact that Japan's share of the U.S. auto market has increased from roughly 15% in the early 1980s to the 25% mark by the mid-1990s. The loss of domestic market share by U.S. producers translated into a loss of profits greater than $6 billion for America's Big-Three automakers in 1995 alone, not to mention related losses in steel and fabricated metals sectors. Industries that were once considered to be the very backbone of the U.S. economy are now struggling to survive.

The erosion of the U.S. industrial base has been identified with the process of **deindustrialization**—*a sustained or marked decline in a region's traditional manufacturing and export sectors due to restructuring and/or the loss of competitiveness.*

American consumers have unknowingly contributed to further losses and leakage by buying (and justifying the purchases of) "made-in-America" Japanese vehicles. Note that the majority of the parts which go into the assembly of American-made Hondas, Toyotas, and Nissans actually originate in Japan. As a larger share of value added is associated with the production of parts in comparison to the assembly of parts, Japanese auto manufacturers and parts producers continue to benefit at the expense of the American economy. Value added is retained in the Japanese production network, and value added is lost in the U.S. production network. In addition, as the aging fleet of Japanese cars and trucks in America

mature, additional imported parts will be required to service and maintain those vehicles. The second flurry of imported parts will serve only to enlarge this sector-specific trade deficit.

Trade statistics from the U.S. Department of Commerce show that the United States imported $14.3 billion in automobile parts in 1994, up from less than $10 billion in 1990, a figure that has increased substantially since the mid-1980s after the invasion of Japanese automobile manufacturing plants. In 1994, the U.S.-Japanese-auto-parts trade deficit stood at $13 billion, up from $9 billion in 1991. It is a deficit that is expected to rise toward the $20 billion mark by the year 2000. To reverse this disturbing trend, the United States must adopt a multidimensional trade strategy to help U.S. producers gain access to restricted geographic markets (like Japan's automobile and automotive parts market), while ensuring that outside competitors engage in free and fair trade practices. As such, trade policies must be formulated on a sector- and product-specific basis, imposing "quid pro quo" (that is, something in return) measures to counteract existing trade imbalances.

The proliferation of Japanese and other foreign-owned manufacturing plants on American soil (or "transplants" as they are sometimes called) is undeniably a by-product of Americans' voracious appetite for foreign goods. The potential to reduce trade-related leakage by "managed trade"—through import and export control—has been greatly diminished as transplant-driven leakage is a self-reinforcing phenomenon. In 1990, foreign-owned manufacturing plants accounted for 13% of U.S. manufacturing value added, up from 10% during the mid-1980s. Today, foreign-owned production facilities accounted for over 15% of value added in U.S. manufacturing sectors. Note that the majority of these foreign-owned facilities rely on imports as production inputs, a condition that furthers transplant-driven leakage. As in the case of automobiles, commodities produced by those facilities and sold in the U.S. economy will rely on imported replacements, parts, and service. In short, two decades of buying foreign goods will sustain and possibly increase leakage due to imports.

Many critics of managed trade are quick to challenge this point, arguing that even U.S. firms import many of their production inputs. While it is true that globalization of production has meant a greater reliance on outside suppliers in virtually all regions and sectors, large geographic variations do exist in the dependence on foreign parts. Note that on average, foreign firms operating in the United States tend to import more than twice as much per worker than do American firms. Japanese firms in the United States, however, tend to import nearly three times as much as the average foreign firm. Certainly, this is a testimony to Japan's unwillingness to support American industry and job growth (Krugman, 1994).

While Japanese production in the United States has led to new jobs in localities absorbing Japanese transplants, these jobs have displaced similar jobs elsewhere in the U.S. economy. For every job displaced by a Japanese automobile assembly plant in the United States, there is at least one additional job lost in other sectors from the loss of value added in parts production. The very fact that Japanese producers acquire the majority of parts and production inputs from other Japanese-owned manufacturers suggests that job losses due to trade imbalances and transplants will continue. From a bilateral standpoint, one may argue that an imbalance in investment flows and the imbalance in market access constitute a hidden source of leakage. The lack of U.S. investment in Japan, largely due to Japan's unwillingness to absorb foreign investment, is yet another contributing factor in America's deindustrialization. From the U.S. perspective, the investment gap has worsened the trade deficit. Consider the fact that well over two-thirds of U.S. imports from Japan are associated with shipments from Japanese corporate parents to Japanese-owned subsidiaries (that is, transplants) in the United States.

Many regional economic analysts agree that persistent bilateral trade deficits are outcomes of protectionist trade policies and not something which is simply due to fluctuations in exchange rates. Consider that the U.S. dollar has been relatively weak over the last decade in comparison to the Japanese yen, a situation that would under normal circumstances make U.S. products more attractive in overseas markets and especially in Japan under a free and fair trade environment. The nationalistic tendencies of the Japanese, along with the inability of U.S. suppliers to penetrate the Japanese market, have contributed to an international trade pattern that goes against expectations and logic. Furthermore, one-sided trade policies and practices are detrimental to all nations of the world economy as they promote network inefficiencies, lessen competition amongst industry and regions, and create a biased distribution of the benefits of trade.

One possible solution to counter the effects of excessively large bilateral trade deficits would involve the imposition of tariffs—taxes—on imports. Tariffs could be levied against imported products in sectors responsible for creating the trade imbalance (for example, manufactured goods). Provisional measures, such as those recently proposed during the Clinton administration (for example, import taxes on luxury cars from Japan), are simply not enough to reverse the tide. Initiating an all-out trade war with a nation like Japan, however, is problematic in the sense that it could lead to problems elsewhere (for example, losses in export income should Japan retaliate and effectively prohibit American goods from crossing its borders). Note that a trade war with Japan could prove costly as the Japanese continue to lend enormous sums of money to the American government to help pay the interest amassing on the U.S. national debt of over $6 trillion.

Trade negotiations with Japan continue to focus on trade barriers and impediments to market access while the import deluge continues to worsen trade-related leakage. Leakage has caused a slowdown in the growth of U.S. economy and is perhaps the leading source of missed employment and investment opportunities in U.S. manufacturing and services sectors. Imports as a share of U.S. domestic demand are reaching record-high levels. In the mid-1970s, imports of nonpetroleum-based goods accounted for less than 10% of all sales in the United States. By 1989, imported goods accounted for roughly 20% of U.S. sales. In the late 1990s, imports now exceed 28% of the sale of nonpetroleum-based goods in the U.S. economy.

Despite overwhelming evidence of bilateral trade problems, results of promotional campaigns aimed at the American consumer, such as "Buy American . . . Your Job May Depend on It," have been largely ineffective. American consumers continue to show little regard (or assume any responsibility) for the problems facing their economy. U.S. consumers tend to be complacent about matters of economy and are far less supportive of domestic industries in comparison to most other nations in the world. Despite the fact that American jobs are being lost or displaced at an alarming rate due to import penetration, deindustrialization, trade deficits and imbalances in high-vale-added sectors, not to mention a national debt that is approximately six times the total debt of all underdeveloped nations, most Americans seem to be under the impression that the late 1990s U.S. economy is strong and stable. U.S. government leaders continue to seek temporary solutions to trade problems, settling on reactive rather than proactive strategies to combat leakage.

Consider the fact that China continues to impose heavy import tariffs on U.S. goods crossing its border (typically raising the price of U.S. imports by 15% to 40%). The U.S. government has done relatively little to reverse this one-sided trade advantage. While some critics have suggested revoking China's "most-favored-nation" status, engaging in an all-out trade war with China may prove futile. Plagued by persistent bilateral trade deficits and discriminatory taxation, U.S. policy makers could find themselves contemplating a retaliation-in-kind approach, including the imposition of identical tariffs on imported goods from regions engaging in one-sided trade. The adoption of such drastic measures is unlikely as such action would increase the cost of imported goods in the United States and reduce consumer purchasing power. Tax increases of the last few decades have already reduced the disposable incomes of most Americans, making further erosion of consumer income a very unpopular idea.

Furthermore, protectionist measures instituted by individual nations may counter the actions of the newly established **World Trade Organization (WTO)**—*an internationally regulated authority to oversee international trade matters and promote free trade worldwide.*

Although it is doubtful that the WTO will be any more effective in dealing with (or solving) bilateral trade problems than individual nations, the existence of the WTO makes it unlikely for nations like the United States to engage in controversial measures that would rock the boat. Enforcing WTO directives to counteract trade problems would also prove difficult. Penalties for noncompliance are likely to take a back seat to diplomacy as the WTO attempts to promote cooperation amongst the world's governments and trading regions.

While the U.S. trade predicament requires action on the national and international economic policy fronts, it is doubtful that the U.S. trade deficit will be a big priority to the WTO, as it is of little concern to the international community. Nevertheless, America's ongoing trade problems have caused irreversible damage to its economy and diminished its future economic development potential. It is a problem that does have global ramifications. As a production and export leader in the world economy, the United States maintains multiple roles as a propagator of innovation and technology and as a significant source of investment capital. Hence, one could argue that reducing the economic growth and development potential of the United States, could result in reducing the growth and development potential of the global economy.

Many experts in regional development argue that international trade policy cannot be viewed as separate from industrial policy. Decisions made by one nation regarding trade can ultimately affect all regions in the globalizing economy. Responsibility rests with national governments to prescribe and implement trade policies that will bring about resolutions that are fair to all parties involved. Nevertheless, the U.S. economy continues to suffer from two decades of irresponsible trade policies and the long-standing impact of deindustrialization. As discussed by Markusen (1993),

> Job loss, and the more invisible foregone production and employment owed to a yielding of market share in internationally expanding industries, is a thoroughly political phenomenon. . . . Over and over, the case studies show how U.S. government practices shape, hinder, or nurture industries. Ineffective trade restraints in the face of grossly "unfair", or at least different, trade practices abroad have contributed to severe losses in U.S. steel and autos. (p. 298)

American workers and industries have suffered irreversible losses, but not necessarily from a lack of competitiveness in most circumstances, but from misguided national economic policy. Many trade analysts feel that supporting free trade (citing the assumed gains of economic growth from exchange and integration) in the absence of a clear-cut industrial policy or national industrial development agenda is something which is not in the best interest of America, nor is it in the best interest of the international community. It is unclear, however, if policy

makers can formulate measures to level the playing field when so little opposition to the current system exists by the power-elite in the political, commercial, and financial sectors of the world economy. The challenge is great as U.S. policy makers must forge reindustrialization strategies to confront deindustrialization, while juggling the complex set of domestic and foreign policy instruments that are increasingly sensitized to trade issues. This effort may be increasingly difficult in a world economy where production is becoming more and more regional interdependent and **multinucleated**—*spread about multiple centers, zones, or regions.*

Trade revisionists have opted to back managed trade policies in order to

> slow the pace of disruptive change and complement a broader industrial policy. . . . in essence, national economic development strategies where each country would anticipate future sectoral specializations and seek to enhance job creation and stabilization. The revisionists, . . . favor targeting industries on the basis of growth potential, job preservation, and strategic importance on economic or foreign policy grounds. (Markusen, 1993, p. 300)

Yet, Markusen notes that it is unlikely that the U.S. trade policy—one so adamantly committed to free trade and opposed to industrial policy—will ever truly succeed in stabilizing the employment or reindustrializing America. Globalization of production has promoted a **new international division of labor**—*the division and resulting specialization of work and production tasks between laborers, economies, and geographic regions.* As such, trade and development policy formulation may benefit greatly from an international dialogue of how jointly managed trade, in line with the industrialization and reindustrialization agendas of rival production regions, can lead to greater regional and global economic stability. A far-too-rigid free-trade ideology may backfire without attention toward the complex issues concerning division of labor and competing industrialization.

Note that sector- or product-specific trade issues are also of great concern to policy analysts, as changes in the world economy have created new opportunities and markets for products and technology. For example, the growth of information service industries and exports worldwide has been astounding. Despite trends in the manufactured goods trade, it is expected that the U.S. trade deficit will lessen as globally integrated production regimes and international information networks begin to dominate the world economy. The competitiveness of U.S. industry in information-related sectors is likely to remain strong in markets where the demand for information and information services are skyrocketing.

The growth of these industries is likely to bring about increasing productivity in the workplace as electronic commerce—*the use of computer networks for accessing information and data for the purposes of carrying out research or business transactions*—becomes increasingly more commonplace and necessary as a tool of convenience and business. Undoubtedly, the rising demand for information and information services and increasing activity on the world's information superhighway will result in the United States securing a leadership position in the information and communications industries.

The phenomenal growth of **information technology (IT)**—*innovations and high-tech products for information and data transfer*—and **information services (IS)**—*enterprises involved in the collection, management, distribution, and sales of information and data to consumers or business*—has been largely supported by companies seeking to gain a competitive edge utilizing information about changing market conditions, consumer preferences, and sales potential of new products and technologies. The era of IT and IS has not only brought about a wave of new and exciting digital and advanced information processing and telecommunications devices, as well as a host of innovations in microelectronics and information transmission (including the use of satellites and fiber-optic cables), but it has changed the very nature in which firms do business and the way in which people interact. Information and information flows have become vital to the development process as firms and industry look for a competitive edge in regional and global markets. The information and telecommunications technology revolution has been championed by the core economies of the NORTH.

Recently, the growth rate of information service exports has exceeded the growth rate of merchandise trade (something that is likely to continue as core economies embark on their journeys into the information age). The U.S. Department of Commerce reports that worldwide, information service exports grew an astonishing 24.8% from 1989 through 1993, while the growth rate of merchandise trade for that same period was only 6.1%. World expenditures for information services are expected to grow at a rate exceeding 12% per year over the next several years.

By 1994, the U.S. dominated the global information service market, capturing 46% of the world's information service exports. Professional and producer services accounted for the largest share of information service exports. The United States is also the world's largest producer of electronic databases by roughly a two-to-one margin. Note, however, that the so-called global market for information services has been largely associated with places that maintain the largest production capacity (that is, the world's core economies). It is also interesting to note that well over 75% of the world's information service market resides in just two geographic regions: the United States and Europe (Table 3.3).

TABLE 3.3	**World Information Services (by Region) Share of World Market in 1994**

Region	Percent of World Market ($282 billion)
United States	46.0
Europe	32.5
Asia/Pacific Islands	18.1
Canada	1.5
Latin America	1.5
Middle East	0.4
Africa	0.0

Source: U.S. Department of Commerce (1995).

REGIONAL TRANSACTIONS AND THE MULTIPLIER EFFECT

Regardless of the source of export income, whether through the exportation of agricultural commodities or manufactured goods or information services, regional economic growth is a direct outcome of the expansion-related impacts of economic transactions and the competitiveness of a region's production network. For example, if exports are high (in terms of value) and there is a high degree of vertical integration within the region, the regional economy will expand rapidly provided that leakage is minimal. *The expansion of the industrial or employment base from the recirculation of money, wages, and income is known as the* **multiplier effect**.

Export-oriented industries bring money into a regional economy. Incoming dollars support a myriad of industries within the region, which in turn pay wages and buy production inputs from suppliers within the region. These transactions allow dollars to circulate and recirculate as export-oriented industries support local industries and local industries support other local industries. Thus, the production of an exportable commodity in a given region supports not only the workers in that industry but workers in other sectors, as producer services are demanded and wages are paid to workers who spend their earnings on goods and services to sustain them and their families. All along the way, dollars are exchanged and recirculated and jobs are created in association with production to satisfy the region's total internal demand for goods (food, clothing, furniture, etc.) and services (accounting, lawn maintenance, auto repair, haircuts, dry cleaners, etc.).

Note that the overall multiplier effect may be diminished if the region imports production inputs or imports more goods and services than it exports (in terms of value). Note also that the loss of export-oriented income could lead to a **reverse multiplier effect**—*the contraction of a regional economy in response to the loss of* *export-oriented jobs and additional job losses in sectors reliant upon the recirculation of export income.* Sluggish regional economic growth is often the result of either the loss of export-oriented income or the inability of the region to sustain its industrial and export bases.

The total overall economic gain or loss to a regional economy can be approximated by estimating a region's **multiplier**—*a number which describes a region's expansion or contraction potential should its export base be increased or decreased.* Multipliers are generally expressed as either employment multipliers or income multipliers. Let us examine each in turn.

Employment Multipliers

TE- Total Employment
BE - Basic Employment
NBE- Non-Ba Employment

The economic growth potential of a region may be quantified by examining a region's employment base and its propensity to engage in production-for-export activities. **Employment multipliers** *highlight the relationship between a region's total employment (TE) and the degree to which total employment is defined as either "basic" employment (BE) and "nonbasic" employment (NBE).*

Basic employment refers to employment in region-forming or export-oriented activities, employment that is basic to the sustainability of the regional economy. Any and all employment related to production-for-export activities would be classified as basic employment. By contrast, nonbasic employment refers to employment associated with region-serving or nonexport-oriented activities. Nonbasic employment includes such things as consumer and producer services, retail trade, government and public administration, and other tertiary sector activities.

Total employment may be defined in terms of its basic and nonbasic components; namely,

$$TE = BE + NBE. \qquad (1)$$

A region's economic base multiplier (m) describes the expansion potential of basic employment and may be defined as

$$TE = m * BE. \qquad (2)$$

By rearranging terms, a crude estimate of a regional economic base multiplier may be obtained by isolating m, where

$$m = TE/BE$$
$$= [BE + NBE]/BE. \qquad (3)$$

Hence, we may express the regional economic base multiplier as

$$m = BE/BE + NBE/BE$$
$$= 1 + NBE/BE \qquad (4)$$

The ratio NBE/BE describes the expansion potential of additional basic employment. It is assumed that each new basic job added to the regional economy will produce a gain equal to the basic job added plus additional number of jobs as defined by the ratio NBE/BE.

For example, consider a regional economy where total employment TE = 250,000 as defined in terms of full-time equivalent (FTE) employment—full-time year-round jobs. Assume that the number of full-time workers employed in export-oriented activities is 100,000 (that is, one out of every 2.5 workers is employed, on average, in a production-for-export operation). It follows that nonbasic employment is NBE = 150,000. Using equation (3), an estimate of the regional economic base multiplier

$$m = 250,000/100,000 = 2.5.$$

Alternatively, the regional economic base multiplier may be calculated from equation (4), where

$$m = 1 + 150,000/100,000 = 2.5.$$

The interpretation of a multiplier of 2.5 is simple and straightforward: Every one basic job (on average) supports 1.5 nonbasic jobs in the regional economy. Thus, for every one basic job added to the regional economy, 1.5 nonbasic jobs are likely to be created on average.

Note that there are several caveats. First, the estimate of a regional economic base multiplier may be inflated due to the impact of income transfers. Note that income transfers also contribute to an increase in the amount of nonbasic employment in a regional economy. Total employment may, thus, be broken down into basic employment BE, linked nonbasic employment LNBE (that is, employment supported by basic employment), and income-transfer-related nonbasic employment (ITRNBE)—jobs supported strictly by income transfers. Subsequently, nonbasic employment may be expressed as NBE = LNBE + ITRNBE, thereby allowing us to rewrite total employment as

$$TE = BE + LNBE + ITRNBE. \qquad (5)$$

Using this approach, a more precise estimate of the regional economic base multiplier m^* would involve calculation of an adjusted regional economic base multiplier of the form

$$m^* = 1 + LNBE/BE. \qquad (6)$$

Note that as LNBE < NBE, the ratio LNBE/BE will be less than NBE/BE, and thus $m^* < m$.

Suppose that it is estimated that 25,000 (full-time equivalent) nonbasic jobs can be attributed to income transfers, and that ITRNBE = 25,000. Hence, linked nonbasic employment LNBE is defined as NBE less ITRNBE. In this case, LNBE = 150,000 − 25,000 = 125,000. The adjusted regional economic base multiplier would then be

$$m^* = 1 + 125,000/100,000 = 2.25.$$

The interpretation of this multiplier is straightforward: Every 1 basic job added to the regional economy is likely to support an additional 1.25 linked nonbasic jobs (accounting for the effects of income transfers). Conversely, one may utilize this framework to estimate the impact of job losses—the negative multiplier effect—estimating the total number of job losses due to the loss of basic jobs. For example, a multiplier of 2.25 would indicate that for every 1 basic job lost in a regional economy, 1.25 nonbasic jobs are lost in linked and supporting sectors. Hence, export-oriented job losses due to trade-related leakage have a much greater (negative) impact of a regional economy than just the loss of the jobs in question. For the loss of basic jobs also means the loss of jobs in linked nonbasic sectors.

Alternatively, we may express equation (2) as follows:

$$TE = [1/(1 - \mathbf{a})] * BE \qquad (7)$$

where $\mathbf{a}$ is the parameter that describes the underlying functional relationship between basic (export-oriented) employment and nonbasic employment. In simple terms, the parameter $\mathbf{a}$ is the proportion of total employment that is nonbasic; that is

$$\mathbf{a} = \text{NBE/TE (or LNBE/TE if accounting}$$
for the impact of income transfers).

Hence, the regional employment multiplier may be written as

$$m = [1/(1 - \mathbf{a})]. \qquad (8)$$

The overall change in a regional economy from expansion of export-oriented employment is part of the process of *circular and cumulative growth* (Pred, 1966). The entry of new export-oriented industry has an "initial multiplier effect" as it both establishes new linkages and reinforces already established linkages in the regional production network. As the network matures and flourishes, it brings about additional investment in infrastructure, increased efficiencies in transportation, communication, and the distribution of producer services, increased public expenditures to support industrial growth, and a boom in wholesale and retail trade.

An expanding industrial base enhances the possibilities of invention and innovation. In addition, regional economic growth will promote efficiency gains as the size and scope of the production network increases and as the regional economy diversifies. This growth will also intensify competition amongst rivals and increase cooperation amongst business and government. The economic gains and benefits associated with these changes may take time to realize, as the lag between the creation or entry of new industry and the hidden benefits of growth from additional investment in infrastructure, increased efficiency and competitiveness, etc. combine to produce a subtle yet important "secondary multiplier effect."

Secondary impacts notwithstanding, the economic base multiplier provides a framework for assessing total job losses due to leakage. For example, consider that roughly 10% of the value added in production is returned to labor in the form of wages (the remaining 90% is accounted for by the cost of raw materials, capital and other production inputs, transportation cost, retail mark-up, etc.). Thus, for every $100 billion in trade-related leakage, a region would lose the equivalent of 300,000 export-oriented jobs (assuming that the average job paid approximately $33,000 per year). A regional economic base multiplier estimate of 2.5 would mean that the loss of 300,000 basic jobs would translate into a loss of an additional 450,000 in nonbasic sectors.

A U.S. merchandise trade deficit of $170 billion is a figure that translates roughly into $17 billion in lost wages—a loss of approximately 480,000 jobs in basic (export-oriented) industries. Using a multiplier estimate of 2.5, this would mean that 1.2 million total jobs were lost as a result of a merchandise trade deficit of this magnitude on an annual basis. Note, however, that the total economic losses associated with this negative multiplier effect is much greater than just lost jobs and wages. Consider that the government would have to support unemployed individuals and the families of nonworking heads of household through social programs, extended unemployment benefits, welfare payments, etc. The loss of jobs means that the government must also contend with billions of dollars of lost tax revenue. Moreover, these figures do not include the negative secondary multiplier effects brought about by the nonrealization of efficiency gains in American industry, lost profits, and lost investment opportunities.

Income Multipliers

Alternatively, it is easy to characterize the expansion or contraction of a regional economy in monetary terms using the "income multiplier" approach. Income multipliers, first introduced by Charles Tiebout in the late 1950s, are used to describe the circulation and expansion potential of incoming dollars in association with consumers' and producers' propensity to spend or invest in the local/regional economy. The income multiplier (K) is typically defined as the total income increase for a regional economy for a given increase in export-oriented income, or

K = $ Total Income Increase / $ Export Income Increase.

An **income multiplier** *describes the circulation and expansion potential of additional income introduced into a local/regional economy by accounting for consumption and investment related leakage.* In other words, how many additional dollars can an incoming dollar generate before it eventually leaks out of the economy. Note that the intensity of leakage is inversely related to consumers' and producers' tendencies to purchase goods and services from suppliers within the regional economy. Hence, the greater the propensity to purchase locally or regionally, the less severe the leakage and the greater the circulation and expansion potential of a dollar in that economy.

More formally, the income multiplier may be expressed as

$$K = 1 / \{1 - [(P_c * m_c) + (P_I * m_I)]\}, \quad (9)$$

where P_c is the propensity of individuals to consume within the region; m_c is the consumption multiplier—the income generated per dollar of sales within the region; P_I is the propensity of individuals to invest within the region; and m_I is the investment multiplier—the income created per dollar invested within the region.

Suppose that for a given local/regional economy, it is estimated that

P_c = .5 (consumers tend to purchase 50% of what they buy from suppliers in the local economy);

m_c = .4 (typically, each dollar in local sales generates an additional 40 cents of local income);

P_I = .2 (roughly 20 cents on every dollar is invested within the local economy); and

m_I = .5 (typically, each dollar invested in the local economy will generate an additional 50 cents in local income).

Using these values, it can be shown that the income multiplier

$$K = 1 / 0.7,$$

where K is approximately equal to 1.43. This outcome suggests that each new dollar brought into the economy will generate an additional 43 cents in income based on current consumption and investment trends. For example, an additional \$1,000,000 in export-oriented income associated with the basic sector would create an additional \$430,000 of income in nonbasic sectors.

The greater the propensity of individuals to consume and invest within the region, the greater the overall multiplier effects. As P_c and P_I increase, so does K. In other words, the greater a region's ability to retain value added by supporting local producers, the larger the expansion capabilities of an additional incoming dollar. The larger, more diversified, and more dense the production network of a regional economy, the larger the values of P_c and P_I and the greater the consumption and investment multipliers m_c and m_I. Note that m_c and m_I will increase (and so will K) with the expansion of the secondary and tertiary sectors.

With the exception of agricultural products, peripheral economies tend to have a relatively low propensity to consume locally produced commodities. They are highly reliant upon imports and outside suppliers, having little capacity to produce the many products which they import. In addition, most peripheral economies have a relatively low propensity to invest in the domestic economy. A lower propensity to invest means less economic and industrial growth. These economies exhibit a high degree of leakage and an overall inability to retain and attract capital. As a result, peripheral nations have smaller consumption and investment multipliers and relatively smaller income multipliers in comparison to nations of the semiperiphery or core.

INTERNAL GROWTH THEORY

Regions may experience growth in response to internal forces. **Internal growth theory** *suggests that a region will experience economic growth in response to increasing population and/or the growth of its urban areas.* While we will sidestep the issue of the quality of growth in relation to internal-driven population increases for the moment, it is important to draw some conclusions concerning economic growth in response to **urbanization**—*the process of becoming urban or urbanized.*

The degree to which a region is experiencing growth in existing urban areas and the rate at which a region is creating new urban areas (that is, the rate of urban growth) is a reflection of that region's internal growth potential. While it is inviting to draw parallels between urbanization and industrialization as linked processes, given examples of 19th- and 20th-century Europe and North America, urbanization is not necessarily driven or sustained by industrialization, nor does it insure that industrialization will continue to take place in areas where an established link had been observed. Hence, it is difficult to generalize about economic development status of a region by the degree to which that region is urbanized or by its rate of urbanization. Population growth in the world's fastest growing cities (for example, Lagos, Nigeria, and Cairo, Egypt) is not necessarily driven or accompanied by industrialization. The concentration of people in towns and cities throughout the world's peripheral regions and the increasing rate of urbanization in the SOUTH is in many ways a reflection of the distorted perceptions people have about employment opportunities of cities.

Urbanization is a process of population concentration and growth. It is an outcome of the increase in population from births in the urban area and from the influx of people as they move from rural areas to urban areas. Movement from rural to urban areas is typically related to job opportunities (real or perceived).

An **urban area,** utilizing the U.S. Census Bureau definition, is *a settlement of 5000 people or more in a continuous built-up urban environment.* While the appropriateness of this definition is certainly nonuniversal given that the urban environment has taken on new shape, form, and meaning over the past several decades, it does provide a useful bench mark for identifying significant population clusters or settlements on the economic landscape. Note that the percentage of world population residing in urban areas has increased dramatically over the last two hundred years. By contrast, the **urbanization rate**—*the rate at which urban areas are growing*—is slowing down (see Table 3.4).

Urbanization may also be viewed as a process that is linked to the tendencies of firms and industries to cluster in space. The availability of employment opportunities in and around an industrial cluster increases the demand for labor, which in turn creates more jobs and an even greater demand for labor via the multiplier effect. It is a self-reinforcing system.

Historically, urbanization of the core was precipitated by the formation and proliferation of industrial clusters in key locations. The "clustering" of producers and suppliers was tied to the availability of resources, transportation and production cost considerations, and the geography of established production linkages. Production clusters arise as firms seek to minimize transport and interaction costs and maximize production and efficiency gains associated with proximity to suppliers and/or markets. Cost savings may also arise from "scale economies" as production levels and output increase at established centers, driving down the cost per unit of output.

> There are costs to transactions across space; there are economies of scale in production. Because of economies of scale, producers have an incentive to concentrate production of each good or service in a limited number of locations. Because of the costs of transacting across distance, the preferred locations for each individual producer are those where demand is large or supply of inputs is particularly convenient—which in general are locations chosen by other producers. Thus concentration of industry, once established, tend to be self-sustaining. (Krugman, 1991, p. 98)

The localized growth of industry and the regional development process are supported, in part, by the growth of markets. Thus, the meteoric growth of demand in the emerging markets of the SOUTH as a by-product of rapid population growth has encouraged urban growth and industrialization. While the trend toward higher rates of urbanization is a worldwide phenomenon, regional variations in urban growth and its net benefits can be found.

Consider the data in Table 3.5, highlighting the percentage of population classified as urban for various selected groups and geographic regions. Note the glaring differences in the urbanization of North America and Europe in comparison to that of Africa or Asia. It is tempting to conclude that the more urbanized a region is, the higher its economic development status. Yet the

TABLE 3.4	The Percentage of the World's Population Residing in Urban Areas for Selected Years (rounded to nearest %)
Year	**Percent of World Population**
1800	3
1850	6
1900	14
1950	30
1965	36
1970	37
1980	41
1990	43
1991	43
1993	44
1995	45
2000	47 estimated
2050	50 estimated

Source: Rubenstein (1992), p. 455; *World Resources* (1994), p. 286; de Souza and Stutz (1994), p. 63; *World Development Report* (1995), p. 223.

TABLE 3.5	Urban Population as a Percentage of Total Population by Income-Group and Selected Geographic Region, 1970–2005

Region/Group	1970	1993	2005*
Low-income economies	18	28	34
Low-income economies (excluding China and India)	19	27	33
Middle-income economies	46	60	65
High-income economies	74	78	78
NORTH	66	71	75
North America	70	77	77
Europe	64	73	76
Commonwealth (former Soviet Union)	57	66	71
SOUTH	30	37	40
Latin America and Caribbean	57	71	76
Asia	25	34	37
Africa	23	34	39
Sub-saharan Africa	19	30	35
Middle East and N. Africa	41	55	58
WORLD	37	44	48

Source: *World Development Report 1995*, pp. 222–223.
* Projection based on trend.

nations of Latin America and the Caribbean stand as an exception to this rule, as much of their populations are concentrated in large cities (a trend that will be discussed in greater detail in a forthcoming chapter). Note that by the year 2005, the world's population is still expected to be less than 50% urbanized.

The largest urban growth rates, as implied by the data in Table 3.5, are associated with regions that would be characterized as least able to absorb more urban growth; in particular, many Asian and African nations. Many of the larger cities on these two continents are already experiencing difficulties in providing adequate levels of public services and other basic provisions to people in the burgeoning urban areas. Note also that even as the rate of urbanization in the world's underdeveloped regions has been higher than the rate of urbanization in the more advanced regions of the world, it is unlikely that the underdeveloped SOUTH will be more than 40% urbanized by the year 2005.

Urbanization has created a potentially dangerous situation for the residents of cities in underdeveloped regions, as adequate access to safe drinking water, sanitation, and health care, not to mention police protection, are always in question. In fact, many of the rapidly growing urban areas of the SOUTH are experiencing **hyper-urbanization**—*a situation where urban population is growing faster than the city's ability to absorb that population in terms of providing adequate housing, services, and opportunities.* The sprawling squatter-settlements and appalling living conditions on the outskirts of Mexico City, where each year a growing number of people join the ranks of the impoverished, provide a grim reminder of the impact of hyper-urbanization. The growing sea of urban poor in the SOUTH represents one of the greatest challenges to governments and policy makers as they try to contend with this most serious global development problem. As the poor flock to large, nearby cities in the hope of finding income-generating opportunities, the cities of the SOUTH must now deal with a growing

incidence of urban poverty, crime, and congestion. Once thought of as centers of power, culture, and production, many of the SOUTH's largest cities now suffer from an image problem as they contend with the unwanted side-effects of hyper-urbanization. These oversized cities may find it increasingly difficult to supply the necessary infrastructure to accommodate new industry (to meet the growing demand for employment opportunities). More disturbing is the fact that many of these cities will find it increasingly difficult to provide a level and mix of public and producer services that is sufficient to meet both the demands of industry and its residents (Figure 3.2).

According to internal growth theory, population growth will lead to an increase in the size of the regional labor force. More labor means more production, more output, and potentially more exchange. On the surface it seems as if internal growth should translate into economic growth and development. However, rapid population growth in age groups that include children (ages 0 through 14) or the elderly (age 65 or higher) can undercut regional development efforts if the working labor force is unable to provide a decent and nondeclining standard of living for the average member of that society. There are many concerns associated with an aging world population. In particular, questions arise as to the ability of an enlarging labor force to support not only a larger world population but a potentially more dependent one.

Consider the labor force statistics shown in Table 3.6, where labor force participants are described as workers between the ages of 15 through 64 for accounting purposes. The **dependency ratio** (DR), is defined as

$$DR = [\% \text{ of pop. } 0\text{–}14 + \% \text{ of pop } 65+] \\ / [\% \text{ of pop } 15\text{–}64] \qquad (10)$$

describing *the relative size of the dependent population in comparison to the size of the labor force.* Ratios for selected years are highlighted in bold (Table 3.6).

Figure 3.2 Hyper-urbanization in developing countries has brought new meaning to the word congestion. The rapidly growing cities of the SOUTH continue to pose major development problems for urban planners and policy makers.

The increase of the world's population in the 15–64 group relative to the dependent groups (ages 0–14 and 65+) has many analysts optimistic over the production potential of the world's labor force itself. Yet, it is uncertain whether an adequate number of employment opportunities can be created outside of the agricultural sector and sustained by the expected economic growth in emerging regions of the SOUTH to generate a sufficient number of opportunities to keep active the world's growing labor force. While a larger world labor force can potentially produce more output to meet a growing demand, the rapid expansion of the labor force in the SOUTH can lead to exploitation of labor, as it is likely to be both inexpensive and abundant. Concerns have also surfaced over the likelihood of an unprecedented movement of labor from underdeveloped to developed regions as a result of the perceived difference in economic opportunities between regions, as in the case of juxtaposed core and peripheral economies (for example, the United States and Mexico).

As with most geographic phenomena, regional variations in the dependency ratio are also evident (see Table 3.7). Despite the evidence of a general decline in dependency ratios at the global level over the last twenty years,

with the greatest decline occurring in peripheral regions, the persistence of higher dependency ratios in less-developed regions is unsettling. Note that the dependency ratios in Africa are considerably higher than the world average. As life expectancies are expected to continue to rise over time in many underdeveloped regions, there is a growing concern over further increases in dependency ratios in the early part of the next century. Regions with relatively high dependency ratios are more prone to be deficit-producing regions than those with relatively low dependency ratios. A high DR may translate into a dependency on outside suppliers or regions.

While dependency ratios give only a rough estimation of how a population may be dependent upon a labor force, they do not address issues of labor use or productivity. Specifically, dependency ratios do not account for differences in a labor force's ability to support people within the region as well as people outside the region.

For example, a subsistence farmer in rural Appalachia would not contribute as much to a regional economy's overall ability to support a dependent population as, say, a design engineer at BMW's production facility in Stuttgart, Germany. The Appalachian farmer may contribute nothing in terms of that region's export base, so no direct benefits accrue to the region's economy as a result of his labor. The engineer, however, not only pays taxes to support the internal infrastructure of his region, buys numerous goods and services and contributes to the recirculation of dollars, but is also directly involved in a production-for-export, which affects the flow and circulation of products and money in the world economy. While the subsistence farmer's crops may only feed himself and his family, the BMW "crop" will be traded in the global marketplace and generate currency to drive economic growth and development in the production region and throughout the global sales network.

TABLE 3.6	Distribution of the World Population by Major Age Groups for Selected Years

| Population Group | Percentage of Total Population by Year | | |
	1980	1995	2025 (projected)
0–14	37	31	24
15–64	57	62	66
65+	6	7	10
Dependency Ratios (DR)	**.754**	**.613**	**.515**

Source: UN *Global Outlook 2000* (1990); *World Resources* (1994), p. 270.

TABLE 3.7	Geographic Variations in Labor Force Dependency Ratios (1980, 1995)

| Region | Dependency Ratios (DR) | |
	1980	1995
Europe	.567	.494
Former USSR	.552	.563
N. and C. America	.642	.576
Asia	.785	.605
South America	.779	.612
Africa	.923	.919

Source: *World Resources* (1994), pp. 270–271.

TRADE AND DEVELOPMENT

International trade may be viewed as both a means to growth and as a means to development. As the world economy expands and the world's population swells to new heights, many regional economies will undoubtedly experience the benefits of economic growth in relation to both external and internal factors. Some of that economic growth will inevitably translate into money that will be useful to fund development efforts. Yet, there will be some regions of the world where the benefits associated with economic growth will not be realized, as population growth will act as a countervailing force.

Consider the data shown in Tables 3.8 through 3.11. The statistics in these tables reveal marked variations in economic growth rates by region and economic orientation. The differences are easily explained by the opposing forces of external and internal growth. Note that the classification system used in this section is summarized in the appendix at the end of this chapter.

The data in Table 3.8 demonstrate that positive GDP growth rates have occurred around the globe, yet the benefits of this economic growth have not necessarily translated into increasing prosperity, as some places have experienced a decline in GDP per capita. In other words, regional population growth rates in many places of the world have outpaced the growth of the regional economy. If successful economic development is contingent upon economic growth, it is difficult to imagine how progress can be made in the human condition when population growth exceeds economic growth and there is less income to go around on a per person basis. Particularly hard-pressed are most peripheral regions of the world: areas of low economic status and high population growth; those with a reduced potential to fully realize the gains of economic growth despite the presence or evolution of a globally competitive export base.

Table 3.9 demonstrates that "developing nations" engaging in the exportation of raw materials or agricultural products have experienced less growth in their GDP in comparison to those who engage in the exportation of manufactured products. This is no surprise given the fact that manufacturing-for-export activities are typically associated with larger regional economic base multipliers than nonmanufacturing activities and given that a greater amount of value added comes from the secondary sector in comparison to other sectors. As petroleum prices have remained fairly low and flat over the last few decades (notwithstanding minor fluctuations due to short-term disruptions in supply), petroleum-exporting nations have fared somewhat worse than other primary commodity exporters in terms of realizing the benefits of export growth in association with oil exports. Note also that nations exporting mostly primary sector commodities and importing manufactured goods run the risk of assuming a trade deficit. Leakage of this variety continues to have a dampening effect on economic growth in the SOUTH (see Table 3.10).

If we accept the premise that economic development status may be improved by increasing a region's income-generating ability or export orientation, then it is essential that a successful development strategy promote secondary sector activities and industrialization. In particular, regional economic development strategies must be

TABLE 3.8	GDP per Capita Levels and Growth Rates by Region		
	Growth Rates in %		
Region	**GDP per Capita 1985**	**GDP 1985–1990**	**GDP per Capita 1985–1990**
WORLD	$ 2700	3.3	1.6
Economies/Regions with Consistently High Growth			
Developed market economies	11,100	3.0	2.5
North America	12,750	2.9	2.0
Western Europe	10,840	2.9	2.6
Eastern Europe and the former Soviet Union	3,650	2.7	1.9
*Regions with Declining Economic Development Status**			
North Africa	1440	2.7	–0.6 (–0.3)
Sub-saharan Africa	450	2.3	–0.4 (–2.6)
Latin America and the Caribbean	2100	1.4	–0.1 (–1.1)
Western Asia	3100	0.3	–2.6 (–4.3)
*Regions Showing Improvements in Economic Development Status**			
South and East			
Asia	500	6.6	4.4 (3.7)
China	430	8.0	6.6 (7.5)

Source: *Global Outlook 2000* UN publications (1990), p. 11.

Note: Figures in 1980 dollars and exchange rates (rounded to nearest tenth dollar).

* Estimated average annual growth rate of GDP per capita from 1980–1990 is shown in parentheses.

devised to foster industrial change that will allow underdeveloped regions to integrate themselves into the world economy and compete with core nations. It is paramount that such regions secure industries that export high-value-added commodities, as these types of industries hold the greatest potential for bringing in the much-needed currency necessary to fuel industrialization and development (as argued by Esfahani, 1991). Moreover, the greater the value added, the greater the overall benefits of production and export. Economic base theory suggests that the recirculation and expansion potential of the incoming dollars from the export and sale of finished manufactured products in the world market is high relative to semifinished materials or nonprocessed commodities from the primary sector. The empirical evidence in Table 3.9 vigorously supports this contention.

Note that developing nations that were not manufacturing-for-export oriented tended to have below-average growth rates; whereas developing economies that exported manufactured products tended to grow at a rate that was above the world average from 1985 to 1990. Note that developing nations engaging in the exportation of primary sector commodities have experienced negative per capita economic growth rates (negative GDP per capita growth rates) over the same period; whereas those oriented toward manufacturing exports have a GDP per capita growth rates that exceed the world average. Nations characterized as "highly indebted" have experienced relatively lower rates of economic growth than those engaged in manufacturing for export. In sum, developing nations that are primary sector exporting or highly indebted have economies that are growing much slower than their human population base. By contrast, developing nations that export manufactured commodities

have experienced an economic growth rate that exceeds their population growth rate.

Projections indicate that the growth rates of GDP and GDP per capita for manufacturing-for-export regions are likely to both increase and remain well above the world average throughout the 1990s. Manufacturing-for-export regions are expected to post a 5% or more increase in GDP and a minimum 3% increase in GDP per capita up until the year 2000. By contrast, primary sector exporters and highly indebted nations are expected to improve their economic positions, although economic growth rates will remain considerably less than the world average (United Nations, *Global Outlook 2000*, p. 11).

It is expected that the world economy (and world production output) will continue to grow at a fairly healthy clip as population growth increases the demand for goods and services, regional and global markets expand, and regional economies become more highly integrated. An examination of the long-term trend in economic growth per capita reveals that growth rates have been slowing, although growth remains strong in several regions (see Table 3.11). Despite the slowdown in economic growth, it is the developed market economies and the emerging nations of South and East Asia that have experienced the greatest gains from past economic growth. This is largely due to their rapid transition from agrarian to industrial, and in turn, to postindustrial economies (despite their population size).

The largest economies and nations in South and East Asia are those that are undergoing the most rapid "industrialization." The level of industrialization and related economic changes in South and East Asia are partly fueled by the dramatic increases in internal demand from a region undergoing rapid population growth and the

| TABLE 3.9 | GDP per Capita Levels and Growth Rates of Developing Nations by Export Orientation |

	Growth rates in %		
Orientation	GDP per Capita 1985	GDP 1985–1990	GDP per Capita 1985–1990
Petroleum-exporting nations	$1630	1.3	–1.3
High-income oil exporters	7550	1.5	–2.5
Primary commodity exporters[*]	640	2.0	–1.1
Major exporters of manufactured products	**1,325**	**4.9**	**3.0**
Least–developed countries (LDCs)	240	3.5	0.7
Highly indebted nations	1790	1.9	–0.8
WORLD		**3.3**	**1.6**

Source: *Global Outlook 2000* UN publications (1990), p. 11.

Note: Figures in 1980 dollars and exchange rates (rounded to nearest tenth dollar).

For a breakdown of the nations included in each of the categories listed above, see the appendix at the end of this chapter.

[*] Also exporters of services.

TABLE 3.10	External Debt, Export and Import Structure, and Balance of Trade for Selected Nations of the SOUTH (1994)

	$ External Debt		Balance of Trade			
Nation	**$ billions**	**as % of GDP**	**$ billions***	**as % of Exports**	**Major Exports**	**Major Imports**
Nations with Trade Deficits						
Argentina	73.0	39.0%	–5.7	36.3%	agricultural products and meat	machinery, fuel, chemicals
Bangladesh	13.5	56.3%	–1.6	67.6%	garments, leather, seafood, jute	capital goods, food, petroleum, textiles
Egypt	31.2	87.1%	–8.1	261.2%	crude oil and petroleum products, cotton, and textiles	machinery and heavy equipment, fertilizer, consumer goods
Ethiopia	3.7	64.9%	–0.8	363.9%	coffee, leather goods, gold	capital and consumer goods, fuel
Ghana	4.6	76.9%	–0.7	70.0%	cocoa, minerals	consumer goods, food, petroleum
Honduras	4.0	139.8%	–0.2	16.4%	bananas, coffee, seafood, minerals, meat, lumber	transportation equipment, machinery, chemicals, manufactured goods
India	89.2	39.5%	–1.1	4.5%	clothing, gems, jewelry, leather products, cotton yarn and fabric	crude oil, machinery fertilizer, chemicals
Jamaica	3.6	94.2%	–1.0	83.3%	alumina and bauxite, sugar, bananas, rum	transportation equipment, machinery, chemicals, food, construction materials, fuel
Kenya	7.0	149.2%	–0.4	27.5%	tea, coffee, and petroleum products	transportation equipment, machinery, chemicals
Nicaragua	11.0	611.0%	–0.5	138.9%	meat, coffee, seafood, sugar, cotton, bananas	consumer goods, machinery, transportation equipment, petroleum
Pakistan	24.0	51.8%	–2.8	41.8%	cotton, textiles, leather, apparel	petroleum, chemicals, machinery, and equipment
Panama	6.7	102.0%	–1.7	324.0%	bananas, seafood, sugar, clothing	capital and consumer goods, crude oil, chemicals, foodstuffs
Peru	22.4	54.5%	–1.0	24.4%	metals and minerals, crude petroleum, fish meal	transportation equipment, machinery, foodstuffs, chemicals, iron, and steel
Philippines	40.0	73.9%	–7.9	58.9%	textiles, fish, coconut products, electronics	raw materials, consumer and capital goods, petroleum
Sudan	17.0	78.8%	–1.3	30.5%	gum arabic, cotton, livestock, and meat	foodstuffs, petroleum, manufactured goods
Tanzania	6.7	321.2%	–0.9	20.3%	coffee, cotton, tobacco, tea, cashews, sisal	manufactured goods, machinery, foodstuffs, crude oil
Thailand	64.3	51.5%	–6.6	14.3%	agricultural products, consumer products, and fish	capital and consumer goods, intermediate and raw materials
Uganda	2.9	95.7%	–0.5	203.0%	coffee, tea, cotton	transportation equipment, food, machinery, petroleum products
Vietnam	8.5	66.2%	–0.6	16.7%	rice, agricultural and marine products	petroleum products, grain, machinery, steel, fertilizer
Zambia	7.3	198.1%	–0.1	11.8%	copper, tobacco, metals, and minerals	transportation equipment, machinery, foodstuffs, manufactured goods
Nations with Trade Surpluses						
Brazil	134.0	30.2%	+10.4	23.8%	minerals and agricultural products	oil, coal, foodstuffs
China	80.0	3.0%	+5.3	4.4%	textiles, apparel, toys, machinery, electronics	rolled steel, oil products, motor vehicles, aircraft
Indonesia	87.0	60.1%	+9.9	24.0%	manufactured goods, fuels, foodstuffs	intermediate and raw materials, consumer goods
Malaysia	35.5	55.1%	+1.4	2.5%	electronics, textiles, rubber, and wood	machinery, chemicals, petroleum
South Africa	18.0	17.0%	+3.9	15.4%	gold, metals, minerals	transportation equipment, machinery, chemicals, scientific equipment
Venezuela	40.1	66.8%	+7.6	50.0%	oil, minerals, steel, manufactured goods	transportation equipment, machinery, and raw materials

Source: *CIA Worldfact Book* (1994, 1995).

* Value of exports less value of imports (1994) in $ billions rounded to the nearest tenth of a percent.

TABLE 3.11	Average Annual Rate of Growth of GDP per Capita for Selected Countries and Regions: Regional Disparities between the Core and the Periphery

Average GDP per Capita Growth Rates (%)

Region/Countries	1960–70	1970–80	1980–90	1990–95ᵉ
WORLD	3.2	1.9	1.3	1.1
Developed market economies	3.9	2.4	2.1	1.7
North Africa	8.2	1.2	–0.3	–0.8
Sub-saharan Africa	1.8	–0.4	–2.6	–3.0
Western Asia	4.1	1.0	–4.3	–4.0
Latin America and the Caribbean	2.7	2.4	–1.1	–1.9
South and East Asia	2.6	4.1	3.7	3.4
Least-developed countries* (LDCs)	1.1	–0.2	–0.3	–0.5

Source: *Global Outlook 2000* UN publications (1990), p. 10.

*Note: LDCs are low-income economies, excluding India and China, with an average 1990 per capita GDP of $240 (in 1980 dollars). These nations account for approximately 9% of the world's population in 1990.

ᵉ Means estimated.

growth and proliferation of overseas manufacturing in areas of high population concentration. It is the large appetite of the developed world (and the core economies of the NORTH) for inexpensive goods from Asia that has largely contributed to the growth of export-oriented manufacturing in this region. Of all the developing regions of the world, only South and East Asia have shown consistently higher-than-average economic growth rates over the last three decades, growth rates that have exceeded even those of the developed market economies of the NORTH.

Figures 3.3 and 3.4 highlight the production-for-export orientation and economic growth characteristics of nations in the SOUTH. Of the Big-Five economies of the SOUTH: Mexico, Brazil, India, South Korea, and China, only the Asian nations are part of the top-20 fastest-growing economies from 1980–1993. As established earlier, there exists a moderately strong and positive correlation between economic growth rates and the presence of a strong manufacturing-for-export base whose advantage lies in the availability of an inexpensive and abundant labor force.

UNEVEN DEVELOPMENT AND POLICY

The paradox of uneven growth and development is rooted historically in the established interdependencies between core and peripheral economies. Many claim that these interdependencies have set the stage not only for the exploitation of the periphery by the core, but are largely responsible for maintaining, and in some cases worsening, regional economic disparities. The dependency problem has plagued development efforts in the SOUTH since the age of colonialist expansion. The domination of the SOUTH by the economies of the core, has given the NORTH a competitive edge. Today, regions cannot compete by simply having a **comparative advantage** in production—*a regional specialization in the production of one or more commodities for which that region has a particular advantage as tied to favorable resource endowments or physical/environmental conditions.* Regions must seek to gain a "competitive advantage" in the global economy, by increasing the sophistication and diversification of their economies.

Competitive advantage may be thought of as *a regional production advantage tied to economic variables and human resources.* For example, favorable economic conditions combined with a cost-effective, productive, and multiskilled labor force is likely to support the expansion of a diverse, globally competitive, and highly integrated production network, which can retain value added and stimulate regional economic growth. This can be done with or without favorable natural resource endowments. Consider that the nation of Japan has very limited natural resources. This limitation has not hampered the rise of Japan's post-World War II economy. Japan gained a competitive advantage by utilizing modern production technologies, a highly focused and productive labor force, and an internally supportive production network, which looked almost exclusively to Japanese suppliers for production inputs and services. As a result, the Japanese economy was able to produce a wide-ranging mix of end-products to meet both the demands of its diverse production network and the demands of consumers in a diverse global marketplace.

There are several problems with development policies that promote specialization through comparative advantage. Once countries move toward a comparative advantage with the hope of improving their trade position and strengthening a specialized export base, they have a tendency in the long run to reduce their overall self-reliance by supporting and sustaining only those industries that are linked to their comparative advantage. They become dependent on outside sources and suppliers for various products outside their comparative advantage, and are sometimes left to import products which they are able to produce should resources be more evenly distributed. These nations then begin to rely on the currency gained through trade to import that which they have no comparative advantage to produce. Countries specializing in raw material production, processing without fabrication, or the production of agricultural commodities for export (that is, cash crops) tend to get a disproportionately small share of

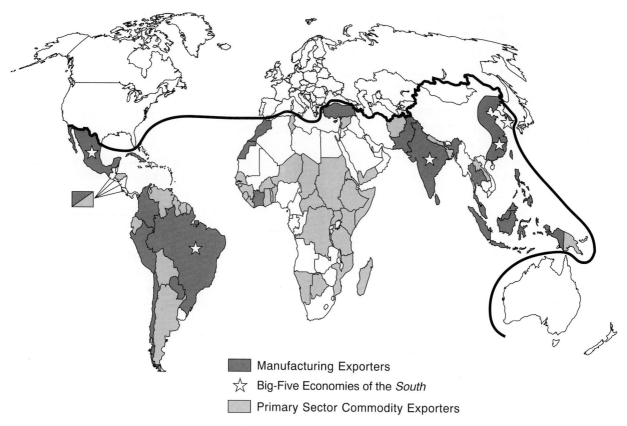

Figure 3.3   Manufacturing and commodities exporters of the SOUTH.

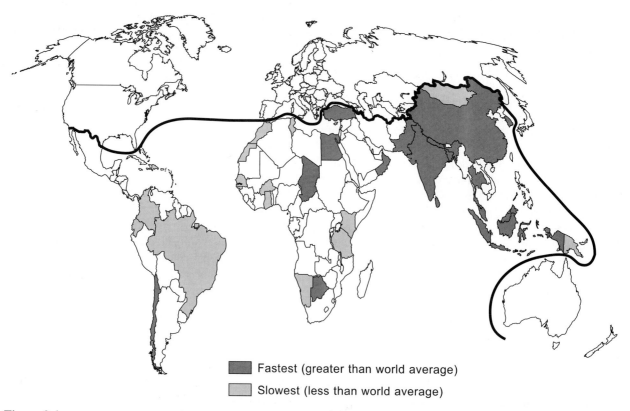

Figure 3.4   Fastest- and slowest-growing economies of the SOUTH.

the overall benefits of utilizing their so-called comparative advantage in production and trade as these products are of low value added. As market prices or the exchange value of finished products, especially manufactured items, have risen faster than prices of raw materials and agricultural commodities in the world economy, the periphery is continually placed at a overall disadvantage relative to the core; its comparative advantage lies mostly in the production of primary sector commodities. Moreover, the core economies of the NORTH control most of the world's manufacturing capacity and value added. Over time, incoming dollars from primary sector exports in the periphery buy fewer (high value) imports. As price levels for industrial output rise faster than price levels of primary sector commodities, trade imbalances, deficits, and leakage money promote the flow of export-related income away from the poor economies of the SOUTH and toward the more-affluent economies of the NORTH.

In short, most developing nations of the world are highly dependent upon the industrialized core as a final destination or market for their exports, despite the emergence of several high-growth semiperipheral economies in the SOUTH.

> While most of the exports of the developing countries still go to the industrial market economies of the western nations, . . . to an increasing extent, developing world exports are being absorbed by developing countries. Despite these developments, primary commodities have remained the dominant exports in the least industrialized low-income and lower middle-income countries. The attendant disadvantage with the persistence of adverse trading balances has led many developing countries to feel that the international trading situation is working to their continuing disadvantage. (Simpson, 1994, p. 131)

Since high population growth in the underdeveloped regions of the world continues at a blistering pace, so too does the growth of large regional labor forces. Although the presence of such cheap and abundant labor in the underdeveloped world is a condition that will ultimately ensure that regions in the SOUTH will attract investment capital, incoming dollars are most likely to flow to production-for-export sectors where low-valued goods are produced. It has been argued that the level of wages determines the type and prices of goods produced. As the majority of nations in the underdeveloped world are characterized as having cheap and abundant labor pools of which to draw upon, the goods produced there are characterized by low value and low prices, resulting in "an unequal exchange in trade if they trade with the rich, industrialized nations" (Simpson, 1994, p.134).

As if this were not a great enough challenge, the underdeveloped regions face yet another obstacle in gaining a global competitive advantage: market access. Questions continue to arise over whether or not peripheral economies can gain access to markets within the core as the world's leading nations begin to align themselves within large continental trading blocs. Regional trade blocs or alliances may make it increasingly difficult for peripheral economies to penetrate established markets should members within the trading blocs promote protectionist measures.

Disadvantages in trade and the lack of a competitive advantage may force poor nations to become even poorer, while the rich nations retain or amass wealth, and improve their economic positions and leverage in the world economy. Not only are many of the less-developed nations of the world falling farther behind in terms of their abilities to produce and consume in a system where the forces of regional and global supply and demand place them at a disadvantage, but they must also contend with high population growth (negating any of the benefits of growth and integration).

Over the past several decades, economic conditions have deteriorated for many peripheral nations in the SOUTH. As most productivity gains in core economies are returned to the workers in the form of higher wages and income, productivity increases in semiperipheral and especially in peripheral nations (which emphasize primary sector production) are usually reflected in lower overall price levels of the products that they export. Hence, the core not only retains the benefits of productivity increases within the core, but is allowed the luxury of importing low-cost commodities from the periphery. Wage-price spiraling within the core and the relatively low-value-manufacturing production orientation of nations outside of the core (especially those opting to exploit their comparative advantage in labor-intensive manufacturing activities) have left the periphery at a large trade disadvantage as prices of high-valued manufactured products rise faster than other commodities. Developing nations have also encountered radical short-term fluctuations in export prices and subsequent losses in export-related earnings—the so-called "export instability" problem. Many analysts contend that this condition may further threaten economic growth and development efforts.

In an attempt to industrialize, many developing regions have tried to move away from their comparative advantage in production and export of primary sector commodities (agricultural products, raw materials, and minerals) in favor of pursuing alternative strategies to economic growth and development. Many regional economies are enacting policies which help encourage industrialization and self-reliance in production. On the international economic policy front, industrialization of semiperipheral economies has been largely supported by policies which promote (1) "export-oriented industrialization"; and (2) "industrialization through import substitution" (Salvatore, 1987a).

Export-oriented industrialization (EOI) refers to *an industrialization process supported by the growth of an export base and a regionally concentrated production* *network.* EOI policies encourage the expansion of export-oriented industry, typically in labor-intensive or low-value-added manufacturing. This is accomplished by directing financial resources toward industries in which the economy may hold a competitive advantage as based upon its human and economic resources. Government involvement comes mainly in the form of subsidies to its domestic export-oriented industry. The role of government may also be expanded to oversee the creation of a stable regional economy and a good "business climate." Energy is focused not only on building a globally competitive export base with respect to one or several key industries, but on creating a production environment that will attract investment capital from outside sources (that is, money from foreign investors). Once establishing a viable export base, a regional concentration of production linkages, and a favorable business climate, the region will experience economic growth, facilitating the formation of a highly integrated production network that can be potentially self-sustaining. Once a self-sustaining production network is in place, it can easily transform a simple center of production into a center of innovation should the business and cultural atmosphere be favorable to support such a change.

Industrialization through import substitution (ITIS) refers to *an industrialization process in which the growth of domestic industry and increased industrial output is supported by import restrictions.* Instead of relying on imports to meet the domestic demand for certain types of products, domestic industry in the regional economy is encouraged to form and produce those products itself. Hence, imports are substituted with output from domestic industry; allowing domestic producers to take advantage of an already existing internal market. In other words, ITIS policies tend to support the replacement of imports with domestically produced products.

The flow of imports into a region is generally restricted by the implementation of **import tariffs**—*a tax or duty on imported goods*—or through the imposition of **quotas**—*a fixed number or quantity of goods allowed into a region or a fixed percentage of a market that is opened to outside suppliers.*

The advantages of EOI policies are threefold. First, EOI policies tend to allow industrializing nations the chance to take advantage of **economies of scale** in production—*savings which accrue from producing a larger overall output level.* This allows some of the smaller industrializing nations the chance to overcome the limitations of their domestic market by pursuing sales abroad. In other words, export sales revenues allow a region to overcome the restrictions associated with the limited amount of income that can be earned from its own (sometimes small) domestic market. Exports allow industry to produce at higher output levels, driving down the average cost of production. First, generating production surplus for export increases the efficiency of industry as it re-

duces "slack" or downtime in a region's production system, allowing expansion and fuller utilization of a region's production capacity. Second, the expansion of industry is not limited by the growth of domestic demand for industrial outputs. In fact, the expansion of an emerging production network is contingent upon the growth of demand in outside markets. Third, **economies of network**—*network-level efficiencies (anticipated and unanticipated) brought about as a result of the formation and strengthening of regional production linkages and efficiency gains in directly or indirectly related industries or sectors*—are also likely to emerge. In short, export production will help solidify and augment various segments of the existing production network, promote efficiency, and stimulate economic growth.

The ineffectiveness of EOI policies are, for the most part, related to the difficulties in overcoming two major obstacles: (1) setting up production-for-export industries in regions that are currently unable to compete with established and more-efficient producers in the core; and (2) obtaining a sufficient degree of import penetration in outside geographic markets to sustain these newly formed industrial clusters. Note that successful import penetration can arouse interest in protectionalism, leading to the imposition of trade barriers or other restrictions on imports. Protectionalist policies may be employed to limit market access, minimize leakage, and shield producers in core economies from the negative effects and losses related to an import deluge. Hence, EOI can fail if external markets become protected and impermeable.

There are three major advantages of ITIS policies. First, ITIS reduces the risk of promoting the growth of industry in a region where a domestic market for one or more products is already known to exist (as evidenced by a given volume of imports). In other words, the observed demand for a given product ensures that sales are inevitable. Second, it is relatively easier for a developing region to protect its own market from imports than it is to establish an export base that must contend with the wide array of trade barriers, which limit market access in external geographic markets. Third, ITIS policies entice producers in the core to relocate production facilities to ITIS-promoting regions to sidestep tariffs and other trade barriers. Relocation of production facilities in response to ITIS policies leads to a breakdown in the resistance of the flow of capital from the core to the periphery and an increasing propensity for both capital and technology transfers from the NORTH to the SOUTH.

There are, nonetheless, several disadvantages to ITIS that require elaboration. Domestic industries falling under protectionalist ITIS policies have no incentive to become more efficient, more globally competitive, or more quality-control conscious. In addition, import substitution may lead to inefficiencies if the domestic market is not large enough to take advantage of economies of scale in production. The larger the domestic market, the higher the probability of successfully supporting the growth of a domestic industry to serve that market. Finally, it is both difficult and costly to replace many manufactured imports with products from newly formed domestic industries. This is especially true for industries which are capital-intensive, technology-driven, or associated with high-value-added production (vehicle and heavy machinery production, microcomputers, advanced telecommunications industries, etc). As a result, many peripheral regions have abandoned ITIS frameworks and have returned to the more traditional approach of finding a comparative advantage. With little hope of ever developing a globally competitive industrial cluster, many of the world's more peripheral regions continue to look to primary sector exports as their only option.

As discussed by Poon (1994), two opposing schools of thought have emerged on the role of exports in the economic growth scenario since the 1950s. There are those who believe that economic growth is a consequence of favorable production conditions and various internal and supply-related factors, which combine to make a region globally competitive. Under this "supply-side" scenario, exports promote growth because they stimulate the efficient use of factors and resources. In turn, efficiency allows a region to establish and enhance its export base and remain competitive over an indefinite period of time. Alternatively, there are those that argue that export-led growth is a by-product of a favorable global trade climate as largely determined by the prosperity of core economies—the largest demand sinks or markets in the world economy. In this "demand-side" scenario, the expansion of any region's export base is not (on average) expected to exceed the growth of demand in the core.

The rate at which exports are expected to lead to economic growth has also been a source of debate between the so-called "export optimists" and "export pessimists." **Export optimists** *tend to place great confidence in markets as ways in which to efficiently allocate production factors and resources, and hence, see exports as a way in which to grow all regions as markets grow in response to increasing population and wealth.* Nations that are able to compete for a share of world exports as a result of production specialization and efficiency gains will not only establish a trade advantage but will also experience increased productivity-driven and growth-driven development. These nations will be most effective in utilizing their production capital, labor force, natural resources, and entrepreneurial skills and know-how, and are likely to experience growth over an indefinitely long period of time. Moreover, growth is expected to spill over to regions with which they interact or trade. The speedy rise of the four **Asian Tigers**—*Hong Kong (prior to 1997), Singapore, Taiwan, and South Korea*—supports the contention that economic growth is a direct result of superior export performance. The rise of the Asian Tigers provides us with a good example of just how effective export-

advancing EOI policies can be in promoting progress on an economic front. It is no surprise, therefore, why this type of development strategy is so widely endorsed by export-optimistic institutions such as the World Bank and the International Monetary Fund.

By contrast, **export pessimists** *tend to be more cautious about the role of markets as mechanisms which promote efficiency and positive change; arguing that the growth of a regional export-base is largely constrained by external conditions (things that lie outside the control of most developing nations).* In short, export pessimists are skeptical about export-led growth and development strategies, as these strategies are far too reliant upon a weak assumption: that there will always be sufficient demand (or growth of demand) in world and regional economies to sustain and expand production-for-export industries in all developing regions. Certainly an established trade position may translate into further export-led growth if external conditions are favorable. Yet export pessimists are quick to point out that there is much risk involved in relying on external conditions and markets. Export pessimists recommend that developing nations promote inward-looking development strategies, such as industrialization through import-substitution, in order to make them less susceptible to fluctuations in external demand, protectionalist measures, or international crisis.

Recent studies suggest that an integrated approach may offer the best of both worlds, where trade-induced growth is brought about by a combination of inward-looking/import substitution and outward-looking/export promotion strategies (that is, a combination of ITIS and EOI policies). Each strategy will serve to compliment and/or supplement the other so that the maximum positive benefits can be achieved in the export-growth relationship. Poon (1994) writes that

> . . . export pessimists have not been wrong to highlight the influence of the trade environment in determining the success of export-led growth. But they have also overlooked the fact that export-led growth has been successful for a few countries even during adverse economic conditions. Such countries have continued to maintain a relatively strong demand for their exports over this period by engaging in higher-value manufactured exports.
>
> A favorable trade environment alone is therefore not sufficient to ensure higher growth via exports. The internal conditions which enable a country to be competitive such as entrepreneurship, labor skills, resources and institutions need also to be present. Similarly, export optimism cannot be over-emphasized in the presence of weak demand. (p. 19)

All in all, the developing regions of the world face many impediments and challenges in their attempt to industrialize and integrate into a globalizing production-and-trade network. Economic theory may tell us that international trade is an engine for economic growth, yet development policies designed to establish production-for-export industries in an attempt to improve the status of an underdeveloped region suffer from several major problems. First, the present global economic system tends to benefit the core at the expense of the periphery (not unlike a "king of the mountain" scenario where those trying to climb up the ladder continue to be kicked back down by those at the top). Second, with less favorable supply and demand conditions, barriers to production efficiency in high-value-added sectors, and a general lack of capital (in comparison to nations in the core), underdeveloped regions are unlikely to experience the same benefits of trade and integration as the core given that they are dependent upon growth of the core to support their own growth. The distribution of future benefits is likely to favor those regions which have already enjoyed the greatest overall cumulative benefits and growth. Hence, uneven growth, deteriorating terms of trade, and the difficulties in overcoming the one-sided aspects of commodity exchange in the world economy continually place the more peripheral regions of the SOUTH at a greater and greater disadvantage. Third, rapid population growth in underde-

veloped regions has precluded many of these regions from realizing any gain from trade and economic growth. Last, problems and setbacks of a trade-related nature (including the possible return of protectionalism) have been overshadowed by other economic and social problems in various regions of the SOUTH.

With a worsening poverty situation, excessive population growth, critical food shortages, civil unrest and chaos, economic and political instability, inadequate access to other necessities (medical care, clothing, shelter, safe drinking water, proper sanitation, etc.), and a mounting international debt, it is only a matter of time before the nations of the SOUTH revert to a short-sighted economic survival mentality. Many have come to recognize that progress on the economic front (including industrialization and export-base development) has sometimes occurred at the expense of progress in the human condition. The opportunity costs are high and rising. Export-led growth and development may continue to bring prosperity to the core and semiperiphery, yet it does not represent a solution that applies to the ailing periphery.

KEY WORDS AND PHRASES

Asian Tigers   65
brain drain   45
capital accumulation   43
capital flight   45
comparative advantage   61
competitive advantage   61
deindustrialization   48
dependency ratio   56
dumping   47
economies of network   64
economies of scale   64
electronic commerce   51
employment multiplier   52
export base   43
export optimists   65
export-oriented industrialization (EOI)   63

export pessimists   65
external growth theory   44
globalization of production   43
hyper-urbanization   56
import tariffs   64
income multiplier   53
income transfer   44
industrialization   44
industrialization through import substitution (ITIS)   64
information services (IS)   51
information technology (IT)   51
internal growth theory   54
leakage   45
multinucleated   50
multiplier   52
multiplier effect   51

new international division of labor   50
quotas   64
region-forming transactions   44
region-serving transactions   44
reverse multiplier effect   51
sustained economic growth   43
trade   43
trade deficit   45
trade imbalance (bilateral trade deficit)   46
urban area   54
urbanization   54
urbanization rate   55
World Trade Organization (WTO)   49

REFERENCES

Alexander, J.W. "The Basic-Nonbasic Concept of Urban Economic Functions." *Economic Geography* 20 (1954), pp. 246–261.

Balassa, B. *The Theory of Economic Integration*. London, U.K.: Allen & Unwin (1962).

Balassa, B. *The Newly-Industrializing Countries in the World Economy*. New York: Pergamon Press (1981).

Balassa, B. "Exports, Policy Choices, and Economic Growth in Developing Countries after the 1973

Oil Shock." *Journal of Development Economics* 18 (1985), pp. 23–35.

Bernstein, H. (ed.) *Underdevelopment and Development*. London: Penguin (1978).

Bhagwati, J.N. *Dependence and Interdependence.* Cambridge, MA: MIT Press (1985).

Bluestone, B., and B. Harrison. *The Deindustrialization of America: Plant Closings, Community Abandonment, and the Dismantling of Basic Industries.* New York: Basic Books (1982).

Blumenfeld, H. "The Economic Base of the Metropolis." *Journal of the American Institute of Planners* 21 (1955), pp. 114–132.

Cairncross, A.K. (ed.). *Factors in Economic Development.* London: Allen & Unwin (1962).

Chilcote, R. *Theories of Development and Underdevelopment.* Boulder: Westview Press (1984).

Chisholm, M. *Modern World Development.* London: Hutchinson (1982).

Clawson, D.L., and J.S. Fisher (eds.). *World Regional Geography: A Development Approach* (6th edition). Upper Saddle River, NJ: Prentice Hall (1998).

Cohen, S.S., and J. Zysman. *Manufacturing Matters: The Myth of the Post-Industrial Economy.* New York: Basic Books (1987).

Cooper, J.C., and Madigan, K. "What's Sabotaging Growth? The Stubborn Trade Gap." *Business Week* (September 4, 1995), pp. 29–30.

Daniels, P. *Service Industries in the World Economy.* Oxford, U.K.: Blackwell (1993).

Dasgupta, A.K. *Economic Theory and the Developing Countries.* London: Macmillan (1974).

deSouza, A.R. and F. P. Stutz. *The World Economy: Resources, Location, Trade, and Development* (2nd edition). New York: Macmillan (1994).

Dicken, P., and P. Lloyd. *Location in Space: Theoretical Perspectives in Economic Geography.* New York: Harper & Row (1990).

Dodaro, S. "Exports and Economic Growth: A Reconsideration." *Journal of Developing Areas* 27 (1993), pp. 227–244.

Durkheim, E. *The Division of Labor in Society* (translated by W.D. Halls). New York: Free Press (1984).

"General Alarm." *The Economist* 3 (January 1992), p. 80.

Ellsworth, P.T., and J.C. Leith. *The International Economy.* New York: MacMillan (1984).

Emmanuel, A. *Unequal Exchange.* London: New Left Books (1972).

Esfahani, H.S. "Exports, Imports, and Economic Growth in Semi-Industrialized Countries." *Journal of Development Economics* 35 (1991), pp. 93–116.

Findlay, R. "Growth and Development in Trade Models." In R.W. Jones, and P.B. Kenen (eds.) *Handbook of International Economics, Volume 1.* New York: North-Holland (1984).

Foust, D., D. Harbrecht, McNamee, and E. Updike. "Showdown Time for U.S.-Japan Trade." *Business Week* 24 (April 1995), pp. 48–49.

Gwyne, R. *New Horizons? Third World Industrialization in an International Framework.* New York: J. Wiley & Sons (1991).

Haggett, P., A.D. Cliff, and A. Frey. *Locational Analysis in Human Geography I.* New York: J. Wiley & Sons (1977).

Harrod, R.F. *Towards a Dynamic Economics.* London: Macmillan (1948).

Hewings, G.J.D. *Regional Input-Output Analysis.* Newberry Park, CA: Sage (1985).

Hirschman, A.O. *The Strategy of Economic Development.* New Haven, CT: Yale University Press (1958).

Ho, D.T. "Developing Countries Fast-Growing Exporters of Manufactures: An Analysis of Competitiveness of Selected Countries during the 1970's." *Trade and Development* 5 (1984), pp. 121–140.

Hoover, E.M. *The Location of Economic Activity.* New York: McGraw-Hill (1948).

Howes, C., and A.R. Markusen. "Trade, Industry, and Economic Development." in II. Noponen, J. Graham, and A.R. Markusen (eds.) *Trading Industries, Trading Regions.* New York: Guilford Press (1993).

Isard, W. *Methods of Regional Analysis: An Introduction to Regional Science.* New York: J. Wiley & Sons (1960).

Kravis, I. "Trade as a Handmaiden of Growth: Similarities between the Nineteenth and Twentieth Centuries." *The Economic Journal* 80 (1973), pp. 850–872.

Krugman, P. "Trade, Accumulation, and Uneven Development." *Journal of Development Economics* 8 (1981), pp. 149–161.

Krugman, P. *Geography and Trade.* Cambridge, MA: MIT Press (1991).

Krugman, P. *The Age of Diminished Expectations.* Cambridge, MA: MIT Press (1994).

Kurtzman, J. *The Decline and Crash of the American Economy.* New York: W.W. Norton (1988).

Kuznets, S. "Economic Growth and Income Inequality." *American Economic Review* 45 (1955), pp. 1–28.

Lewis, W.A. *The Theory of Economic Growth.* London: Allen and Unwin (1955).

Lewis, W.A. "The Slowing Down of the Engine of Growth," *American Economic Review* 70 (1980), pp. 555–564.

Linder, S.B. *An Essay on Trade and Transformation.* New York: Wiley (1961).

MacBean, A.I. *Export Instability and Economic Development.* Cambridge, MA: Harvard University Press (1966).

Maizels, A. *Exports and Economic Growth of Developing Countries.* London, U.K.: Cambridge University Press (1968).

Malecki, E.J. *Technology and Development* (2nd edition). New York: Wiley (1997).

Markusen, A.R. "Trade as a Regional Development Issue: Policies for Job and Community Preservation." In H. Noponen, J. Graham, and A.R. Markusen (eds.) *Trading Industries, Trading Regions* (1993), pp. 285–302.

Martin, R.L., and B. Rowthorn (eds.). *The Geography of Deindustrialization.* London, U.K.: MacMillan (1986).

Morton, K., and P. Tulloch. *Trade and Developing Countries*. London: Croom Helm (1977).

Myint, H. "The 'Classical' Theory of International Trade and the Underdeveloped Countries." *The Economic Journal* 67 (1958), pp. 317–337.

Myint, H. *The Economics of Developing Countries*. London, U.K.: Hutchinson (1964).

Myrdal, G.M. *Economic Theory and Underdeveloped Regions*. London: Duckworth (1957).

Nurske, R. *Patterns of Trade and Development*. New York: Oxford University Press (1961).

Perloff, H., E. Dunn, E. Lampard, and R. Muth. *Regions, Resources, and Economic Growth*. Baltimore: Johns Hopkins (1960).

Pfouts, R.W. (ed.). *The Techniques of Urban Economic Analysis*. New York: Chandler Davis (1960).

Poon, J. "Effects of World Demand and Competitiveness on Exports and Economic Growth." *Growth and Change* 25 (1994), pp. 3–24.

Prebisch, R. *Towards a New Trade Policy for Development*. New York: United Nations (1964).

Pred, A. *The Spatial Dynamics of U.S. Urban-Industrial Growth*. Cambridge, MA: MIT Press (1966).

Richardson, H.W. *Regional Growth Theory*. London: MacMillan (1973).

Riedel, J. *Myths and Reality of External Constraints on Development*. Brookfield, VT: Gower (1987).

Rodwin, L., and H. Sazanami. *Deindustrialization and Regional Economic Transformation*. Winchester, MA: Unwin Hyman (1989).

Romer, P. "Growth Based on Increasing Returns Due to Specialization." *American Economic Review* 77 (1987), pp. 56–62.

Rubenstein, J.M. *An Introduction to Human Geography* (3rd edition). New York: Macmillan (1992).

Salvatore, D. *International Economics* (2nd edition). New York: MacMillan (1987a).

Salvatore, D. (ed.). *The New Protectionalist Threat to World Welfare*. New York: North-Holland (1987).

Sen, A. *Growth Economics*. Harmondsworth, England: Penguin Books (1970).

Simon, J.L. *The Economics of Population Growth*. Princeton, NJ: Princeton University Press (1977).

Simpson, E.S. *The Developing World* (2nd edition). London: Longman (1994).

Streeten, P. (ed.). *Trade Strategies for Development*. London: Macmillan (1993).

Stutz, F.P., and A.R. de Souza. *The World Economy: Resources, Location, Trade, and Development* (3rd edition). Upper Saddle River, NJ: Prentice Hall (1998).

Theberge, J.D. *Economics of Trade and Development*. (New York: J. Wiley & Sons (1968).

Thirlwall, A. *Growth and Development* (2nd edition). New York: J. Wiley & Sons (1977).

Tiebout, C.M. "Exports and Regional Economic Growth." *Journal of Political Economy* 64 (1956), pp. 160–169.

Tiebout, C.M. "The Urban Economic Base Reconsidered." *Land Economics* 32 (1956), pp. 95-99.

Tiebout, C.M. "Community Income Multipliers: A Population Growth Model." *Journal of the Regional Science Association* 2 (1960), pp. 75–84.

Tyler, W.G. "Growth and Export Expansion in Developing Countries, Some Empirical Evidence." *Journal of Development Economics* 9 (1981), pp. 121–130.

Tyson, L.D., W. Dickens, and J. Zysman. *The Dynamics of Trade and Employment*. Cambridge, MA: Ballinger (1988).

United Nations. *Global Outlook 2000*. Washington, D.C.: UN Publications (1990).

Wheeler, J.O., P.O. Muller, G.I. Thrall, and T.J. Fik. *Economic Geography* (3rd edition). New York: J. Wiley & Sons (1998).

Wilson, A.G. *Urban and Regional Models in Geography and Planning*. New York: Wiley (1974).

World Development Report 1995 (published for the World Bank). New York: Oxford University Press (1995).

World Resources 1994–1995 (World Resources Institute). New York: Oxford University Press (1994).

Developed Market Economies—North America, southern and western Europe (excluding the Mediterranean nations of Cyprus, Malta, and the now fractured former Yugoslavia), Australia, New Zealand, Israel, Japan, and South Africa.

Developing Nations—the nations of Latin America (Central and South America) and the Caribbean, Africa (other than South Africa), Asia (excluding China and other centrally planned economies of Asia—Vietnam, North Korea, and Mongolia—and also excluding Israel and Japan), Oceania (excluding Australia and New Zealand), and Cyprus, Malta, and the now fractured Yugoslavia.

North Africa—Algeria, Egypt, Libya, Morocco, Tunisia.

Sub-saharan Africa—nations of Africa (excluding the nations of Northern Africa and the nation of South Africa).

Latin America—Developing nations in the Western Hemisphere (excluding North America and the Caribbean) or Central and South America.

Oil Exporting Countries—Angola, Algeria, Bahrain, Brunei, Cameroon, Congo, Ecuador, Egypt, Gabon, Indonesia, Iran, Iraq, Kuwait, Libya, Nigeria, Oman, Qatar, Saudi Arabia, Syrian Arab Republic, Trinidad and Tobago, United Arab Emirates, Venezuela.

Major Exporters of Manufactured Products—Brazil, the Four Tigers of the Pacific Rim (Hong Kong, South Korea, Taiwan, and Singapore), Coastal China. Note: Other "manufacturing-oriented" and developing countries not included in this category are Argentina, Chile, Columbia, Costa Rica, Côte d'Ivoire, Cuba, Cyprus, El Salvador, Guatemala, India, Indonesia, Malaysia, Malta, Mexico, Morocco, Nicaragua, Pakistan, Paraguay, Peru, Philippines, Sri Lanka, Swaziland, Thailand, Turkey, Uruguay, Zambia, Zimbabwe.

Least-Developed Countries (LDCs)—Afghanistan, Bangladesh, Benin, Bhutan, Botswana, Burkina Faso, Burundi, Cape Verde (islands), Central African Republic, Chad, Comoros, Ethiopia, Equatorial Guinea, Gambia, Guinea, Guinea-Bissau, Haiti, Lao People's Democratic Republic, Lesotho, Malawi, Maldives, Nepal, Niger, Rwanda, Somalia, Sudan, Togo, Uganda, United Republic of Tanzania, Vanuatu, Yemen.

Primary Commodities Exporters—LDCs plus Bolivia, the selected nations of the Caribbean and Central America, Fiji, Ghana, Guyana, Jordan, Kenya, Lebanon, Liberia, Madagascar, Mauritius, Mozambique, Namibia, Papua New Guinea, Senegal, Solomon Islands, Surinam, Tunisia, Zaire.

Highly Indebted Nations—Argentina, Bolivia, Brazil, Chile, Colombia, Costa Rica, Côte d'Ivoire, Ecuador, Jamaica, Mexico, Morocco, Nigeria, Peru, Philippines, Uruguay, Venezuela, former Yugoslavia.

4

ECONOMIC DEVELOPMENT AND MODERNIZATION: PROCESSES AND PERSPECTIVES

OVERVIEW

There are numerous viewpoints on how regional economies progress and develop. Some economic geographers see regional development as a process which occurs in "stages," as a sequence of changes linked to economic growth or integration to the global production network. They believe that each region or nation occupies a position on the global economy's development ladder, with core economies at the top and peripheral economies toward the bottom. The more optimistic of the lot envisage the nations at the bottom as proceeding upward, each at its own pace. Other economic geographers contend that economic development is something that occurs all too slowly in the absence of remedial policy or intervention. They argue that the lack of development policy forces many regional economies to remain at a given level. Critics of this viewpoint argue that internal and/or external market forces will ultimately catapult all regional economies upward. Still others view unequal development as an unrelenting consequence of a world economic system that sustains inequality and unequal access to capital. They see economic disparity and underdevelopment as a direct outcome of the exploitation of labor and resources, with disadvantaged people and regions paying the price for the maintenance of the ladder itself. In essence, the ladder is viewed as a stable and unchanging entity, with economic and political conditions favoring rich and powerful nations at the expense of the poor and powerless. Both rich and poor nations are viewed as part of one defective world system that sustains both development of the core and the advancement of underdevelopment in the periphery.

No matter how one views or characterizes the development status of the world's nations, there is little contention over the fact that additional progress needs to be made. Economic history has taught us that development is something that can occur rapidly if and when the preconditions for development are just right. Although the pace of economic change has hastened somewhat since the Industrial Revolution, all nations have not achieved a satisfactory level of development or integration into the global production network. Many argue that the forces of "political economy" are largely to blame.

As a modern-day exercise in social science, *studies in* **political economy** *focus on regional development and change, highlighting and analyzing the interrelationship of political and economic processes.* Regardless of one's regional development perspective, it is well recognized that the interrelationship between political and economic processes has contributed to marked geographic variations in the rate of economic growth and economic development.

ECONOMIC SYSTEMS: FORM AND FUNCTION

The contemporary post-1950s world economic system is made up of three fundamental components. First, there is a single and predominant world market that is "capitalist" in its orientation. **Capitalism** refers to *an economic system where ownership and control are in the hands of private or individual economic agents whose actions are motivated almost entirely by personal economic gain or profit in response to changing market conditions and investment opportunities.* Capitalism is "a mode of production and social organization in which the material conditions of existence are dominated by production for profit" (Knox and Agnew, 1994, p. 421). The capitalist tendency of the world economy implies that regional growth and development are inextricably linked to the myriad of external forces and factors that influence economic change. In short, regional economic change is known to be affected by **markets**—*spheres within which price-setting forces of supply and demand tend to operate to determine the production, consumption, and exchange (and ultimately the physical movement) of goods and services.* Second, since the end of World War II, two superpowers (as defined in terms of political orientation) have dominated the global political scene. The balance of power in world political economy is thus viewed as "bipolar" in the sense that it is organized around two contrasting systems as best represented by the United States and the former Soviet Union. The conflict and tension between these two opposing superpowers, and the ongoing drama of power politics and economics, short of violent confrontation, has been referred to as the Cold War. Third, in matters of economy, there is a hierarchy of regions and nations that vary in terms of their control or dependence on other regions and/or nations. The relative position of the regions or nations within that hierarchy, and their subsequent influence, is a direct result of both a capitalist global economy and the political and economic realities that shape the dynamic relations between the core, semiperiphery, and periphery.

Recent trends suggest that a significant shift has occurred in the preference of how economies are structured. With the political disintegration of the Soviet Union and its empire, the privatization of industries in Latin America, Asia, and Eastern Europe, and the global-wide diffusion of Western ideologies on matters of economy, there has been a dramatic change in the way individuals and government envision the development process. Increasingly, the many paths to development have involved policies and actions (on the part of government and individuals) that have distinct capitalist overtones. In particular, the end of the Cold War has been marked by a general movement away from "centrally planned" systems toward systems which are "market-based."

Centrally planned systems *are characterized by the public ownership and control of capital and an economy which is entirely regulated and managed by the "State" or centralized governing authority.*

Market-based systems *are founded on the principle of "free-enterprise" and characterized by the individual or private ownership of capital and a decentralized economy in which the production and distribution of goods and services are guided by markets.* Markets are viewed as instruments of efficiency, bringing buyers and sellers together. Through the forces of supply and demand, markets help determine the price of commodities and the quantity produced and exchanged.

Although a large portion of the world nations may be characterized as having attributes of both centrally planned and market-based systems, depending upon the degree to which their government regulates or manages the economy, most nations have been moving away from central planning and toward free-market systems. Currently, the world economy is composed of regions or nations which are typically categorized as falling under one of four distinct types of economic systems: (1) command economy; (2) capitalist economy; (3) mixed economy; or (4) traditional economy.

Command economy *is a centrally planned system, most commonly identified with regions that are under the control of socialist, Marxist, or communist regimes.* It is a form of "totalitarianism," an oppressive system in which the "means of production" (that is, production capital) is owned and controlled by a central governing authority. This authority is commonly referred to as "the State." The State controls virtually all aspects of the economy including production and price levels, resource and investment flows, and the distribution of goods and services. Although command economy rarely exists in a pure form, good examples of this type of system would include the former Soviet Union, present-day People's Republic of China, and the economies of North Korea, Cuba, Vietnam, and Nicaragua (under the former Sandinista regime). Before the break-up of the Soviet Union and the end of the Cold War, approximately 2 billion people (roughly 37% of the world's population) were living in command economies. In the mid-1990s, only about 1.3 billion people (roughly one-quarter of the world's population) reside in command economies (mostly in China) (Figure 4.1).

The shift away from command economy has been largely due to the dismal economic performance of command economies in general. It has been established that the economic growth rates of nations under command economy have severely lagged behind the growth rates of their market-based system counterparts. Regions and people that lack **economic sovereignty**—*the power to influence and guide economic change*—have begun to push for alternative ways in which to organize their economic systems, create economic opportunities, and stimulate growth. The desire for economic sovereignty has brought about a push for more personal and economic freedoms and the need for people to have a stronger voice in matters of political economy.

Figure 4.1 A constant military presence in China's command economy keeps a watchful eye on people's activities. Here border guards monitor the flow of pedestrians crossing a bridge.

Market-based economies are firmly rooted in a belief in capitalism. **Capitalist economy** *is characterized by (a) private or corporate ownership of capital goods and the means of production; (b) investments that are determined by private decision-making bodies or individuals (rather than a central planning authority); and (c) the determination of price levels, production outputs, and the distribution of goods and services in a free-market setting.* The capitalist free-market system is composed of both a "resource market" and a "product market" in which businesses and households can be thought of as both agents of production and agents of consumption. The circular flow of goods and services and money in the resource and product markets are depicted in Figure 4.2. Examples of capitalist economy include the United States, Canada, most of western and central Europe, Japan, as well as most Latin American and Asian nations.

In reality, most economies are **mixed economies—***hybrid economic systems, which have characteristics of both command and capitalist economies to varying degrees.* The degree to which an economy is "mixed" depends on the role of government, the nature of the ownership of production capital, and the extent to which the government is involved in economic decision making. Note that mixed economies are found in just about every corner of the globe. When a mixed economy leans toward free-market activities and private ownership of capital, it is generally referred to as a capitalist economy.

Although predominantly capitalist in its orientation, the U.S. economy could be characterized as mixed given

that the U.S. government is actively involved in regulating industry and intervening in most economic matters. For instance, the U.S. government assumes an active role in stabilizing the economy by taxing income, redistributing wealth, providing basic services, and even producing goods and services that would not be otherwise available given that they are not profitable (for example, the U.S. Post Office (historically), the Environmental Protection Agency, and the Army Corp of Engineers). To a large extent, the market-base socialist economies of Canada, Britain, France, and Scandinavia are also good examples of mixed economic systems.

Consider the fact that while more than 90% of Sweden's land and resources fall under the domain of private ownership, the government maintains an active role in the distribution of income by taxing individuals who earn more than $50,000 a year at around the 80% level. A large share of Sweden's tax revenues go to fund social programs such as a national health-care network that provides lifetime medical care for each member of society. While Japan and many European nations support the free-market system, as well as private ownership and decision-making in matters of economy, the Japanese and European governments subsidize and protect many of their domestic industries. Government involvement buffers these industries from outside competition by reducing the shock of import penetration on domestic companies from drops in market prices and lost revenue. Modern-day China could also be labeled as a mixed economy, given that it has attributes of both market-based and centrally planned systems.

Despite the fact that most nations could be labeled as either capitalist, centrally planned, or mixed, roughly half of the world's people could be classified as belonging to regions that function under **traditional economy**—*a system where participating economic agents are engaged in the most basic of life-sustaining activities (that is, subsistence agriculture).* This is true despite the supposed orientation of their national economy or the degree to which their government regulates and manages the economy. In geographical terms, the concept of traditional economy may be applied to regions covering large portions of the earth's surface. Traditional economy is associated with places where a majority of the active members of the labor force are engaged in agrarian or subsistence endeavors such as agriculture and animal husbandry. Hence, traditional economy is commonly associated with rural settlements in peripheral and semiperipheral areas, where people are, for all intents and purposes, isolated from the workings of modern economy. Although some may live just outside of a large urban area, their lives and livelihood are neither affected by nor dependent upon the economics of the urban system. The lives of people in traditional economies go largely untouched by the forces which affect the ebb and flow of secondary and tertiary sector production in the capitalist world economy. In

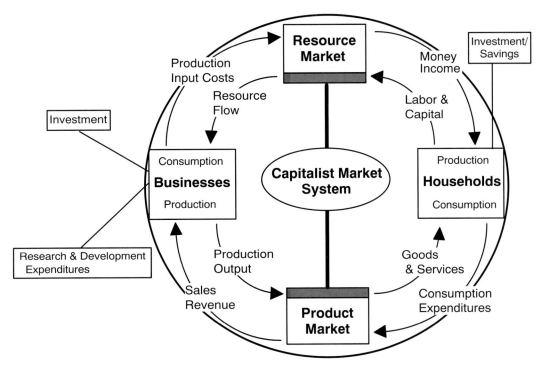

Figure 4.2 Free-market economy.
Source: Adapted from McConnell and Brue, 1993 (Circular Flow Market Systems).

many ways, national and global economic development policies, commerce and trade, and big business are of no real concern to them as their focus is on subsisting.

Transition and Change

Over the past few decades, the large majority of mixed economies have been shifting further away from elements of command economy and toward a more pronounced form of capitalism (that is, away from the public ownership and control of capital toward the private ownership and control of capital). Ironically, it is a trend that is in direct opposition to that which was predicted long ago by political economist Karl Marx—the so-called founding father of Marxism. Marx viewed capitalism as a system flawed and doomed for failure. He envisioned that a conflict would inevitably arise between workers and the owners of production capital in capitalist society (that is, a class conflict between the poor and rich). Subsequently, the poor would revolt and rise up to abolish a system that perpetuated a class distinction. This revolution would lead to an abandonment of capitalist ideals, and a movement toward the "communal" ownership and control of capital would follow.

Socialism was viewed as *the transitional stage between capitalism (in a post-revolution period) and communism.* It is a stage in which an economy is transformed into a system that emphasizes the needs of all members of society. **Communism** *is a system distinguished by the*

communal or community ownership of production capital and the unequal distribution of goods and payments to individuals according to their work and need. While the underlying features of a communist society may seem good in theory, its ideological framework renders it virtually useless as a foundation for a truly operational economy. The very notion of a sustainable socialist or communist system in a world economy that is largely guided by capitalist ideals and market forces defies logic.

Furthermore, it is especially true that large centrally planned systems (that is, those covering large geographic areas) are simply too impractical to manage. Consider the case of the former Soviet Union, a region whose central planning authority found it virtually impossible to control such a large and complex economy, which covered a geographically vast and diverse empire. The price-setting aspects of centrally planned economies alone make them increasingly less feasible as all national economies and production regimes become more interdependent and globalized and as commodity flows are sensitized to price signals in the global marketplace.

The transition from central planning to a market-based system, however, has not been smooth. For example, attempts at restructuring the economy of Russia have been challenged on several fronts. Expectations of consumers over opportunity and economic growth have not been met, and as a result, the effectiveness of political leaders (such as Boris Yeltsin) continues to be questioned. The economic transition period in Russia has been marred by a serious banking crisis. As the Russian citizenry find it increasingly difficult to trust in their fledgling economy and their shaky currency (the ruble), the Russian people will continue to experience the growing pains of economic reform and rebirth. Macroeconomic instability and currency devaluation have led many to panic. Consumers continue to place pressure on the banking system, threatening withdrawal of their savings with the intent of converting rubles to dollars, a situation that has only served to exacerbate the financial crisis.

This type of predicament leaves people to ponder whether they were better off under the old regime. It is likely, therefore, that many regions which attempt the transition from command economy to market-based economy will revert to the more familiar practice of heavy-handed government. A backlash of command economy supporters will inevitably allow political hardliners to regain control of many vulnerable regions. As nations like Russia, Poland, and others combat transition-led instability and crisis, supporters of capitalist ideals and market-based economy will be severely castigated or ostracized. Many pro-reformists will lose their political power base in the process. Although it is unlikely that former command economy regions will return to the rigid, old-style central planning given the prevailing trend toward free-market activity, economic crisis in many regions will undoubtedly bring about new forms of mixed economy as regions experiment with ways in which to contend with the growing pains of reform and transition.

Despite rapid market reforms, slowing population growth, an ongoing trade surplus, and low rates of inflation (about 4% per annum over the last decade), China's economy has not been without its share of problems. Consider that China boasts a GDP of approximately $1 trillion, with an economic growth rate of roughly 10% per year since the mid-1980s. The recent reassimilation of Hong Kong into the Chinese economy has also elevated China's position in the world economy. China's export surge and export-oriented industrialization has been unparalleled. Exports have been responsible for generating $40 billion of revenue in 1987 and over $180 billion in 1997. Revenues from exports will exceed $250 billion by the year 2000.

Note that while the dollar value of Chinese exports has been impressive overall, most exportable commodities tend to be of relative low value (clothing, footwear, toys, nondurables, etc.). China's current economic quandary, however, is not necessarily export related, but attributed to limitations imposed by its centrally planned economy. State-owned industries such as steel, machinery production, petrochemicals, telecommunications, and high-technology continue to pile up billions of dollars in losses each year. The Chinese government continues to subsidize these industries at a high opportunity cost. Bad loans made to many state-owned firms have placed the Chinese banking system in jeopardy. Annual losses of

state-owned industries approached $100 billion per year in the late 1990s.

In response to these problems, a restructuring of state enterprises is currently under way. This includes the planned conversion of some 300,000 state-owned businesses, which account for approximately one-third of China's output, into corporations owned by private shareholders. The privatization of production capital would mean that government's role in the economy will be minimized. Many faltering or bankrupt state-owned industries will be sold to foreign investors or Chinese tycoons in Hong Kong.

As an economic superpower, China intends to export its way out of this dilemma. The modernization plan set into action in 1978 by Deng Xiaoping continues as the Chinese production-for-export network expands and as foreign investment remains strong. In just two decades, a tired command economy was converted into a fast-paced, investment-absorbing, market-driven socialist economy. This hybrid system has lifted over 200 million Chinese out of poverty since the late 1970s. Yet despite the economic miracle, well over 300 million Chinese remain in poverty.

Shortcomings of central planned systems notwithstanding, governments continue to influence regional economic development policy. While the triumph of market-based economy has been truly extraordinary, bringing growth to regions around the globe, socialist systems continue to survive in one form or another, despite empirical evidence that command economies tend to grow slower than market-based economies (on average). The extent to which governments should become involved in matters which affect individuals, industry, and regional economic growth and development is something that remains controversial. Individuals' preference for government involvement may differ dramatically depending on their political mindset or development perspective.

PERSPECTIVES ON ECONOMIC DEVELOPMENT

Although there are many competing paradigms on paths to economic development, there are three predominant "mindsets" or perspectives on development policy as highlighted and discussed by de Souza (1990).

1. The Conservative Mindset

The **conservative** perspective is inherited from the writings of political economists Adam Smith (1723–1790) and David Ricardo (1772–1823), as exemplified in Adam Smith's 1776 treatise entitled *An Inquiry into the Nature and Causes of the Wealth of Nations*. It is a mindset that works under the assumption that human economic agents need incentives to be productive. Conservatives believe in free-enterprise and free-market economy, where decisions and actions (motivated by profit) are guided by market forces. It is a belief rooted in capitalist ideals, as capitalism furnishes the perfect vehicle for innovative and hard-working individuals to realize their potential and be rewarded for their entrepreneurial spirit, savvy, and genius as they create, market, and sell products and services for financial gain. It is a system that allows creative and industrious individuals to achieve maximum personal worth and satisfaction and accumulate wealth.

The conservative perspective on development is based on the philosophy of **laissez-faire**—*a doctrine opposing government or public-sector interference in markets or in the economic affairs of individuals and/or industry.* As such, conservatives are avid supporters of the notion of "comparative advantage" and "free trade." They see the production specialization of regions and trade as ways to stimulate economic growth. It is the belief of the conservative that the growth will inevitably lead to more growth.

Conservatives are not only advocates of free trade and growth, but are fundamentally opposed to the regulation of industry by government. Conservatives adhere to the belief of finding market solutions to economic problems. They see the **invisible hands of the market**—*the laws of supply and demand as governed by needs, wants, and economic scarcity*—as the ultimate guiding force in the allocation and distribution of human, physical, and economic resources. Conservatives believe strongly in the efficiency of markets. They hold strong to the conviction that economic growth will inevitably translate into gains and opportunities for all members of society, and the benefits of growth are likely to "trickle-down" as growth and prosperity will create jobs and opportunities for people in all income strata. In addition, conservatives believe that eradication of poverty is indeed possible by promoting economic growth.

Conservatives argue that international economic development problems may be eliminated with increasing participation of underdeveloped countries in the world economic system. They see the combination of integration and free trade as the ultimate antidote for what ails most depressed economic regions. The majority of conservatives contend that regional economies should seek their comparative advantage, specialize accordingly, and engage in trade for the purpose of gaining currency to fuel development efforts. This is the very basis behind external growth theory, where nations utilizing their comparative advantage can achieve integration into a world economy by establishing a specialized production-for-export base (an argument originally put forth by David Ricardo as a way to promote production efficiency and the wise use of resources).

Conservatives contend that economic difficulties and problems in underdeveloped regions can usually be traced back to any or all of the following:

1. internal obstacles in the physical environments and environmental limitations associated with poor soils, unreliable climate, natural hazards, and unfavorable resource endowments;

2. internal obstacles in regional economy that create noncompetitive business climates, restrict access to technology and capital, lower productivity levels, and make for insufficient economic incentives due to risk and uncertainty and/or unstable political systems;

3. problems inherent to the prevailing political systems, which causes regional fracturing, civil disorder or chaos, and/or disruption of production and distribution mechanisms;

4. the noncooperative aspects of the indigenous culture as defined by its traditions, values, religious beliefs, and/or its resistance to change; and

5. excessive population growth and the subsequent burden placed on domestic production systems.

Once these impediments are overcome, conservatives are convinced that virtually any region or economy could elevate their economic development status.

Contemporary advocates of the so-called "right-wing" conservative mindset would include such public figures as former Republican presidents Richard Nixon, Ronald Reagan, George Bush, as well as political figures such as Newt Gingrich, Bob Dole, Trent Lott, and Rush Limbaugh. The World Bank, historically, has exhibited a tendency to back conservative economic development initiatives. Consider the 1988 World Bank Report, which focused on "poverty" as a global development problem. In that publication the World Bank outlines a strategy for reducing the incidence of poverty. The stated policy objectives were twofold:

1. promote broadly based economic growth that will generate income-earning opportunities for the poor through the continuation of global market expansion and the proliferation of market-based operations; and

2. ensure economic development by improving access to education, health care, and social services to help the poor take advantage of the benefits of economic growth (once realized).

Opponents to conservative development strategies claim that the conservative perspective is consistent with blaming the disadvantaged for being disadvantaged; the equivalent of "blaming the victim." Critics of conservative economics cite that the cycle of poverty and economic distress (and associated socioeconomic ills) perpetuate unequal access to production and financial capital and resources, precluding many individuals, societies, and regions from realizing their economic growth or development potentials. They point to the stratified nature of the world economy and the widening gap between the rich and poor nations—the increasing polarization of regions—as evidence that economic growth does not necessarily lead to development. Even in the more affluent nations, we find a disappearing middle class, a growing number of people falling below the poverty line (in absolute terms), nagging unemployment problems, the expansion of low-wage service sectors that offer little incentive for those with few options, and trickle-down effects that never really trickle down to those at the bottom of the socioeconomic totem pole. The creation of growth-driven employment opportunities in low-wage sectors does little to remedy poverty, as the economic rewards associated with low-wage employment are marginal. Overall growth of low-wage employment may contribute to the further decline in real wages (observed in the U.S. economy over the last two-and-a-half decades). As government budgets tighten in response to reduced revenue potential as a consequence of declining real wages, the poor will experience reduced access to public services and less in the way of poverty alleviation.

It is wishful thinking to believe that the benefits of regional and global economic growth will be equitably distributed. Even the World Bank's development strategy is suspect as it rests on the assumption that a sufficient number of good employment opportunities will be generated by economic growth to accommodate the growing number of poor in underdeveloped regions; it is the quality of employment opportunities that matters and not simply the quantity.

Undoubtedly, conservative development policies that promote economic growth are most beneficial to core economies (where the lion's share of the world's wealth, production capacity, and power are already concentrated). Growth will allow producing regions and industry in the core to maintain their competitive edge. The conservative agenda, however, does little in the way of providing a remedy for what ails the more peripheral economies.

At present, the poorest of the poor nations have little hope of securing the much-needed investment capital that would allow them to industrialize and compete with production zones in the core and semiperiphery. Moreover, conservative economic development policies promote the exploitation of peripheral and semiperipheral regions, particularly those with favorable resource endowments and/or cheap and abundant labor pools.

Opponents to free-market activity contend that a pure market-based system would fail to bring about the necessary changes in regional economies that would ensure development and progress for people at all levels in society. They argue that it is the continued exploitation of the underdeveloped world's resources (both natural and human) which reinforces economic imbalances and uneven development. The origins of these imbalances can be traced to "colonialism" and "hegemony."

Colonialism (sometimes referred to as imperialism) *is an historically rooted phenomenon marked by the creation, proliferation, and perpetuation of a colonial empire across many geographic regions as influenced by the economic agendas of the controlling forces within a nation that has laid claim to various territories and/or directly established colonies in foreign lands which fall under their authority and control.* Colonialism represents a legacy of domination and control of peripheral areas by core economies. It is a spirit embodied in the early formation of European-controlled settlements in Asia, and elsewhere the use of overseas labor and resources to establish a European-dominated trade-and-export system. Following the discovery of the Cape of Good Hope in 1498 and the voyages of merchant adventurers to the East Indies, commercial interests such as the East India Company (originally established by the Dutch in the early 1600s and later by the English and French) laid claim to various territories across East Asia. European claims and colonies were established in the 17th and 18th centuries by private companies with an active support of government, spanning lands which crossed Indonesia, Malaysia, and India. The political, economic, and administrative influence of these colonies is still felt today as production linkages and trade routes, which emerged hundreds of years ago, still affect the flow of money and goods throughout this region.

Hegemony refers to *the domination of a region by one state or nation in a particular historical epoch through economic, military, political, financial, and/or cultural means.* A classic example is the former Soviet Union's post-World War II political grip on eastern European nations. The military presence of the former Soviet Union in countries such as Poland, Hungary, and Romania helped them to maintain territorial domination and expand the economic depth and geographic bounds of their production network. As a ruling and influential entity, the Soviet presence in these lands also preserved and diffused socialist ideas and helped spread communist propaganda, while simultaneously allowing them to exploit the area's physical and human resources.

2. The Liberal Mindset

While conservative economic development policies work well for highly advanced economies, there are those who argue that conservative policies overlook the ramifications of market failures. Opponents of conservative ideals charge that conservative economic growth initiatives due little more than maintain regional imbalances in power and wealth in the world economy and further preserve socioeconomic imbalances. Alternatively, the **liberal** perspective advocates government intervention to correct imbalances when and where observed.

Liberals tend to support a "regulated" economic system, one that allows for the controlled distribution of wealth and the benefits of growth in accordance with nationally or regionally established economic development objectives. Liberals are, in general, staunch proponents of government intervention in matters of the economy. They see the public sector as a mediating instrument that plays a pivotal role in the development process. Liberals not only support government intervention in the marketplace to correct for market failures, but also support the regulation and taxation of industry and the rich to restrain the influence of power structures of the "privileged."

Liberals contend that there must be accountability for excessive or destructive economic growth, and that subsequent damages to society and to the environment must in some way be recovered. They believe that the socioeconomic ills and economic disparities are perpetuated and sometimes exacerbated by free-market activities. It is a belief rooted in the notion that the allocation of resources, the distribution of wealth, the availability of capital, access to services, and the general working of the economy cannot be placed solely in the invisible hands of the market. The liberal perspective is based upon the writings of the economist John Maynard Keynes (1883–1946), forming the basis of modern-day Keynesian economic theory.

Liberals place great emphasis on individual equality and social justice. They believe that government programs and legislation are necessary to attack the problems endemic to the so-called **culture of poverty**—*the disadvantaged people and regions held (or oppressed) within the vicious cycle of poverty, or those chronically hindered by adverse economic conditions and distress and subject to the array of social and psychological problems which stem from feelings of despair and hopelessness* (Figure 4.3).

The liberal viewpoint is marked by the insistence that inequalities can be reduced through the redistribution of wealth. This is easily accomplished by taxing the wealthy and redistributing the wealth to the poor. Hence, liberals tend to support progressive income-tax schemes aimed at imposing higher taxes on those with the greatest ability to pay, while lessening the tax burden on those with more modest incomes. Income-redistribution will, as they see it, ensure that socially acceptable changes occur rapidly and directly, and at a rate that exceeds changes brought about by trickle-down economics.

Opponents of the liberal viewpoint cite the heavy-handed "tax-and-spend" ethic as nothing more than a futile attempt at combating problems which only markets are capable of solving. Conservatives would label income-redistribution policies as destructive, counter-productive, inflationary, and debt-increasing. Conservatives are quick to point out that progressive taxation and income redistribution reduces personal savings and investment and ultimately reduces a region's economic growth potential. Conservatives argue that penalizing the rich and economically powerful (people or groups that have worked hard to achieve success in the capitalist system) serves only to weaken the foundations of the economy and reduce the likelihood of regions sustaining a globally competitive position throughout the next millennium. Preserving a system that rewards hard work and creativity and provides not only monetary compensation for contributing members of society, but the freedom to accumulate savings and wealth, is certainly a worthy goal. Savings and investment mean economic growth and opportunity. Economic growth and opportunity mean innovation and leadership—things that ultimately translate into positive changes for all members of society as an economy flourishes. While liberals see progressive taxation as a means to an end (and as a way to fund social

Figure 4.3   The culture of poverty: urban decay and economic decline continue to trap the inner-city poor.

programs), conservatives see it as a growth-reduction mechanism that punishes the more-productive members of society and dampens entrepreneurial spirit.

Liberals believe that problems of income disparity and poverty are, in part, by-products of rising price levels that stem from the "monopolistic" or "oligopolistic" tendencies of industry—where markets come under the control of one or a few leading sellers who ultimately raise prices above a level that would emerge under a more "competitive" environment, as producers attempt to increase profits. Conservatives counter by arguing that government intervention and regulation of industry lessens the efficiency of producers. Penalizing large and productive firms will not allow them to pass on the savings associated with efficiency to consumers. Reducing the competitive position of large and efficient producers who have gained a large or controlling share of a geographic market, on the basis of promoting competition, is counter-intuitive. Conservatives question the justification of a government that decides to support smaller and less-efficient producers at the expense of promoting production inefficiencies. Large, few, and efficient producers may actually offer the consumer savings as efficiency gains due to size—"economies of scale"—are passed on to the consumer in the form of lower prices (as production costs are lowered on a per-unit-output basis).

When it comes to matters of the environment, the liberal perspective is less apt to be criticized because there are many cases in which no market exists to control the negative effects of economic growth. Some economic activities bring about **externalities**—*spillover effects from production or consumption activities that go unpriced or unaccounted for in the market.* In the case of a "negative" externality, such as damage to the environment, liberals clearly support restrictions to limit damages or the imposition of impact fees or taxes to recoup the cost of the damage inflicted on the environment (Figure 4.4).

In matters of regional and global development, liberals believe that things cannot be left up to the market alone. Whenever market mechanisms fail to provide basic human needs (for example, housing, health care, adequate food and income, and security), the government must step in and right the wrong by providing aid and assistance to needy individuals or regions. This includes the distribution of foreign aid in international crisis situations, humanitarian aid in times of war or natural disasters, or simple gestures of good will. Contemporary followers of the liberal perspective include Bill Clinton, Al Gore, Jimmy Carter, and economist John Kenneth Galbraith.

Critics of liberal development policy in the United States, while usually in full agreement that the desire to help those in need is an admirable one, may show concern for any of several reasons. First, economic assistance is only a partial and temporary solution to economic development problems, one that does not allow people of a

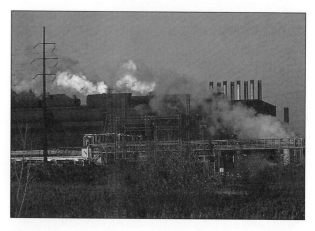

Figure 4.4 The drawbacks of industrialization: smoke-stack industry emitting pollutants into the air.

given region to help themselves by eliminating the underlying cause of the development problem itself (if one can indeed be identified). Second, if economic assistance comes in the form of loans (which must be paid back) it further reinforces the dependency of the periphery on the core. Should underdeveloped regions default on those loans, they run the risk of earning a reputation as economically volatile regions, further reducing their long-term development potential and their ability to attract investment capital. When economic assistance from a core region becomes commonplace or expected, developing regions become dependent and not self-reliant. In addition, a "bail out" of a peripheral economy from a crisis situation is a large financial burden for those supporting the income transfer.

Economic development loans to regions threatening to default on prior loans are not uncommon. Consider the fact that the United States continues to offer financial assistance in the form of loans to Mexico for development purposes or as a means to stabilize the Mexican economy when crisis situations arise (despite Mexico's lackluster economic performance and loan repayment history). In addition, the opportunity costs of economic assistance are high as it usually means a loss of investment money (and growth) somewhere else in the world economic system.

Despite these concerns, liberals view the need to intervene and assist in development matters as a moral obligation and an economic necessity. As the United States and Mexico further advance trade relations, preserving the health of the Mexican economy is of paramount importance to U.S. industry, because roughly 70% of Mexico's imports come from U.S. producers. But there are also concerns over who ultimately pays the bill. Financing social programs, development projects, loans, or humanitarian aid is problematic because money for such endeavors must come from somewhere. It usually means that the working middle-class taxpayer, already overburdened with taxes and subject to eroding real incomes, will foot the bill.

External growth theory suggests that the opportunity costs of funding many international development programs, loans, or income transfers may be high, as that money may not be used to support economic growth and development efforts on the domestic front once the transfers are made and leakage occurs. From an economic efficiency standpoint, there is little incentive for the nations of the core (when viewed either individually or collectively) to redistribute wealth amongst the world's more peripheral regions, as they do so at the core's expense.

3. Radical Mindsets

Two significant schools of **radical** thought have surfaced in response to the prevailing political and economic climates in the capitalist world economy; namely, the libertarian and Marxist perspectives.

Libertarians *are advocates of the doctrine of "free will" and extreme supporters of personal freedom in matters of economy, opposing or rejecting restrictions placed upon them by a governing authority.* The libertarian philosophy is founded on the principle of individual liberty. In an extreme case, libertarians are promoters of "anarchy," denying the very existence of any ruling power or prohibiting authority. The libertarian mindset has been dismissed outright as a viable economic development perspective given that it essentially supports the breakdown of order in society and/or economy, an outcome that certainly cannot promote progress in the human condition. Liberty in action amongst individuals uninterested in the collective good is problematic, as the total social and economic cost of externalities (that is, spillover effects) of individual action would most likely be greater than the total benefits associated with the realization of personal freedom. Unlike the libertarian position, the Marxist perspective can be taken more seriously, as it offers a critical look at capitalism, rooted in social and economic theory.

Marxist ideology may be traced back to the writings of the political economist Karl Marx (1818–1883)—the founding father of **Marxism**—*a radical philosophy on political economy that emphasizes the dialectic approach in the analysis of development problems, and the ongoing "class struggle" between the working poor and the rich or powerful.*

Marxists *view conservative and liberal economic policies as part of a built-in agenda through which the controlling agents of the capitalist system, those that own the "means of production," seek to perpetuate the status quo and exploit the working class.*

The spoils of economic growth, including those that trickle down to the poor or the periphery, are viewed as a way in which to draw attention from the underlying causes of development problems. Marxists believe that it is the nature of capitalism itself that is responsible for the oppression of the world's labor force and uneven development. The control of the global economy by corporate interests, governments, or economies of the capitalist core is viewed as the primary development obstacle.

Marxists argue that under capitalism, human needs are subordinate to the needs of the marketplace. In other words, only commodities with "exchange value" are produced, while the demand for commodities with "use value" remains unmet. The exchange value of a commodity refers to its prevailing market value or selling price, whereas use value refers to the utility of a commodity from the perspective of those in need. For example, as a surplus producer and exporter of agricultural products, the United States attempts to sell much of its food surplus to countries that cannot afford to purchase it at market prices (rather than distribute food to countries strictly on the basis of need or use value). Only in the case of crisis situations does the U.S. government, for example, offer such food assistance to nations teetering on the brink of famine.

Marxists believe that the capitalist economy and the market serves the interests of the ruling classes and the rich, yet does little to benefit the working class or the poor. The very existence of **class**—*subdivisions of a population or society defined by wealth or access*—forms the basis of a theorized class conflict between **labor**—*the working poor*—and **capital** (as a class)—*those who control the means of production.* Labor supplies its labor power to capital in exchange for wages, payment that is substantially less than the value that labor contributes in production. This class distinction serves as a basis by which capital can exploit labor and reproduce the very conditions that perpetuate exploitation, inequalities, regional disparities, and unequal access to factors of production and resources. Labor adds value to products by changing their form and place utility. As a class, labor is exploited as it produces more wealth in the value-added sequence than is returned to it in the form of wages. This situation allows capital to realize profits, expand their base of wealth, and continue to exploit labor through the further "accumulation of capital" and control of the means of production. Hence, "surplus value" is said to be extracted from labor.

Surplus value *is the difference between the value of labor embodied in a product and wages returned to labor for their production input.* The concept of surplus value was central to Marx's **labor theory of value.**

From the radical perspective, the owners of the means of production will, if unchecked, move to maximize profits or increase surplus value by raising the prices of goods and services as high as possible while paying the lowest wages possible. Organized attempts at increasing wages or benefits (to recover surplus value) will occur periodically in this class struggle. When conflicts arise or pressures are introduced to increase wages beyond what is deemed as reasonable, capital may respond in a variety of ways. Employers may

1. meet the demand of labor by reaching an agreement through compromise or arbitration, thereby allowing some of the surplus value to be returned to labor;
2. introduce labor- or cost-saving equipment to ensure profit levels are maintained;
3. lay off or terminate those involved in organized labor activities and tap into the "reserve army" of the unemployed and those willing to work for the same wage or even less; and/or
4. simply pack up and move their operations to more hospitable business climates, particularly those with a significant labor cost advantage (Figure 4.5).

Marxian doctrine contends that labor exploitation sets the stage for a relentless class struggle between the working poor and the rich. Inevitably, the class struggle would make workers unite and revolt against oppressive elements in society. The transition to a nonoppressive society (void of class distinction) would occur only after a revolution, where the working class would rise up to overthrow the ruling class. The post-revolution formation of a "socialist party" would then enable society to carry out its transition to communism. The end result of this transition is a state of social and economic equality for all: a classless "communist society." This society is characterized by communal ownership of all property (including the means of production). Though Marx was somewhat vague on the specifics of this transition and the actual make-up of post-revolutionary communist society, it is most likely to be similar in structure to that of the "kibbutz" (plural kibbutzim) or the "moshav" (plural moshavim).

The **kibbutz** is an Israeli settlement (typically rural and agricultural) where land is communally owned and all economic and social functions are collectively organized and directed, with activities and behavior directed toward improving the common good. It is a system in which the role of the family in society becomes secondary to primary focus on the "collective." This differs from the **moshav,** a settlement with an organizational structure that is "cooperative" rather than collective. Here the family remains as the basic social and economic unit, where all production and marketing activities are done for the mutual benefit of the families within the settlement.

Although impressive in their own right, the success and longevity of the kibbutz and the moshav are most likely attributable to the cultural homogeneity within these systems, the relatively small size of these settlements, and the limited geographic area under consideration. Viewed in this light, the kibbutz and moshav are models which are not so easily applied elsewhere. It is unlikely that they offer a real solution to the many and varied problems associated with underdevelopment in the socially, politically, economically, and culturally heterogenous (and much-divided) world economy.

Contemporary critics of Marxist and extreme left-wing rhetoric often cite the recent trend away from centrally planned systems toward free-market-based economy as proof of the general failures of socialist systems. In addition, critics are quick to point out that the overwhelming majority of conflicts in the world economy are not class oriented, but rather ethnic, cultural, or religious based. Moreover, the universal appeal for "democracy" is a sign that regions, economic agents, and people prefer having a voice in matters of their own economic destiny. The push toward economic freedom, free-market activity, privatization of industry, and democracy are in direct contradiction to the predictions of Marx.

According to Slater (1995), the general failure of Marxian economics as a guide to regional or global development policy is attributed to three interrelated problems:

Figure 4.5   Organized labor "on strike".

1. the supposition that capitalist economy and subsequent economic development determine the outcome of social and political processes (something which is not a given);
2. the belief that all sociopolitical problems and changes are "class" related;
3. the assumption that a "revolution" had to be engineered by an insurgent working class, rising up to overthrow the power base of the privileged.

In reference to this last point, it is only necessary that people in a given region share a common political imagination or a collective will to overthrow an established and oppressive system, a vision that does not have to manifest itself in a working-class revolution.

In what may be the supreme irony, the "democratic revolution" in the former Soviet Union was largely a result of policies initiated in the upper-echelon of government as opposed to the result of a "grassroots" movement. Mikhail Gorbachev's **glasnost** (in reference to the openness and outward-looking nature of the economy) and **perestroika** (in reference to economic restructuring and the transition from a military- to consumer-oriented production system) were changes in economic ideology championed by leadership in government. Transformations toward capitalist economies have also been widely supported by the poor in most regions, although to varying degrees and with varying degrees of success. At the global level, it is difficult to envision a world society with a common vision. Moreover, it is difficult to imagine a vision that is brought about from a common experience, which is based solely in terms of "class relations."

STAGES OF MACROECONOMIC DEVELOPMENT

Rostow's Theory of Modernization

Contemporary "modernization" theory was first introduced in 1960 by Walter Rostow. Rostow postulated that nations proceed (sequentially) through five stages of development. His observations were based upon the modernization of European nations from the dawn of the Industrial Revolution to the modern post-Industrial era. In Rostow's model, global integration through export-led growth is an outcome of internal changes, which are brought about by the public and private sectors (that is, by government and industry). More specifically, modernization is a process that must be supported by a national commitment to development policies which promote domestic investment and industrialization.

Stage one is characterized by a **traditional society**—an economy oriented around subsistence activities and a social and economic system that is virtually closed or isolated. Production activities utilize low-level technological inputs (as gauged by Western standards), revealing a society that is seemingly unconcerned with modernization. Current-day examples of regions might include the subsistence herding cultures of the Sahel and Saharan regions of Africa. In general, stage one regions are associated with fairly remote areas and/or the rural agricultural zones of the periphery and semiperiphery.

The second stage is known as "preconditioning." **Preconditioning** comes about when a society favors economic change and modernization to bolster national prestige, personal profit, entrepreneurial climate, and a belief in a better way of life. Many sub-Saharan African nations and large portions of Latin America and the Caribbean would fall into this category even today. Note that preconditioning involves the organization and cooperation of political and economic elements and institutions to pursue the goals of economic progress (that is, development through economic reform). Although development goals are established internally, development efforts do not exclude the involvement or assistance of outside entities. Hence, it is not uncommon for nations in the preconditioning stage to seek development loans or outside investment transfers. Initially, investments are made to promote surplus-for-export agricultural production and possible production in low-value-added export industries. Exports are viewed as a way in which to promote foreign exchange and the formation of a production network. The bulk of investment is usually targeted toward exploiting a nation's comparative advantage in primary sector or "entry-level" manufacturing activities such as clothing and textile production.

Stage three is known as **take-off.** Take-off is marked by the breakdown of resistance to modernization across all economic sectors and regions. Heavy investment flows pour into the various target sectors to promote economic growth and industrialization. Economic transfers would include the reinvestment of national income, equivalent to at least 10% of the nation's Gross National Product. Monetary flows to key production-for-export activities are supported by industrialization-through-import-substitution (ITIS) policies. These measures are accompanied by a transformation of portions of the agricultural system to promote commercial agricultural products that can be used as exports. Emphasis is placed on increasing the efficiency of the agricultural system (where possible) to allow for the production of surplus and permit excess labor in agricultural sectors to redirect their efforts toward the secondary sector.

Hence, major shifts are likely to occur in the composition of the labor force as portions of that labor force move out of primary sector activities and into entry-level manufacturing. A large, inexpensive, and productive labor force can help expedite the take-off process, as corporations from abroad relocate production facilities to take advantage of low-cost production environment. Take-off becomes fully realized when a nation's comparative advantage shifts from agricultural exports to labor-intensive manufacturing exports. Big winners here are the sectors which produce clothing and textiles, shoes, toys and trinkets, and standardized consumer electronics.

At the outset, measures are commonly employed to protect newly formed industries from outside competition. Later, the production network takes on a life of its own. Regional "growth poles" begin to emerge on the economic landscape, particularly in and around urban areas. **Growth poles** *are production centers or zones associated with one or two key fast-growing industries.* As the growth of labor-intensive industries supports other activities such as producer and transportation services, the region's evolving production network becomes self-perpetuating. The agglomeration of industry and the formation of linkages help to diversify and modernize the economy. Signs of modern culture diffuse out to even the most remote geographic areas. It is not uncommon for economic output or national income (as measured on a per capita basis) to double during this period. China, over the last two decades, fits well into this category as it has undergone rapid take-off. The rapidly industrializing cities along China's coast have earned the reputation of being growth poles of the Pacific Rim.

Though the world economy remains in a state of transition, analysts who study the rate of economic growth have remarked on the increasing speed at which change is occurring in various corners of the globe. Empirical evidence reveals that it is now possible for countries to develop more rapidly than ever before. Indeed, the pace of change has accelerated. The time required to move through the "take-off" period has steadily decreased over time (see Table 4.1).

The globalization of production and the growth of world export markets have allowed nations to double their income per capita at unprecedented rates. Both the United Kingdom and the United States, for example, achieved a doubling of output per capita (as well as substantial improvements in the overall quality of life) in

about 50 years. Japan's take-off period took a little more than 30 years. In more modern times, the economies of Turkey, Brazil, South Korea, and China have moved even more rapidly through the take-off period, doubling output per capita in 20 years or less. Relatively short take-off periods are now being observed in economies such as Mexico, Chile, India, and Pakistan.

There are several reasons for the shortening of the take-off period. First, there is a higher propensity for regional interaction as evidenced by the intensification of trade flows amongst producing nations as the production-for-export regimes of the core economies go global—the so-called transition to "advanced capitalism." **Advanced capitalism** *is a period marked by the globalization or internationalization of production networks due to the increasing mobility of capital and the intensification of competition amongst firms for multiple and diverse geographic markets.* The push to go global and expand operations overseas is, in part, an outcome of the search for low-cost production environments to achieve a global competitive advantage. Note that internationalization of production has also allowed firms increased access to consumers in foreign markets. Second, the rate of take-off has increased due to **space-time convergence**—*reductions in, and the virtual elimination of, the friction of distance in commodity and information transport.* The convergence of space and time has been credited to dramatic improvements in transportation and communication technology and the increased investment in transport infrastructure worldwide. These advancements have allowed people, products, and information to be transported around the globe at a speed unheard of 30 years ago. Third, **techno-driven development**—*progress and changes brought about by new production processes and the application and marketing of new technologies*—has allowed the nations of the core and semiperiphery the opportunity to advance their competitive positions in the world economy. Technology has led to the creation of new products, new markets, new forms of corporate organization, not to mention expanding export bases for "high tech."

Stage four of Rostow's model is the **drive to maturity.** The drive to maturity is characterized by the diversification of a nation's production and export bases and the redirection of national income toward many "capital-intensive" industries. Benefits of increased productivity and trade are then steered toward social welfare, investment abroad, private and public sector expansion, transportation and communication infrastructure, etc. High technology is applied in various production sectors, transforming some urban growth poles into **techno-poles**—*industrial districts which support high-tech industries.*

As a manufacturing base becomes increasingly diverse, high-value-added manufacturing exports begin to flourish and foreign direct investment (both outgoing and

TABLE 4.1	The Take-Off of Selected Economies: Periods in Which Output per Capita Doubled

Nations	Period in Which Output per Capita Doubled
United Kingdom	1780–1838
United States	1839–1886
Japan	1885–1919
Turkey	1857–1877
Brazil	1961–1979
South Korea	1966–1977
China	1977–1987

Source: *World Development Report* (1991), p. 12.

incoming) accelerates. The benefits of economic growth spread swiftly through the entire national economy. Consumers and producers become receptive to new products and technologies. Examples of stage four economies include Japan (just a few short years ago) and the four tigers of the Pacific Rim (Hong Kong, Singapore, Taiwan, and South Korea), and recently Malaysia—the "fifth tiger."

Note also that an emerging China is now entering into this stage. The rapid diffusion of modern technology is testimony to its commencing drive to maturity. In a nation with a GNP per capita of less than $800 per person, it is not uncommon to see consumers in urban areas using cell phones, pagers, fax machines, laptop computers with Internet access, and electronic mail.

Once the drive to maturity is complete, a nation enters into the fifth and final stage of Rostow's model: **high mass consumption.** High mass consumption is characterized by the internal restructuring of a regional economy and the general movement away from manufacturing activities and toward services. In this stage, both the mix of production outputs and the structure of imports and exports change. Relatively less emphasis is placed on necessities, as the propensity to consume luxury items increases. This stage is marked by the heavy utilization of high technology by consumers and producers. Consumers become reliant on the latest technology. The emphasis on personal consumption translates into advancements in food preparation, packaging, and marketing, the commercial sale of information and financial services, enhanced education services, diverse product offerings, and improvements in transportation. Spending in the public and private sectors accelerates, with increasing expenditures on leisure-time activities, health care, social programs, and national security. Import penetration becomes widely accepted and welcome by consumers.

Stage five economies experience a rapid expansion of the tertiary sector due to the overall increase in demand for services. The United States, most of western and central Europe, and recently Japan are prime examples of economies that have reached this final stage. In reference to Rostow's model, the nations of the "developed" world can be described as falling into either the drive-to-maturity or the high-mass-consumption stages. It is unclear, however, that the poorest and most peripheral regions of the world will ever exit the preconditioning stage or truly experience the benefits of take-off given their debt structure, lack of globally competitive industries, and rate of population growth. Perhaps economic development and modernization require internal and external preconditions that no longer exist, particularly, as the world economy is no longer in its formative stage of development.

The global-wide conditions that worked for many of the core and semiperipheral nations over the last century may not work for the world's peripheral regions, as the dominant core-periphery relations have already been established. W. A. Lewis (1977) contended that peripheral nations must be able to expand their export bases and capture significant shares of the markets of the developed world if they are to have a fighting chance at raising their economic development status. Yet even if they are able to capture a significant share of the world's leading markets, many poor nations may never really experience the benefits of export-led economic growth given their population growth rates and the increased competitiveness within the global production system. The world economy is now experiencing a glut of product output. There are a multitude of competing nations, companies, and product lines. Historically, the take-off of what now would be labeled as semiperipheral nations involved the exportation of low-valued, entry-level manufactured goods. Today the world is flooded with such low-cost products/exports, making it extremely difficult for the preconditioned nations of the periphery to earn a sufficient amount of export income given that they must compete with established low-cost producers in the semiperiphery.

Consider the data on per capita exports of manufactured goods to the high-income economies of the NORTH as shown in Table 4.2. When measured on a per capita basis, the export-led growth potential of the core is much greater than the periphery. High- and upper-middle-income economies tend to dominate exports from high-value sectors such as transportation equipment, whereas low- and lower-middle-income economies tend to emphasize exports of lower value like textiles and apparel. As low- and low-middle-income nations are expected to undergo the greatest increases in population, it is likely that per capita exports will shrink through time. This will preclude many of them from adequately reinvesting in their economies and diversifying their manufacturing base. In short, many poor nations will most likely be prohibited from ever moving through the take-off stage. Note the stark contrast between the per capita exports of high-income nations and low-income nations and the variations by region. Nations of sub-Saharan Africa and South Asia, such as Bangladesh, Afghanistan, Bhutan, Nepal, India, Pakistan, and Sri Lanka currently have the lowest per capita export earnings, exporting less than $15 per person to the high-income economies of the NORTH. This is a rather small figure in comparison to Japan, a nation that exports over 100 times that amount (on a per capita basis).

Per capita export measures, however, mask the importance of the textile and apparel industries to developing economies. As such, one may consider alternative measures, which expose the relevance of various export-oriented industries. To examine the relative strength of a region's export base on a product-specific basis, one must uncover that region's "revealed comparative advantage."

Revealed comparative advantage (RCA_{ij})—*is an observed export advantage of a particular industry or*

TABLE 4.2	Per Capita Exports of Manufactured Goods to Leading High-Income Economies of the NORTH (1993)

In 1993 U.S. $ per Capita

Region or Income Group	Electrical Machinery and Electronics	Transportation Equipment	Textiles and Apparel	Total Exports to High-Income Economies
WORLD	41.96	**54.95**	31.30	333.06
Low	2.83	0.35	**12.12**	31.49
Low (excluding China and India)	0.12	0.33	**8.46**	12.86
Low-middle	8.48	2.54	**26.25**	79.33
High-middle	**69.79**	26.89	46.91	286.04
High	211.05	**339.84**	98.37	1788.61
Severely-indebted nations	0.07	0.12	0.32	1.39
sub-Saharan Africa	0.29	0.54	2.47	14.09
South Asia	0.15	0.19	10.47	16.99
Middle-East and North Africa	2.76	2.15	19.42	43.94
East Asia and Pacific	20.26	2.10	25.64	99.97
Europe and Central Asia	9.93	7.92	37.71	118.20
Latin America and Caribbean	26.49	18.87	17.25	124.97
United States	122.95	158.82	20.49	853.87
Japan	306.32	453.98	1.57	1570.88

Source: *World Development Report* (1995), pp. 162–163 (Table 1. Basic Indicators) and pp. 192–193 (OECD Imports of Manufactured Goods by Origin); based on Hanink, 1994, p. 336.

Note that the top contributing production-for-export sector is shown in bold (by income group).

activity (j) in a given region (i) as measured by the following index:

$$RCA_{ij} = [x_{ij}/ X_i] / [x_{wj} / X_w] \qquad (1)$$

where x_{ij} is the value of exports of product j (or exports from sector j) from a region i, X_i is the total value of all exports from region i, x_{wj} is the value of world exports of product j (or world exports from sector j), and X_w is the total value of all world exports. This index reveals the relative strength of a region's production-for-export base in comparison to the world average. An $RCA_{ij} > 1$ indicates that a region is exporting a greater share of product j in comparison to the nations of the world as a whole. An $RCA_{ij} < 1$ indicates that a region is exporting a relatively smaller share of product j in comparison to the world as a whole. In short, the higher the value of RCA, the greater that region's revealed comparative advantage in a particular export activity (see Hanink, 1994, pp. 341–342).

The revealed comparative advantages of exports for selected regions and income groups are highlighted for selected manufacturing-for-export industries in Table 4.3. Note that only exports to the leading high-income economies of the NORTH are considered (as opposed to world exports) in the calculation of the RCA index. The statistics suggest that peripheral regions hold the greatest revealed comparative advantage in low-value manufacturing; a trait that is undoubtedly linked to the labor-intensive nature of these industries and the availability of an inexpensive and abundant labor force in these areas. As the degree of periphery decreases, the revealed comparative advantage in high-value-added manufacturing, such as chemicals and transportation equipment, increases. The greatest revealed comparative advantages in electrical machinery and electronics are held by upper-middle-income (mostly semiperipheral) nations located in East Asia and the Pacific Rim and in Latin America.

TABLE **4.3**	Revealed Comparative Advantages in Industry-Specific Exports to Leading High-Income Economies of the NORTH (1993)

Revealed Comparative Advantage (RCA) by Industry

Region or Income Group	Textiles, Clothing	Electrical Machinery and Electronics	Chemicals	Transportation Equipment
Low-income	**4.095**	0.714	0.322	0.066
Low-income (excluding China and India)	**7.000**	0.071	0.245	0.157
Lower-middle-income	**3.521**	0.849	0.593	0.194
Upper-middle-income	**1.744**	**1.936**	0.492	0.570
High-income	0.585	0.936	**1.129**	**1.152**
sub-Saharan Africa	**1.862**	0.167	0.720	0.230
East Asia and Pacific	**2.734**	**1.611**	0.262	0.127
South Asia	**6.553**	0.071	0.288	0.067
Latin America and Caribbean	**1.468**	**1.683**	0.585	0.915
Middle East and N. Africa	**4.702**	0.500	**1.212**	0.297
Europe and Central Asia	**3.394**	0.667	0.814	0.406
China	**3.425**	**1.000**	0.310	0.048
Philippines	**3.021**	**2.841**	0.119	0.042
Malaysia	**1.140**	**3.738**	0.186	0.078
S. Korea	**2.106**	**1.968**	0.345	0.375
United Kingdom	0.511	0.794	**1.593**	0.885
Germany	0.510	0.841	**1.331**	**1.273**
United States	0.255	**1.142**	**1.076**	**1.127**
Japan	0.106	**1.548**	0.347	**1.752**

Source of data used in calculating RCA indices: *World Development Report* (1995), pp. 192–193.

Note: RCA values using exports to leading high-income nations of the NORTH only. Revealed comparative advantages by industry (for RCA's > 1) are highlighted in bold.

In addition to geographic variations in revealed comparative advantage, there is much empirical evidence to suggest that the current patterns of global trade, especially amongst members of the core, is a persistent one. It has been argued that the stability of international trade is an outcome of historical inertia. In many ways, international trade linkages are characterized as "self-reinforcing" (Hanink, 1994). It is unlikely, therefore, that Rostow's modernization theory is applicable in a world economy whose dominant core-(semiperiphery)-periphery relations have already been ensconced by the geography of production and trade and reinforced or perpetuated by the very nature of advanced capitalism. As such, one could posit that the development of any one given region must come at the expense of another region. For instance, the export bases development of Asian nations has displaced the exports from economies outside this region. Moreover, there is doubt as to whether many underdeveloped nations of the periphery can build and

sustain a globally competitive export base for a period of time necessary to secure enough national income (for investment purposes) to foster a self-perpetuating production network. Despite this doubt, there is evidence that structural changes are occurring in low-income economies (see Table 4.4). Furthermore, long-term growth in GDP per capita and exports has been observed throughout the SOUTH (see Tables 4.5 and 4.6). Questions and concerns remain, however, as to whether these trends provide sufficient proof to support the contention that the underdeveloped world is "developing."

PORTER'S GLOBAL COMPETITIVE ADVANTAGE

Many economic geographers believe that the concept of promoting comparative advantage has outlived its usefulness as an economic development policy goal. As the

TABLE 4.4	Structure of Production of Low-Income Economies 1970 and 1993

	Distribution of Gross Domestic Product (%) in Low-Income Economies		
Year	Primary	Secondary	Tertiary
1970	37	29	33
1993	28	35	38

Source: *World Development Report* (1995), p.166.

Note: Percentages may not sum to 100% due to error introduced by rounding.

TABLE 4.5	Historical Trends in GDP per Capita

	Average Annual Growth Rate of GDP per Capita (%)	
Region	1913–1950	1950–1989
Asia	–0.1	3.6
Latin America	1.2	1.4
sub-Saharan Africa	n/a	.8
Eastern Europe	1.4	2.0
Developing Economies	1.0	2.7
O.E.C.D.	1.1	2.3

Source: *World Development Report* (1991), p.14.

Note: "n/a" means not available.

world becomes increasingly interdependent and interconnected, regions must seek to establish and exploit not a comparative advantage, but a global "competitive advantage." Porter (1990) contends that a region's **global competitive advantage** is *an outcome of four major considerations: factor endowments, internal demand conditions, density of industrial linkages, and corporate structure.*

1. **Factor endowments**—*the internal condition of a region's infrastructure and its physical, human, capital, and knowledge-based resources, which includes the way in which economic agents organize themselves, utilize information, skills, tools and production inputs, and create and combine knowledge of how to produce.* Factor endowments may be viewed as either "basic" or "advanced." **Basic factors** include resources, climate, location, population size (and the size of the region's labor force), and land. **Advanced factors** include education attainment levels, the competitiveness of a region's labor force in relation to skills and productivity, research and development (R&D) capabilities as tied to a region's propensity to invest in the development of new products and technology, and the quality and characteristics of its communication and information networks.

2. **Internal demand conditions**—*the size, health, vitality, and characteristics of a region's internal market (domestic demand).* The greater the number of domestic buyers, the greater the potential for industries to achieve "economies of scale" in production. The more demanding, discriminating, and sophisticated the consumer, the more pressure is placed upon competing firms to produce high-quality goods and services and make improvements in design when possible.

3. **Density of industrial linkages**—*a recognition that production efficiency and competitiveness is directly proportional to the breadth and depth of a region's industrial linkages.* Consider the case of South Korea, a nation that has emerged over the past few decades from a simple exporter of entry-level

manufactured products to a premier ship-building, automobile-manufacturing, and steel-producing giant. Although the rise of South Korea's domestic steel industry was initially supported by entry-level manufacturing, solid linkages to ship-building, heavy manufacturing, and automobile production have been responsible for the fortification of this industry.

4. **Corporate structure**—*highlighting the importance in the way in which firms are organized and managed, things which have a direct bearing on the competitiveness of a firm (in a business environment that has become sensitized to customer service and satisfaction).* Efficient organization is known to enhance a firm's decision-making processes, increasing both its competitiveness and its longevity.

The geography of the corporation has undergone marked changes due to the forces of competition in the global economy. Companies have reengineered and streamlined their operations in order to promote efficiency (the attainment of maximum output while minimizing the cost of labor and capital inputs). Workers have been displaced by machines, others have been up-skilled, and still others have assumed new and expanded responsibilities in the production process. Not only are leading firms cognizant of external market conditions as they try keeping up with competitors, monitoring signals from consumers, and anticipating the latest trends in regional and international markets, but they have become more internally "flexible." Flexibility means reorganizing themselves in ways that allow them to make necessary adjustments or production shifts in a less costly or time-consuming manner. By remaining flexible, firms can meet the ever-changing needs and demands of consumers in the diverse regional and global marketplace.

Traditional top-down corporate structures (such as the organization of the firm as a pyramid with executive-level management at the top and workers below) have been replaced by the "horizontal model," leading to significant increases in productivity and efficiency. According to Byrne (1993), the **horizontal model** of corporate structure has seven key attributes.

1. The firm is organized around the entire production process and not simply a series of individual tasks.
2. The firm is arranged in such a manner as to reduce supervision, combine fragmented tasks, eliminate steps that fail to add value, and allow for greater flexibility through interchangeability (as team members are encouraged to develop multiple skills).
3. Self-managing teams serve as the basic building blocks of the corporation, overseeing all aspects of the production process, with teams and team members held accountable for their performance and productivity.
4. Performance evaluations are customer-driven, based on customer satisfaction and not simply profit or quarterly earnings.
5. Rewards are based upon team results and performance rather than individual productivity.
6. All employees (sales representatives, management, and team leaders) are placed in direct contact with suppliers and customers to form a better understanding of the needs of the company and the market, and how to better improve product lines or the production process.
7. Employees and management are informed and updated on all changes in the production process and not just on a need-to-know basis. (Figure 4.6)

The structure and organization of today's corporations have been influenced by increased competition in globalizing production and the search for production cost advantage. Transport costs have become less important in the production equation as major advancements in shipping, packaging, and distribution as well as communication and information technologies have reduced the time it takes for inputs to arrive at a production facility or finished products to reach the market. It is, however, still costly for firms to remain responsive to changing market signals in their attempt to stay on the competitive edge. Globally-minded corporations constantly need to be aware of the many forces that shape and influence production advantage, marketing, and sales, not to mention conditions that affect firms' abilities to compete internationally. Producers must be sensitized to the capacities of local and regional suppliers, changing market conditions, and the changing needs and whims of consumers in the diverse regional markets.

While transportation cost considerations are of little importance to many industries, information and the application of new organizational technologies have become

TABLE 4.6	Historical Trends in the Growth of Exports	
Nations, Region or Income Group	**Average Annual Growth Rates in Exports (%)**	
	1970–1980	**1980–1993**
Low-income economies	2.7	6.4
sub-Saharan Africa	1.0	2.5
Severely-indebted nations	2.0	2.4
South Asia	4.2	7.3
East Asia and Pacific	9.0	10.8
Middle East and N. Africa	−0.8	−1.0
Selected Low- and Middle-Income Economies		
China	8.7	11.5
India	5.9	7.0
Bangladesh	−2.4	9.8
Pakistan	3.1	10.0
Malaysia	3.3	12.6
Thailand	8.9	15.5
Latin America and the Caribbean	0.9	3.4
Upper-middle-income economies	2.7	4.2
Selected Upper-Middle-Income Economies		
Brazil	8.6	5.2
Mexico	5.5	5.4
Venezuela	−6.8	1.7
S. Korea	22.7	12.3
Malaysia	3.3	12.6
South Africa	7.9	5.4
High-income economies	6.0	5.1
Selected High-Income Economies		
Hong Kong	9.9	15.8
United States	7.0	5.1
Germany	5.6	4.2
Japan	9.2	4.2
France	6.8	4.4
Italy	6.9	4.3
United Kingdom	4.3	4.0

Source: W*orld Development Report* (1995), p. 186.

essential ingredients in the modern production equation. The recent movement toward **flexibility** in production— *the adaptability and responsiveness of firms to market signals and the changing needs of consumers and the subsequent utilization of new methods of production, distribution, and inventory control.* Flexibility has not only changed the nature of competition, but it has forced firms to look beyond their production and geographic limitations.

> Flexible structures have allowed . . . firms (capital) more freedom to choose from a world selection of key production locations. As a result, fewer locational decisions are now taken within the context of a single country. Instead, different regions have to compete directly with each other in global markets. (Bennett, 1991, p. 107).

Porter (1990) provides a unique perspective on the development of nations as part of global economic integration. Economic development status is discussed in terms of regions' global competitive advantage. Similar to Rostow's modernization theory, Porter's competitive advantage model ranks a region's development status in four stages.

Stage one is the **factor-driven stage,** where economic growth and development potential is first tied to a region's comparative advantage. This is determined by the availability of "basic factors" of production, including a favorable natural resource base and/or an abundant, inexpensive, and competitive labor force. This stage is associated with nations that are large exporters and/or processors of primary sector commodities. The export of low-value commodities places these regions, however, at a tremendous trade disadvantage. If resource endowments are favorable, but the region is unable to compete in the industries which require those raw materials as production input, it runs the risk of remaining a primary commodity exporter rather than an exporter of processed or fabricated manufactured products. In an attempt to secure foreign capital and support national economic development efforts, the region may rapidly deplete and squander precious resources. Most of the least-developed nations of the world today may be classified as having an economic development status that is "basic-factors related." For example, the nations of the Middle East and many African nations have a reputation of being exporters of fuels, minerals, and metals. In many cases, raw materials are purchased and consumed by developed regions without concern over the opportunity costs of depleting those resources. The outward flow of these commodities greatly reduces the future economic development potential of these regions.

Stage two is the **investment-driven stage.** It is based upon a region's willingness and ability to invest in domestic industries in an attempt to improve its competitive position. It is the stage of modernization of industry and infrastructure. The focus of this stage is the use of "advanced factors" of production in capital intensive industries, which require highly skilled labor, or in industries that have a large labor-cost component. Furthermore, the emphasis is usually on production of standardized commodities using production methods or technologies transferred from the core. Activities generally include the

Figure 4.6 Today's modern corporate environment facilitates the flow of information between employees and management.

manufacturing of textiles, clothing, shoes, trinkets, toys, recreation equipment and sporting goods, and hand-held tools. As domestic industries mature, production emphasis may turn toward commodities of higher value (for example, machinery and electronic devices). Examples of investment-driven transformations include post-World War II Japan, modern-day Mexico, and the high-growth economies of the Pacific Rim (for example, China and Malaysia).

As nations secure a globally competitive export base and a steady source of incoming dollars to fuel industrialization, they stand ready to enter into the **innovation-driven stage.** This third stage is characterized by the creation of new products and new production technologies for new markets, where competitive advantage is no longer tied to basic factors or cost, but is a direct result of technological innovation and techno-driven change. Regions that have entered this stage are likely to give birth to firms and industries which operate across international boundaries—"transnational corporations." These regions will also experience **advanced network economies**—*savings arising from the interconnectiveness of firms and suppliers in or across multiple regions or markets and/or the interregional coupling of complementary production networks which span multiple jurisdictions about the globe.* Consider the numerous firms, networks, and regions joined together in the production of the "airbus" (a short-to-medium-ranged subsonic jet passenger plane). Although a handful of European-based firms lead the way in the development airbus technology, production and assembly of the airbus and its parts extend over many regions and networks in Europe, west Asia, southeast Asia, and the Americas.

The more-advanced nations of Europe (Germany, France, etc.) and Asia (Japan, Singapore, South Korea, etc.) are prime examples of innovation-driven economies.

As nations exert a commanding influence in the world economy and their industries mature and diffuse over space, they enter into the **wealth-driven stage.** This fourth stage is typically associated with the relative decline of a national economy due to its inability to consistently reestablish itself as an innovator in a globally competitive economic system. The focus of the economy changes. Emphasis is placed upon the preservation of status and long-term economic stability, as opposed to entrepreneurial- and innovative-led change. The distribution of old wealth becomes more common than investment in new enterprise. As a result, economic growth rates become sluggish as consumers and industry become complacent. Production of consumer durables shifts to external low-production-cost locations and import penetration intensifies. The growth of exports slows, followed by the contraction of the secondary sector, which in turn leads to a general slowdown in the national economy. Examples of wealth-driven economies might include Great Britain and the United States as they continue to lose ground to foreign competitors.

Consider the case of the U.S. economy, which came fairly close to losing its global competitive advantage during the 1970s and 1980s due to complacency. In particular, the automobile and steel industries buckled (and some almost collapsed) as a result of (a) the effects of intense and unprecedented international competition; (b) the failure of industry to adequately reinvest and/or modernize and upgrade existing production facilities; (c) the inability of domestic producers to make significant gains in productivity and product quality; (d) misguided trade and economic policies fostered by reckless leadership in government; (e) continued pressures from organized labor to increase wages and benefits in the face of shrinking profit margins; and (f) the perceived risks of increasing research and development (R&D) expenditures to stimulate innovation-led industries such as alternative energy, education, science, and high tech.

Fortunately, the recent restructuring of the U.S. steel and automobile industries (by way of downsizing, streamlining, improving quality control, and increasing labor productivity) has allowed these industries and the U.S. economy to buy some time. Many industries that were once in danger of losing their competitive edge have now been given a second chance, but their survival and success are highly dependent upon the mutual cooperation of government, industry, and labor, and their ability to move beyond bottom-line economics toward promoting larger social and development goals (D'Costa, 1994, p. 128).

Advanced production factors, technology, and the diffusion of technology in the capitalist world economy are largely controlled by the core and semiperiphery. Hence, it is unlikely that peripheral economies will ever leave stage one of Porter's model. Much optimism remains, however, as to the future of the semiperiphery, as regions and industry in the wealth-driven core become increasingly complacent. It is likely that significant market shares will be lost to the newly industrialized economies in Asia and Latin America.

GLOBAL DEVELOPMENT: CHOICES AND PROSPECTS

Development strategies have changed with the transforming world economy. Noninterventionalist "laissez-faire" policies, which rely on the invisible hands of the market to guide the economic affairs of regions and nations, have been slowly replaced with strategies that call for limited government intervention (when and where necessary). The role of the government in the development process has certainly undergone profound changes. Governments are now viewed as agents of economic security, functioning to counter market distortions, and ensure price stability, as well as maintain their more traditional roles of providing national defense and maintaining law and order. In addition, governments utilize "fiscal policy" and the power of taxation and spending to generate revenues to support social programs, as well as "monetary policy" and the setting of interest rates to stabilize the economy and sustain economic growth. Perhaps two of the more important functions of government are its ability to establish a system of money and credit and to control the growth of the "money supply." With reference to the latter, a stable money supply is generally viewed as a desirable economic goal; it is well known that an excessive supply may increase consumer spending and fuel inflation. Yet, there is a downside to government regulation and control. For example, placing too much of a tax burden on business and industry is something that is viewed as undesirable by many because it discourages entrepreneurial activity, savings, and investment. Thus, too much government has the potential to limit economic and employment opportunities by limiting economic growth.

The laissez-faire development philosophy has been critically reexamined over the last 30 years. Advocates of change have called for the support of **laissez-passer**—*a doctrine which supports the free and uninhibited movement of goods, people, information, technology, and money in the world economy.* Proponents of laissez-passer maintain that the fast track to economic growth and development involves the unimpeded flow of commodities through the worldwide elimination of tariffs, quotas, and all other barriers to movement or trade. In addition, advocates of laissez-passer call for no immigration quotas or restrictions, thereby allowing people to relocate in response to regional and global market signals based upon their perceptions of which regions offer the greatest opportunities. In this system, the free movement of money and capital are viewed as essential ingredients in the development formula. Opponents of laissez-passer charge that it is an unrealistic platform for constructing national or global development policy because the world economy is composed of heterogeneous regions and cultures, each unique in their own right and each bound by their own agenda and view of what constitutes development. Moreover, freedom in movement does not guarantee that movement will take place, nor does it ensure that desired changes will be observed in all regions.

The lack of a common vision or a single objective reality amongst the world's economic regions or cultures may perpetuate the need for government involvement to help meet region-specific development needs. Consider how the movement of money in the world economy was simplified through the establishment of an international gold standard. The gold standard ensured that the currencies of all countries could be fixed in relation to gold. This was done in an attempt to promote international stability. The Bretton Woods system of fixed exchange rates was later abandoned in the early 1970s (in accordance with the declining economic and political influence of the United States in a post-Vietnam War era), making currencies commodities that could be bought and sold in the international marketplace. Since then, governments have assumed an international finance role. Governments continue to have a necessary preoccupation with exchange rate management, as currencies are known to undergo radical price swings in the international marketplace. Rapid fluctuations in the price of currencies can have a dramatic impact on a nation's purchasing power. Governments generally react in ways which dampen the volatility of a currency's value in the international market. The value of a nation's currency can have a marked impact on production, consumption, and trade flows. If the value of a nation's currency is high relative to others, it sets the stage for declines in production-for-export sectors and opens the door to foreign competitors offering goods at relatively lower prices. For example, a weak U.S. dollar against the Japanese yen would mean that Japanese imports would become less attractive to U.S. consumers because they would increase in price. As noted by Hanink (1994, p. 122), "currency exchange values can be as much of a determinant as they are a result of international trade patterns."

Note also that the flow of production capital in the world economy is becoming an increasingly important regional development tool. The movement of capital was originally regulated by the International Bank for Reconstruction and Development (that is, the World Bank). This organization assumed the responsibility of managing post-World War II capital transactions to ensure the movement of capital toward both war-torn and underdeveloped regions. In addition to development loans from organizations such as the World Bank and the International Monetary Fund, globalization of the world's production network has enhanced the flow of private capital from the NORTH to the SOUTH. Over the past few decades, corporations have also shown an increasing tendency to produce outside of their domestic economies as they respond to growing international markets. In addition, investors continue to look to the international community for investment opportunities. Suffice it to note that the increased mobility of capital continues to improve economic development prospects in many regions of the SOUTH.

Contemporary global development policies have been recast in ways which reflect a lost sense of confidence in the doctrines of laissez-faire and laissez-passer. Critics of conservative development policies argue that greater emphasis has been placed on "planned" investment to break the vicious cycles of underdevelopment and poverty. Policy analysts and development theorists now acknowledge the "equilibrium aspects of underdevelopment" and the supply- and demand-side conditions that place the nations of the SOUTH in a constant disadvantage to nations of the NORTH. This imbalance was originally labeled by Nurske as the "curse of poverty" (Weaver and Jameson, 1981).

On the supply side, the world's poorest nations have the least to save and invest. Without investment there is little growth. Without sufficient growth there can be no savings. And without savings, there can be no investment. Little or no economic growth means a continuation of the status quo and a worsening of the average human condition over time.

On the demand side, poverty and underdevelopment mean that new markets will not be created nor will growth occur in existing markets beyond that which is population driven. Hence, there are few profitable alternatives for capital and little incentive for entrepreneurs to take courses of action that are vital to the interest of the SOUTH.

This situation has been complicated by several factors. First, inadequacies of markets in underdeveloped regions to operate as either efficient clearing houses for both human and physical resources or to account for externalities has led to less-than-optimal use of resources and less-than-maximum profits or outputs in most economic sectors. Missed opportunities and lower levels of efficiency mean less ability to attract further investment capital. Second, to be effective, investment in industry and infrastructure must occur simultaneously. The additional investment requirement signals a lower potential return to investors. Last, the lack of social (overhead) capital and the low profitability of investments in the public sector has created further shortages of capital in underdeveloped regions. These regions find it very difficult to generate sufficient revenues to pay for roads, schools, hospitals, bridges, power-generating facilities, etc., at a time when investments in infrastructure and human capitals are becoming increasingly vital to development.

Development planning throughout the last two decades has centered around a belief that "good governance" is essential to positive economic and social change. As part of the new global development agenda, progress in the human condition can come about through formation and coordination of regionally tailored development policies which are sensitized to the economics and the "geopolitics" of the world economy. Policy makers and analysts continue to extend the terrain of intervention in the formation of policies cast to encompass the needs of a world society. Overcoming the problems of underdevelopment, nonetheless, is a monumental task, especially at a time when markets and economic change continue to be unusually cruel to labor.

Rising inequalities in income and wages over the past several decades is a trend which continues to run counter to global development initiatives. Prosperity has been limited to the skilled workers of the core and relatively less-skilled labor in the more rapidly advancing semiperipheral regions. In contrast, prospects of positive economic change have been reduced throughout the stagnating or declining nations of the periphery (especially in sub-Saharan Africa and the Middle East). While even coordinated national and global development efforts are unlikely to bring about a convergence in income, there is still much potential for generating greater income-earning opportunities worldwide.

Visionaries at the World Bank see two global scenarios. The first entails a continuation of past trends, including the domination of the world economy by the core. This situation would lead to further polarization and widening differences between the core and periphery. It is a scenario that is best described as one of minimal growth and **divergence.** By contrast, the orchestrated integration of strong domestic economic policies, unified under a global development agenda that emphasizes investment in "human capital," could lead to the **convergence** of regions in the world economy. Policy measures to promote integration would increase the potential of all regions to experience robust job creation, increased worker productivity, and improvements in the quality of employment opportunities and working conditions. Investments in human capital and infrastructure are critical to achieving convergence. Education of the global workforce and enhanced labor productivity are key to the development equation.

> The principal determinant of the outlook for workers is domestic investment—in capital, education, infrastructure, and technology. The divergence scenario assumes that recent trends in investment continue or deteriorate, that a sizable share of those already enrolled in schools drop out prematurely, and that the overall productivity of labor does not rise rapidly. The convergence scenario assumes that investment rates pick up, that enrollments stabilize at current levels and dropout rates decline, and that improvements in infrastructure, technology transfers, and improvements in the quality of governance contribute to rising productivity . . . supported by rises in savings rates, lower fiscal deficits in the rich

countries, and reasonable amounts of international capital flows, including development assistance. The effort in Sub-Saharan Africa must be especially strong." (*World Development Report 1995*, p. 118)

Projections of economic growth as measured by GDP per capita and exports by geographic regions are shown in Table 4.7 for each scenario. If these projections are accurate, production and export-led growth in a convergent world economy could improve global integration and facilitate spatial flows.

Governments must meet the challenge and formulate development policies that are proactive rather than reactive. To override the legacies of inequality and dependence, outcomes that have long been associated with core-periphery distinction, it is critical that governments and policy makers (past mistakes and miscalculations notwithstanding) show a renewed commitment to a global development effort. In geographic terms, the formula for success will be different depending upon the region in question. Furthermore, policy makers and governments must be realistic. The changes and improvements brought about by "convergence" may be slow in coming in light of rapid population growth.

Economic Development at the Local and Regional Levels

Even when national economic development objectives are met and national economies show signs of advancement, there will be geographic variations in the distribution of the economic benefits. Furthermore, all development problems cannot be attacked using top-down strategies. Many problems will require a local or regional approach.

The success or failure of local or regional development efforts are highly dependent upon four key components:

1. local or regional investments in human capital as represented by the quality of the local or regional labor force;
2. the ability to secure a long-term commitment for financial support and the initial **venture capital** (also referred to as risk capital) from financial institutions, government, and/or the private sector;
3. an entrepreneurial and management capacity to initiate and oversee local industries and the development of a local or regional production network; and
4. an information and technology environment that effectively links agents and firms within the local or regional production network (Bennett, 1991, pp. 114–115).

Moreover, the success of economic development policy orchestrated at the local or regional level hinges greatly upon its ability to be socially or culturally accepted. Global and regional shifts from command economy to free-market economy or vice versa are indications that social and cultural forces are at work at a larger level. In other words, local and regional economic policy may be greatly influenced by national movements. Distinction should also be made as to the degree to which **individualist culture** versus **collectivist culture** is at work (see Table 4.8).

The degree to which members of a given society accept inequalities will determine how vigorously they pursue development policies which foster varying degrees of progress amongst the various "strata" of society. In general, individualist culture (and the support of individualism) tends to be more polarization tolerant. If polarization of power and wealth is widely accepted or institutionalized, more conservative economic policies will be pursued on the assumption that all members of all socio-

---

TABLE 4.7	World Bank Projections of Growth under Two Competing Development Scenarios: Divergence (D) and Convergence (C)

	Average Annual Growth (%)					
	In GDP per Capita			In Exports		
		Scenario				Scenario	
Region	Actual 1970–1990	D (1994–2010)*	C	Actual 1980–1990	D (1994–2010)*	C
China	4.6	2.3	3.9	11.3	4.7	6.6
East Asia	5.5	3.0	4.4	10.2	5.3	6.5
Former Communist Bloc nations	–3.0	0.9	3.5	2.1	2.2	5.6
Latin America	1.7	1.4	3.3	2.4	3.8	7.0
Middle East and North Africa	0.8	1.4	3.4	4.2	3.3	5.5
South Asia	2.0	2.4	4.0	6.3	6.6	8.9
sub-Saharan Africa	–0.3	–0.5	1.7	3.1	3.6	6.7
Advanced Market Economies	1.9	1.6	2.3	1.5	2.9	3.7

Source: *World Development Report* (1995), p. 120.
Note: D = Divergence; C = Convergence; * projections.

TABLE 4.8	Traits of Individualist and Collectivist Cultures

Individualist Culture	Collectivist Culture
emphasis placed on the individual	emphasis on society at large
free-market oriented	central-planning oriented
polarization tolerant	polarization intolerant
atomized labor movements	united labor movements
involvement of individuals with organizations primarily calculative	involvement of individuals with organizations primarily moral
modernist ethic	traditionalist ethic
biased toward conservative development policies	biased toward liberal development policies
policies and practices that allow for initiative and apply to all (universalism)	policies and practices that are based on loyalty and an individual sense of duty and obligation and vary according to social relations (particularism)

Source: After Berry, Conkling, and Ray (1993) pp. 10–11, with modifications.

economic strata will benefit from growth and its trickle-down effects. While this strategy may not eliminate polarization, it does have the potential to improve conditions for the average member of society. This result would essentially lessen the need to draw attention to polarization if people, socioeconomic groups, or regions at the lower end of the spectrum would experience real improvements in their economic conditions.

By contrast, collectivist cultures tend to be less tolerant of economic polarization, and hence, support development policies which promote depolarization. Many argue that an inherent trade-off exists between economic growth and depolarization. Nonetheless, the nature and tone of economic reforms (and their success or failure) must be viewed in terms of the cultural forces which influence those reforms and their traits or objectives.

As part of the collectivist agenda, more attention must be focused on the workings of "informal economy" in the local and regional development process. **Informal economy** refers to *that portion of an economy in which productive and useful functions are performed without formal controls or remuneration (that is, production activities which involve volunteer or unpaid domestic labor or non-market-oriented subsistence).* In particular, the role of women in informal economic sectors of developing regions must be fully explored. Note that women workers tend to be more highly concentrated than men in nonwage employment throughout Latin America, Africa, and Asia. Gender, age, and equality issues abound in reference to informal sector employment. In particular, the undervalued role of women and children in the regional production systems of the periphery needs to be exposed.

Antidiscrimination policies and protection of labor in the workplace have become increasingly important to the development process in all underdeveloped regions. Although intervention in labor markets has helped improve standards, wages, job security, and access to employment opportunities and capital, much more needs to be done to ensure that labor (and especially women and children) are properly treated and not exploited.

CONCLUDING REMARKS

Intervention and managed economic development is costly at any level. While financial or production resources are limited in many underdeveloped regions, monetary and technology transfers from the developed world could prove invaluable. Should we conceptualize world economic development as a zero-sum game, it would inevitably lead us to accept the conclusion that development in the periphery must come at the expense of the core. It follows, therefore, that global collectivism must come at the expense of global individualism. A push, however, toward global collectivism may be difficult in an era where economic transformation has been largely supported by individualist culture and the need for personal freedom in matters of economy. Recent trends toward free-market activity, privatization of industry, and global capitalism are in direct contrast to the principles of collectivist culture. Hence, it may prove difficult to rally support for collectivist-oriented regional and global development policies given the rise and prominence of individualist culture in the capitalist world economy.

KEY WORDS AND PHRASES

advanced factors   86
advanced capitalism   82
advanced network economy   88
basic factors   86
capital (as a class)   79
capitalism   71
capitalist economy   72
centrally planned systems   71
class   79
collectivist culture   92
colonialism   76
command economy   71
communism   73
conservative   74
conservative mindset   74
convergence   91
corporate structure   86
culture of poverty   77
density of industrial linkages   86

divergence   91
drive to maturity   82
economic sovereignty   71
externalities   78
factor-driven stage   88
factor endowments   86
flexibility (in production)   87
glasnost   81
growth poles   82
hegemony   76
high mass consumption   83
horizontal model   87
individualist culture   92
informal economy   93
innovation-driven stage   88
internal demand conditions   86
investment-driven stage   88
invisible hands of the market   75
kibbutz   80

labor   79
labor theory of value   79
laissez-faire   75
laissez-passer   89
liberal   76
liberal mindset   76
libertarian   79
market-based systems   71
markets   71
Marxism   79
Marxist   79
mixed economy   72
moshav   80
perestroika   81
political economy   70
Porter's global competitive
    advantage   86
preconditioning   81
radical   79

REFERENCES

Alchian, A.A., and W.A. Allen. *Exchange and Production Theory in Use.* Belmont, CA: Wadsworth (1969).

Bennett, R. "National Perspectives on Global Economic Change." In *Global Change and Challenges: Geography for the 1990s* (R. Bennett and R. Estall, eds.). London: Routledge (1991), pp. 103–117.

Berry, B.J.L., E.C. Conkling, and D.M. Ray. *The Global Economy.* Englewood Cliffs, NJ: Prentice Hall (1993).

Borts, G.H., and J.L. Stein. *Economic Growth in a Free Market.* New York: Columbia University Press (1964).

Brandt, W. *North-South: A Program for Survival.* Cambridge: MIT (1980).

Braverman, H. *Labor and Monopoly Capital.* New York: Monthly Review Press (1974).

Byrne, J.A., "The Horizontal Corporation." *Business Week.* New York: McGraw-Hill (December 20, 1993), pp. 76–81.

Chisholm, M. *Geography and Economics* (2nd edition). London: G. Bell & Sons (1970).

Crosswell, M. "Growth, Poverty Alleviation, and Foreign Assistance." In D. Leiziger (ed.) *Basic Needs and Development.* Cambridge, MA: Delgsh Lager, Crom and Haines (1981).

de Souza, A.R. *A Geography of World Economy.* Chapter 2: World Views. Columbus, OH: Merrill (1990), pp. 23–35.

de Souza, A.R., and F.P. Stutz. *The World Economy.* New York: MacMillan (1994).

D'Costa, A.P. "State-Sponsored Internationalization: Restructuring and Development of the Steel Industry." In H. Noponen, J.

Graham, and A.R. Markusen (eds.) *Trading Industries, Trading Regions.* New York: Guilford Press (1994), pp. 92–139.

Dobb, M. 1937. *Political Economy and Capitalism.* London: Routledge & Kegan Paul (1945).

Dobb, M. *Studies in the Development of Capitalism.* London: Routledge (1946).

Doti, J.L., and D.R. Lee. *The Market Economy: A Reader.* Los Angeles, CA: Roxbury (1991).

Dunning, J.H. *The Globalization of Business: The Challenge of the 1990s.* London: Routledge (1993).

Eatwell, J., M. Milgate, and P. Newman (eds.). *Marxian Economics.* New York: W.W. Norton (1990).

Frank, A.G. *Dependent Accumulation and Underdevelopment.* New York: Monthly Review Press (1978).

Friedman, M. *Capitalism and Freedom.* Chicago: Chicago University Press (1962).

Galbraith, J.K. *Economic and Public Purpose.* London: Penguin (1975).

Galbraith, J.K. *The Age of Uncertainty.* Boston: Houghton Mifflin (1977).

Galbraith, J.K. *The History of Economics: The Past as the Present.* London, U.K.: Penguin (1991).

Griffin, K.B. *Globalization and the Developing World: An Essay on the International Dimensions of Development in the Post-Cold War Era.* Geneva: UNRISD (1992).

Hamilton, F.E.I. "Industrial Restructuring: An International Problem." *Geoforum* 15 (1984), pp. 349–364.

Hammer, M., and J. Champy. *Reengineering the Corporation.* New York: Harper Business (1994).

Hanink, D.M. *The International Economy: A Geographical*

Perspective. New York: J. Wiley & Sons (1994).

Harvey, D. *The Limits to Capital.* Oxford, UK: Basil Blackwell (1982).

Hirschman, A.O. *The Strategy of Economic Development.* New Haven: Yale University Press (1958).

Howes, C. "Constructing Comparative Disadvantage: Lessons from the U.S. Auto Industry." In H. Noponen, J. Graham, and A.R. Markusen (eds.) *Trading Industries, Trading Regions.* New York: Guilford Press (1994), pp. 45–91.

Johnston, R.J., P.J. Taylor, and M.J. Watts. *Geographies of Global Change: Remapping the World in the Late Twentieth Century.* Oxford, UK: Blackwell (1995).

Kaplinsky, R. *Third World Industrialization in the 1980's: Open Economies in a Closing World.* London, U.K.: Cass (1984).

Keynes, J.M. *The General Theory of Employment, Interest and Money.* New York: Harcourt Brace (1965).

Knox, P., and J. Agnew. *The Geography of the World Economy* (2nd edition). London, U.K.: Edward Arnold (1994).

Krueger, A.O. *Trade and Employment in Developing Countries* (Volume 3: Synthesis). Chicago: Chicago University Press (1983).

Krugman, P.R. *Globalization and the Inequality of Nations.* Cambridge, MA: National Bureau of Economic Research (1995).

Kuhn, J. *The Structure of Scientific Revolution.* Chicago: Chicago University Press (1970).

Lewis, W.A. *The Theory of Economic Growth.* London, U.K.: Allen & Unwin (1955).

Lewis, W.A. *The Evolution of the International Economic Order.*

Princeton, NJ: Princeton University Press (1977).

Lipietz, A. *Towards a New Economic Order.* Cambridge, MA: Policy Press (1992).

Little, I. *Economic Development: Theory, Policy, and International Relations.* New York: Basic Books (1982).

Livingstone, I. (ed.) *Economic Policy for Development.* Harmondsworth, England: Penguin Books (1971).

Maddison, A. *Economic Progress and Policy in Developing Countries.* New York: W.W. Norton (1970).

Marglin, S. *Growth, Distribution and Prices.* Cambridge, MA: Harvard University Press (1984).

Marx, K. 1862–3. *Theories of Surplus Value, Volumes I–III.* London: Lawrence & Wishart (1969–72).

Marx, K. 1867. *Capital: A Critique of Political Economy, Vol. I.* New York: Vintage (1977).

Marx, K. 1894. *Capital: A Critique of Political Economy, Vol. III.* New York: Vintage (1981).

Marx, K., and F. Engels. 1848. *The Communist Manifesto.* New York: Washington Square Press (1964).

McConnell, J.E., and S.L. Brue. *Macro-economic: Principles, Problems, and Policies.* New York: McGraw-Hill (1993).

Myint, H. *The Economics of Developing Countries.* New York: Praeger (1964).

Myrdal, G. *Economic Theory and Underdeveloped Regions.* London: Duckworth (1957).

Peet, R. *International Capitalism and Industrial Restructuring.* Boston: Allen & Unwin (1987).

Peschel, K. "Spatial Structures in International Trade: An Analysis of Long-Term Development." *Papers of the Regional Science Association* 58 (1985), pp. 97–111.

Porter, M.E. *The Competitive Advantage of Nations.* New York: Free Press (1990).

Razin, A., and E. Sadka. *Population Economics.* Cambridge, MA: MIT Press (1994).

Ricardo, D. *The Principles of Political Economy and Taxation.* New York: E.P. Dutton (1912).

Ricardo, D. *Works of David Ricardo* (edited by P. Sraffa). London: Cambridge University Press (1951).

Rostow, W.W. *The Stages of Economic Growth.* Cambridge: Cambridge University Press (1960).

Rostow, W.W. *The Economics of Take-Off into Sustained Growth.* London, U.K.: Macmillan (1963).

Sachar, A., and S. Oberg (eds.). *The World Economy and the Spatial Organization of Power.* Brookfield, VT: Gower (1990).

Sheppard, E.S. "Value and Exploitation in a Capitalist Economy." *International Regional Science Review* 9 (1984), pp. 97–108.

Sheppard, E., and T.J. Barnes. *The Capitalist Space Economy: Geographical Analysis After Ricardo, Marx and Sraffa.* London: Unwin Hyman (1990).

Slater, D. "Trajectories of Development Theory: Capitalism, Socialism, and Beyond." Chapter 5 of *Geographies of Global Change: Remapping the World in the Late Twentieth Century* (R.J. Johnston, P.J. Taylor and M.J. Watts, eds.). Oxford, U.K.: Blackwell (1995), pp. 63–76.

Smith, A. 1776. *An Inquiry into the Nature and Causes of the Wealth of Nations.* (edited by E. Cannan). New York: Random House (1965).

Sraffa, P. *Production of Commodities by Means of Commodities.* Cambridge: Cambridge University Press (1960).

Storper, M.S., and R.A. Walker. *The Capitalist Imperative.* Oxford: Basil Blackwell (1989).

Sweezy, P. *The Theory of Capitalist Development.* New York: Monthly Review Press (1942).

Taylor, P.J. *Political Geography* (2nd edition). London: Longman (1989).

Wallerstein, I. *The Modern World-System: Capitalist Agriculture and the Origins of the European World Economy in the Sixteenth Century.* New York: Academic Press (1974).

Wallerstein, I. *The Capitalist World-Economy.* Cambridge: Cambridge University Press (1979).

Wallerstein, I. *The Politics of the World-Economy.* Cambridge: Cambridge University Press (1984).

Weaver, J., and K. Jameson. *Economic Development Competing Paradigms.* Lanham, MD: University Press of America (1981).

Wheeler, J.O., P. Muller, G.I. Thrall, and T.J. Fik. *Economic Geography* (3rd edition). New York: Wiley (1998).

Wood, A. *North-South Trade, Employment and Inequality: Changing Fortunes in a Skill-Driven World.* Oxford, U.K.: Clarendon Press (1994).

World Development Report 1990: Poverty (published for the World Bank). New York: Oxford University Press (1990).

World Development Report 1991: The Challenge of Development (published for the World Bank). New York: Oxford University Press (1991).

World Development Report 1995: Workers in an Integrating World (published for the World Bank). New York: Oxford University Press (1995).

5

HUMAN SETTLEMENTS AND POPULATION

OVERVIEW

Social scientists continue to study problems of overpopulation and the relationship between the number of people in a region and the ability of that region to sustain that population. Recent population growth trends in the underdeveloped SOUTH have now become a growing development concern. The question is not whether the earth is running out of space for its inhabitants, but rather how long the earth's resources will be able to support its growing numbers. It is interesting to note that if all of the world's people were brought together in one place (at arm's length from one another), they would easily fit into a geographic area less than the size of Texas.

Scarcity of space, however, is a real concern for the large portion of the world's population that live in highly populated zones. More importantly, there is a diminishing "quality of life" in many high population growth areas. **Quality of life** refers to *the environmental and psychological aspects of social and economic well-being as existing or perceived by individuals or group.* Changes in the well-being of the average person in a given society or region can be used as a gauge of economic development performance. Unfortunately, as the populations in most underdeveloped regions continue to grow faster than their regional economies, diminishing quality of life has become the norm rather than an exception.

Science and technology have allowed the world's production capacity to expand over time. Yet uneven population growth and regional variations in human carrying capacity have combined to create serious resource limitations in many underdeveloped regions. The limitations imposed by excessive population growth are beginning to be realized in the world's fastest-growing urban areas, especially in the large urban areas of the SOUTH. In particular, limits to food production have been partly attributed to the increasing scarcity of agricultural land.

Land scarcity in the urban systems of the SOUTH has also become a serious problem. **Developable land—** *land that can be easily converted from rural to urban land uses—* is in short supply in many underdeveloped regions. Severe shortages exist in regions where land parcels adjacent to cities are currently being used for agricultural purposes. Land on the "urban fringe" of cities in East and Southeast Asia, for example, is not easily converted to urban land use as it increases the burden placed on the entire rural agricultural system to produce food for Asia's growing numbers. The loss of rural land translates into a loss of food output; something that is not acceptable in light of the increasing demand for food. Land conversion from rural to urban uses has the added cost of displacing traditional rural farmers and their families to the unfamiliar urban landscape.

Hardest hit by rural and urban land scarcity are those nations and regions least able to deal with the problem. Many of these nations face additional challenges, such as poor soils, dry climates, susceptibility to natural hazards, and/or limited amounts of productive agricultural land. For these and other reasons, **demographics**—*the scientific study of population characteristics and population* *growth*—has become increasingly important to the social science community and to economic geographers. Demographic studies are critical because of four essential facts:

1. more people are alive today than at any point in history;
2. the world's population has been increasing at a very rapid rate since the end of World War II;
3. the vast majority, roughly three-quarters, of the world's population lives in underdeveloped countries; and
4. virtually all of the world's new population growth is concentrated in underdeveloped countries.

Demographic scholars wonder whether the world's population will ever exceed the earth's human carrying capacity. This concern is, of course, founded in the belief that the earth possesses a finite ability to support human, plant, and animal life. The earth's human population threshold has certainly increased due to advancements in technology and efficiency gains in production. At present, it is unlikely that human civilization is in any immediate danger of exceeding the earth's carrying capacity. Many experts agree, however, that technological limits and production ceilings are now being reached. If so, forthcoming limitations in production systems will have many undesirable impacts on underdeveloped regions and contribute further to a diminishing quality of life. Population growth has created an intolerable world poverty dilemma and continues to fracture a world economy that many see as teetering on the brink of disaster.

Will the related problems of hunger, malnutrition, disease, and starvation trigger mass uprisings of global proportions or worldwide revolution? Are we to face famines on a massive geographic scale or will they be restricted to the lands and people of the periphery? Are the world's most pressing development problems likely to be neglected given that these problems are largely confined to the underdeveloped SOUTH, far removed from the wealth and power of core economies? And will there be time enough to react and circumvent human misery and suffering should regional and global food production shortages threaten the very survival of people in underdeveloped regions, those with the least access to surplus food and other basic survival resources? Indeed, these are tough questions to answer.

We do know that world population has doubled in the last 30 years, and that the earth's population is approaching an unprecedented 6.5 billion people. Before effective global development polices can be implemented, it is critical that we gain an understanding of the spatial distribution of population and regional variations in population growth rates. While this chapter focuses on the geography of population, subsequent chapters will discuss the implications of limitations and geographic trends in population change.

The world's population is currently estimated at roughly 6.3 billion people (in 1999). Most of the world's people live in just a few highly concentrated geographic areas. These areas cover only a small portion of the earth's total surface. It is of interest to note that 71% of the earth is covered by water, with the remaining 29% covered by land. Yet only 25% of the land surface is easily inhabitable. In other words, about one-quarter of the earth's land surface is not associated with extreme climates or conditions that place significant limitations on the size of human settlements. Astonishingly, more than 80% of the world's population lives on just 7.5% of the earth's surface!

It is also interesting to note that world population distribution is in close proportion to the size of land area by continent (that is, the larger the continent, the larger the population), despite that fact that the geographic distribution of population by continent is far from uniform. Consider the breakdown of land area by continent as shown in Table 5.1. Note that with the exception of Antarctica, the ranking of continents by size is almost identical to the ranking of their respective populations.

WORLD POPULATION GROWTH

In the book entitled *Global Citizen* (1991), Donella Meadows offers startling revelations on world population growth. She calls for careful assessment of human-environment relations and the impact of global population change in light of regional resource limitations.

> Each day on this planet 35,000 people die of starvation, 26,000 of them are children. This human toll is equivalent to 100 fully loaded 747-jets crashing every day. It is the same number of deaths every three days as were caused by the Hiroshima atomic bomb explosion. And each day, because of population growth, there are 220,000 more mouths to feed. (p. 37)

Note that approximately 2.5 people are added to the earth's population each second. Each day the net growth in the human population is enough to populate a city the present size of Baton Rouge, Louisiana. Each week the total number of people swells to the size of a city larger than Houston, Texas (approximately 1.6 million). In little more than one month, the number of new additions would rival the current population of New York City (a population of more than 6 million people). The vast majority of these people are born in the most impoverished regions of the SOUTH. These statistics highlight a potentially dangerous situation and a problem that requires the immediate attention of all nations and people. Regional resource imbalances should not be tolerated in a world economy where there are more than enough resources to go around.

From 1965 to 1985, roughly 76 million people were added to the earth annually. From 1985 to 1990, more than 87 million people were added to the world annually.

Currently, the world's population is increasing by a little over 95 million people per year. This disturbing global population growth trend has been used by alarmists to bring attention to the problem. The upward growth trend has recently showed signs of reversing, however. It is estimated that over the next 50 to 100 years, the annual incremental growth of the world's population will diminish. Despite the expected slowdown, the annual world population growth comes at a time when fuel and energy, jobs, water, food, agricultural land, and resources are becoming increasingly scarce. It took 2 million years for the earth's human population to grow to one billion people. It now takes slightly more than 11 years for the earth's population to increase by an additional one billion. By the year 2030, the world's population is likely to reach and exceed the 8 billion mark.

Shortly before the year 2100, the earth's population is expected to break the 10 billion barrier. By the year 2100, it is estimated that the earth will be home to twice the amount of people that were alive in 1990. Many see the explosion of this so-called **population bomb,** *a phrase coined by Paul Ehrlich (president of the Zero Population Growth Society) to emphasize the serious nature of global population growth,* as one of the greatest threats to human existence and world economic stability. Whether only a portion of the world's population suffers from inadequate food and resources or if the entire world were to suffer from wars initiated by inequality, all of humanity stands to lose. Finding solutions to population growth related problems remains at the top of the priority list in all development agendas.

The world's population engine continues to run hard. Each year, world population reaches unprecedented levels despite recent declines in the annual growth rates in most regions. Even as the world's annual population growth rate continues to decline, an average of 1.4% shortly after the turn of the century, more people will be added on to the earth per year than in any time in history (see Table 5.2). This latent population expansion effect is due to the expanding nature of the world's population base, as a smaller percentage growth rate on a larger base produces a larger expansion effect in absolute terms. If

| TABLE 5.1 | Land Area Statistics, Size and Population Rankings by Continent |

Continent	Area (square miles) × 1000	Percent of Land Area	Rank[*] Area	Rank[*] Population
Africa	11,700	20.2	2	2
Antarctica	5400	9.3	—	—
Asia	17,300	30.1	1	1
Europe	3800	6.6	5	3
North America	9500	16.2	3	5
South America	6900	11.9	4	4
Oceania	3300	5.7	6	6
World	57,900	100.0		

Source: *Goode's World Atlas,* 19th edition (p. 250).
Area rounded to nearest one-hundred thousandth; percent land area rounded to nearest tenth of a percent.
[*] Ranking from largest to smallest (excluding Antarctica), with population ranks based on projected populations for the year 2025.

| TABLE 5.2 | World Population Growth Statistics (selected years) |

Period	Average Annual Population Change (%)	Average Annual Increment (in millions)	World Population for Years Shown (in millions)
1980–1990	1.75%	82.5	1990: 5295.3
1990–1995	1.68%	92.8	1995: 5759.2
1995–2000	1.59%	95.0	2000: 6380.0[*]
2000–2005	1.43%[*]	91.0[*]	2005: 6840.0[*]
2005–2025	1.25%[*]	84.0[*]	2025: 8470.0[*]
			2100: 11,000.0[*]

Source: *World Resources 1994–95* (pp. 268-269).
[*] Forecast (rounded).

current trends continue, world population is expected to stabilize at approximately 11 billion people by the year 2100.

THE GEOGRAPHY OF POPULATION

The spatial distribution of population on the earth's surface is highly variable. Of importance are the locations of the world's major **population concentrations**—*large geographic areas with the highest concentrations of people as measured on a people per land unit basis* and the locations of the world's major **population voids**—*large geographic areas with the lowest concentrations of people as measured on a people per land unit basis*. Before identifying the world's population concentrations and voids, let us compare some important population statistics by geographic region.

Table 5.3 highlights world population distribution by geographic region for selected years. Also provided are population **forecasts**—*projections based on past, current, and future trends*—for the year 2025 to highlight regional trends in population growth.

Over the period 1950 to 2025, the earth's population will increase by approximately 236%. Note, however, that population growth rates are highly variable over space. For instance, the total population of Africa and South America is expected to increase by approximately 610% and 450% respectively, over the same period. The population growth rates of Africa and South America are extreme when compared to the expected 117% population increase in North America or the 36% increase in Europe over the same time period. The most troubling feature of global population explosion is that the largest increases are expected in regions classified as underdeveloped (many of which are peripheral).

> Between 1950 and 1990, about 85 percent of the growth of the world's population occurred in the less developed regions and about 15 percent in the more developed regions. Over the next 35 years, about 95 percent of population growth will occur in the less developed countries . . . . In 1950 about two-thirds of the world's people lived in less developed countries. In the early 1990's, this figure had increased to about 75 percent and the United Nations projects that by 2025 about 85 percent of the earth's population will reside in the poorer nations. (Seitz, 1995, p. 25)

The "density" of population in a given region can also be used as a rough measure of concentration. A region's **population density** *is measured by the number of people per unit land area (that is, the total population of a region divided by the region's total land area)*. Population density is commonly reported as the total number of people per square mile (or square kilometer), a figure that is typically rounded to the nearest person per square land unit.

Note that the world's population density in 1995 was approximately 109 people per square mile (rounding to the nearest person). This figure is calculated by dividing 5700 million (5.7 billion) people by the earth's total land area, which is approximately 52.5 million square miles (excluding the continent of Antarctica). Using the same approach, it is easy to verify that the world's population density in January 1999, is 6300 million divided by 52.5, or approximately 120 people per square mile.

Population density is a crude measure of the burden placed on a region's resource and production system by that region's population. Small countries or regions, as defined in terms of geographic area, do not necessarily have high population densities, nor do large countries or regions (as defined in terms of geographic area) necessarily have low population densities. Note that the five largest countries in terms of land area vary widely in terms of their population density (see Table 5.4).

A list of the world's most densely populated nations (considering only those nations with 10 million or more people) is provided in Table 5.5. Note that the majority of high population density nations are located in the SOUTH. In addition, most of these highly populated nations are of peripheral status. Other, smaller high-density regions (regions with less than 10 million people) include Hong Kong and Singapore, where population density exceeds 10,000 people per square mile, and the island of Manhattan (New York City), with a population density that exceeds 50,000 people per square mile!

TABLE 5.3	Population by Major Geographic Region		
	Population (in millions)		
Region	**1950**	**1995**	**2025**[*]
World	2516.2	5759.3	8472.5
Asia	1377.3	3407.6	4900.3
Africa	222.5	744.0	1582.5
N. and C. America	202.6	418.6	599.8
(N. America)	(166.1)	(276.6)	(360.4)
South America	111.6	319.7	452.0
Europe and CWIS	572.6	804.6	886.2
(Europe)	(398.1)	(516.0)	(541.8)
Oceania	12.6	28.8	41.3

Source: *World Resources 1994–95* (pp. 268–269).

N. America—United States and Canada;
C. America—Mexico, Belize, Guatemala, El Salvador, Honduras, Nicaragua, Costa Rica, and Panama;
CWIS—Common Wealth of Independent States (former USSR.) less southwestern Asian republics of Kazakhstan, Tajikistan, Turkmenistan, Uzbekistan, and Kyrgyzstan;
Oceania—Australia, New Zealand, Solomon Islands, Fiji, Papua New Guinea.
Note: Figures shown in parentheses are for subsets of major group identified in bold.
[*]Forecasts.

TABLE 5.4	Selected Statistics for the World's Five Largest Countries by Land Area (1994)

Country	Land Area*	Percent of World's Land Area	Population Density**
1. Russia	6,592,000	12.8%	23
2. Canada	3,852,000	6.4%	7
3. China	3,691,000	6.2%	321
4. United States	3,615,000	6.0%	68
5. Brazil	3,286,000	5.5%	46

Source: *Goode's World Atlas* (19th edition) pp. 245–249.

* Area expressed in square miles, data is rounded to nearest one-thousand square miles.

** Population density expressed in terms of people per square mile (January, 1994).

MAJOR POPULATION CONCENTRATIONS

There are four major population concentrations on the earth's surface. These population concentrations represent regions which cover an extensive amount of the earth's land surface and contain a large number of people in a relatively small area. The world's major population concentrations are illustrated in Figure 5.1. They are located in East Asia, South and Southeastern Asia, Euro-West Asia, and North America. A list of the world's most populated nations is also provided in Table 5.6.

1. East Asian Population Concentration

Approximately one-quarter of the world's population currently resides in the East Asian population concentration. China, the most populated nation in the world, accounts for more than 1.2 billion people (slightly more than 20% of the world's total population). Note that one out of every five people in the world are Chinese. Other notable contributors to the East Asian population concentration include Japan and South Korea.

TABLE 5.5	The Most Densely Populated Nations of the World (of size 10 million or more people)

Country	Area (×1000) in Square Miles	Population Density (people per square mile) 1995	2025*
Top Three by 2025			
Bangladesh	55.6	2306	4016
S. Korea	38.0	1189	1324
India	1237.1	752	1127
Other Selected Nations			
The Netherlands	15.9	975	1111
Vietnam	127.4	579	919
Philippines	115.8	598	908
Japan	145.8	864	872
Nigeria	356.7	356	801
Pakistan	339.7	397	764
United Kingdom	94.2	617	640
Germany	137.9	590	608
China	3689.1	335	417
Indonesia	752.4	268	377
Thailand	198.1	294	365
Ethiopia	446.9	130	292
Iran	632.4	106	289
Mexico	759.5	123	181
United States	3787.5	69	85
Brazil	3286.5	49	67
World (excluding Antarctica)	52,500.0	109	152

Source: *Goode's World Atlas* (19th edition), pp. 245–249, and *World Resources* 1994–95, pp. 268–269.

Note: Population density figures are rounded to the nearest person per square mile.

*Forecasts.

2. South and Southeast Asian Population Concentration

This concentration contains approximately one-third of the world's population and is home to the world's second largest nation—India. India holds the largest share of population in this region, with more than .9 billion people. Other big contributors to this population concentration include Indonesia, Pakistan, Bangladesh, Vietnam, the Philippines, Iran, and Thailand (see Table 5.6). When combined, the East Asian and South and Southeast Asian population concentrations account for roughly 60% of the world's population. Not only is the Asian corridor the most concentrated in terms of population, but it is one of the fastest growing as well. No surprise that the average yearly increase in the labor force (ages 15–64) from 1980 to 1990 in Asia was about 2.2% per year, substantially above the world average of 1.9% per year and much greater than the population growth rate for the United States, which stood at slightly less than 1% per year (on average) over the same time period. Note that the average annual increase in Asia's population from 1985 to 1995 was well over 55 million people per year. The nations comprising these population concentrations are expected to add more than 60 million people to the earth per year from 1995 to 2005.

3. Euro–West Asian Population Concentration

The Euro-West Asian population concentration, extending from western and central Europe through western portions of the republics of the former Soviet Union, accounts for 15% of the world's population. It is a region that contains well over 850 million people. Europe accounts for about three-fourths of the population in this concentration, with more than 510 million people.

Before the disintegration of the Soviet Union, the USSR. was the third largest nation in the world, with a population of approximately 282 million people in 1990. At that time, Russia contributed roughly half of this total. As an independent nation, Russia was home to roughly 150 million people in 1995 and is currently the sixth largest nation in the world. Today, Russia is the largest single nation in the Euro–West Asian population concentration. Note that the now united Germany is the second largest nation in this concentration, with more than 80 million people living within the borders of what was once a divided nation. Other large nations in this formidable concentration include Turkey, the United Kingdom, Italy, and France (see Table 5.6).

4. North American Population Concentration

The North American population concentration encompasses the United States and Canada. This region accounts for just less than 5% of the world's population. The large majority of these people are located in the northeastern United States and southeastern Canada. With the break-up of the Soviet Union, the North American population concentration is now home to the world's third-largest nation—the United States—with

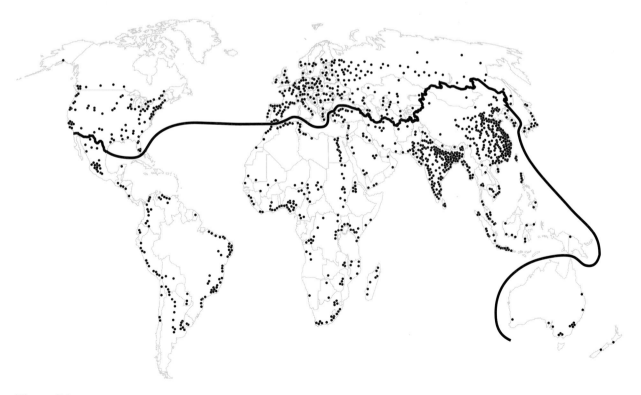

Figure 5.1   Major world population concentrations and voids.

approximately 263 million people in 1995 (about 4.6% of the world's population in that year). In the year 2005, the U.S. population is expected to climb to the 275 million mark.

All in all, the four major population concentrations of the world account for approximately 80% of the world's population. The remaining 20% are scattered about various pockets and enclaves over the earth's inhabitable land surface, typically near waterways, coastal areas, or large and established in-land urban settlements. Although no large and continuous spatial concentration of people exists on the continents of Africa or South America to justify a "major concentration" status, the regions are home to an impressive 744 and 320 million people, respectively, (in 1995). Notable is the fact that many sub-Saharan African nations are vying for the title of world's fastest-growing regions.

Consider the population data in Table 5.6 and the changing ranks of the most populated nations from 1995 to 2025. Twenty-three of the thirty most populated nations are located in the SOUTH. Note also that the world's underdeveloped regions are the fastest-moving in terms of rank. Specifically, the underdeveloped nations of Nigeria, Ethiopia, Zaire, and Iran will show the greatest upward movement in population rank. Should current population growth trends in Asia continue, India is likely to overtake China and become the world's largest nation around the year 2060.

TABLE 5.6	Population Statistics for the World's 30 Most Populated Nations (1995, 2025)		

1995 Rank	Country	Population (in millions)		Top-Twenty Rank in 2025 and (change)
		1995	2025*	
1	China	1238.3	1539.8	1 (–)
2	India	931.0	1393.9	2 (–)
3	United States	263.2	322.0	3 (–)
4	Indonesia	201.5	283.3	5 (–1)
5	Brazil	161.2	219.7	8 (–3)
6	Russia	150.0	173.0	9 (–3)
7	Pakistan	135.0	259.6	6 (+1)
8	Japan	125.9	127.0	13 (–5)
9	Bangladesh	128.3	223.3	7 (+2)
10	Nigeria	127.0	285.8	4 (+6)
11	Mexico	93.7	137.5	11 (–)
12	Germany	81.3	83.9	19 (–7)
13	Vietnam	73.8	117.0	14 (–1)
14	Philippines	69.3	105.2	15 (–1)
15	Iran	66.7	144.6	10 (+5)
16	Turkey	62.0	92.9	18 (–2)
17	Egypt	58.5	93.5	17 (–)
18	Thailand	58.3	72.3	
19	United Kingdom	58.1	60.3	
20	Italy	58.0	56.2	
21	France	57.7	60.8	
22	Ethiopia	58.0	130.7	12 (+10)
23	Ukraine	52.4	55.0	
24	Myanmar (Burma)	46.6	75.6	20 (+4)
25	South Korea	45.2	50.3	
26	South Africa	42.7	73.2	
27	Zaire	43.8	104.5	16 (+11)
28	Spain	39.3	42.3	
29	Poland	38.7	43.8	
30	Colombia	35.1	49.4	

Source: *World Resources 1994–95* (pp. 268–269).
* Forecasts.

MAJOR POPULATION VOIDS

A population void refers to regions of very low population density. These areas are relatively uninhabited or virtually uninhabitable. For our purposes, we will define population voids as large and extensive geographic areas having population densities of two or less people per square mile. There are seven major population voids in all.

1. The African-Asian Void

This void spans a very large and expansive low population area that covers a significant portion of the earth's surface. Covering two continents, this void accounts for the sparsely populated areas that cover roughly 30% of the earth's land surface. This void stretches approximately 10,000 miles from the Atlantic shores of northwest Africa across the great Saharan Desert and large tracts of the Arabian Peninsula, cutting deep into the heart of central Asia, all the way to the southern and outermost confines of the great Siberian Plain (north to Siberia and south to the Mongolian Plateau and highland areas of Tibet). For the most part, this void is composed of extremely arid zones or rugged, cold, and treacherous mountainous areas (regions with low human carrying capacities).

Notwithstanding the limited ability to support life throughout these areas, there are ribbons of minor population concentration contained within this void. Not surprisingly, a number of geographically limited settlements are associated with various sources of water. These include population settlements along the fertile Nile River Valley, the coastal regions of the eastern Mediterranean basin, and pockets south of the Caspian Sea. With the exceptions of the few population clusters listed above, the North African and Northern Euro-Asian voids are characterized by harsh physical and environmental settings, with conditions that present formidable obstacles to the formation of human settlements.

2. The West-Central North American Void

This void is another rather extensive region of low population density. The sparsely populated area associated with this void covers the northern Arctic Circle (in the Western Hemisphere), an area which includes northern Alaska, Canada, and Greenland, with a southern extension that cuts through the western and central United States and down into Mexico (along the rugged terrain of the Rocky Mountain chain, south to the arid plateaus of northern Mexico). This population void accounts for about 20% of the earth's land surface.

3. The Antarctica Void

This void is an isolated land continent of extreme cold, with permanent ice that covers an area of more than 5 million square miles (equal to about 9% of the earth's land surface). With the exceptions of military and research outposts, the continent of Antarctica is virtually uninhabited by humans.

4. The Amazon River Basin Void

This void covers more than 3 million square miles or just over 5% of the earth's land surface. It is an area just slightly less in size than the United States. The majority of this void is a densely forested tropical region in central South America, geographically associated with the floodplain of the Amazonian river system.

5. The Oceania Void

This void is an area covering approximately 5% of the earth's surface, which includes most of the continent of Australia, and large portions of the islands of New Zealand and Papua-New Guinea (with the exceptions of several highly populated coastal cities).

6. The Southwestern African Void

This void is a low population density region that blankets most of the Kalahari Desert and its surrounding areas. This void accounts for roughly 3% of the earth's land surface.

7. The Southern South American Void

This void is an area which covers the southern tip of the South American continent (spanning southern Chile, Tierra del Fuego and Patagonia), portions of the southwestern Argentinean Lowlands (along the Pampas), and along the southernmost stretch of the Andes mountain chain. This void accounts for approximately 2% of the world's land surface.

Note that the seven major population voids account for approximately 75% of the earth's land surface. The low population density in these areas reflects people's preference to live in regions that are not "extreme" in terms of climate or physical setting. There are smaller voids with relatively low population densities associated with the European Alps, the Appalachian region of the eastern United States, and the Altiplano-Atacama Desert regions of Bolivia and Chile.

POPULATION DOUBLING TIME

Indexing a region's population growth can be easily accomplished by calculating its **population doubling time**—*the number of years it will take a region's population, at a current growth rate, to double in size*. In general, the population doubling time (DT) of a region may be calculated using the **rule of 70**—*a method for* *approximating doubling time in years.* To approximate the number of years it would take a base population of a region to double, one may use the following formula:

$$DT = 70 \text{ / annual population growth rate}$$

where annual population growth rate is expressed in percentage terms. Examples of various doubling times for selected growth rates are given in Table 5.7.

The annual growth rate of the world's population in 1999 is approximately 1.6%, indicating that the world's population would double every 44 years if the world's population growth would continue at that rate indefinitely. We have already established, however, that the world's population growth rate has been showing signs of declining, and will continue to fall well below the 1.6% annual growth rate over time. It is very unlikely, therefore, that the earth's population will double in 44 years.

Population growth rates in many underdeveloped regions remain high (in comparison to the world average), and are likely to stay that way for some time. Thus, many underdeveloped regions must contend with a doubling of their populations in less than 40 years. Note the dramatic differences in the population doubling time of the devel-

oped NORTH versus the underdeveloped SOUTH, as shown in Table 5.8. Prior to stabilization of world population in 2100, the underdeveloped SOUTH will bear the greatest burden of the world's population growth (see Figure 5.2).

Figure 5.3 highlights national population growth rates, as measured by the annual rate of population increase in 1994. Notice the spatial regularities that exist in the distribution of those rates in terms of NORTH versus SOUTH. Today, many nations in sub-Saharan Africa as well as the Middle East have annual population growth rates that exceed 3.0%.

These statistics indicate that the populations of underdeveloped nations will become a larger and larger percentage of the total population before the world's population stabilizes. The continent of Africa will exhibit the largest relative increase in its population size. In fact, Africa is the only geographic region that will show an increase in its share of world population by the year 2025 (see data in Table 5.9).

THE ECONOMIC GEOGRAPHY OF POPULATION: PRIMACY AND DUALISM

In many peripheral and semiperipheral regions of the world, population settlements are centered in and around one location, typically the largest urban area or settlement in a given nation. These centers of gravity and high population density are generally referred to as "primate cities." A **primate city** *is an urban settlement which contains a disproportionately large share of the total population of a region or nation.* It is not uncommon for a primate city to contain 20% or more of a nation's population. Not only is the primate city the largest settlement, but it usually contains two or more times as many people as the second-largest city in a given country. The primate city serves as the central activity point for the nation and as a center of gravity for the national economy.

TABLE 5.7	Doubling Times and Growth Rates
Annual Population Growth Rate (%)	**Number of Years to Doubling**
2.8	70/2.8 = 25
2.5	70/2.5 = 28
2	70/2.0 = 35
1.5	70/1.5 = 47*
1	70/1.0 = 70
0.5	70/0.5 = 140
0.2	70/0.2 = 350

*Rounded to the nearest year.

TABLE 5.8	Population Growth Statistics *Developed NORTH versus Underdeveloped SOUTH*			
	Total Population in Millions		**Population Growth Rate**	**Doubling Time (years)**
	1994	**2010**		
World	5602.8	7111.5	1.65%	42
Developed world	1163.8 (20.8%)	1236.5 (17.4%)	.48%	148
Underdeveloped world	4439.0 (79.2%)	5875.1 (82.6%)	2.05%	34

Source: de Blij and Muller (1994), A1—Appendix 1.

Note: Figures in parentheses show percentage of world population.

Primate cities are said to have and maintain "primacy" in virtually all social and economic matters. In addition, primate cities are most representative of a region's history and heritage and they are centers of culture. In most cases, a primate city is the nation's capital. With the exception of a few large cities in the developed NORTH (for example, Paris, France, and Madrid, Spain), the primate city concept is largely associated with the underdeveloped nations of the SOUTH. Many good examples of primate cities are found in Latin America (see Table 5.10).

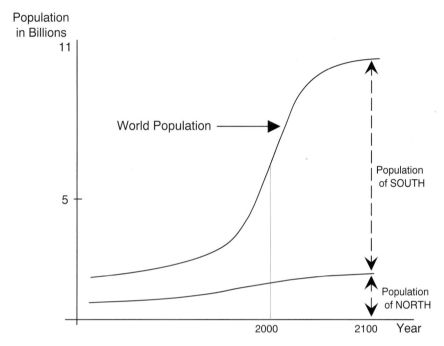

Figure 5.2 Population growth: trends and projections.

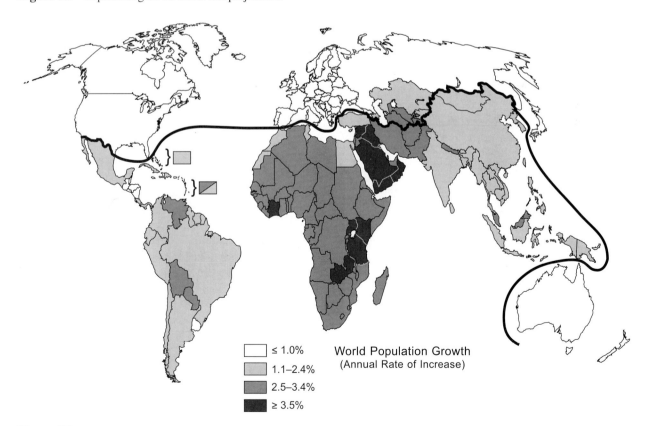

Figure 5.3 World population growth (by percent annual rate of increase).

In general, the percentage of a nation's total population which resides in a primate city in the SOUTH is large in comparison to the percentage of the population which resides in the largest urban areas in the developed NORTH (regions that typically have many large urban centers). For example, the city of Santiago is home to roughly one-third of Chile's population, whereas New York City accounts for only about 5% of the total U.S. population. Note that the metropolitan area populations (the city and its surrounding urban area) would account for an even greater percentage of the nation's population than those listed in Table 5.10. For example, the Mexico City metropolitan area in 1995 was home to more than 21 million people. That is a figure that represents well over 20% of Mexico's population.

TABLE 5.9	Geographic Breakdown of World Population by Percentage Share 1950, 1990, and 2025 (projected)

	Percent Share of World Population		
Region	1950	1990	2025
Africa	8.9	12.3	18.8
N. and C. America	8.2	7.5	6.7
South America	4.5	5.6	5.4
Asia	54.9	58.9	57.8
Europe	15.9	9.7	6.5
Oceania	0.5	0.5	0.5
Commonwealth (former Soviet Union)	7.0	5.4	4.2

Source: Based on *World Resources 1994–95* data (pp. 268–269).

TABLE 5.10	Population Statistics for Selected Primate Cities (1994)

City, Nation	Primate City Population* (in millions)	Percentage of Nation's Population
Mexico City, Mexico	14.1	16
Cairo, Egypt	9.3	16
Managua, Nicaragua	.7	17
San Salvador, El Salvador	.9	18
Caracas, Venezuela	4.0	20
Lima, Peru	4.6	22
Santiago, Chile	4.1	30
Panama City, Panama	.8	31
Kingston, Jamaica	.8	32
Buenos Aries, Argentina	11.0	33

Source: *Goode's World Atlas,* 19th edition (pp. 245–249, 251)

* Figures represent "city" population and not the populations of the entire "metropolitan area."

Note: Population statistics are rounded to the nearest hundred thousand and percentages rounded to the nearest percent.

While most primate cities in the SOUTH show the presence of a highly advanced economy, they may be fairly unrepresentative of the remainder of the development status of the remaining portions of that region or nation. The predominant economic activities of most underdeveloped regions, which occur outside the confines of the primate city, are quite different from the activities found within the primate city. For example, the modern financial, industrial, and government and service districts of Mexico City stand in stark contrast to the surrounding landscapes of poverty which predominate Mexico City's juxtaposed rural areas and the squatter settlements about the urban fringe.

The modern primate cities of the underdeveloped world and the traditional societies of the rural **hinterland** *(an area falling under the political or economic influence of a given settlement or city of primacy)* are testimony to the "pluralistic" nature of society and the "dualistic" nature of regional economies.

Pluralism refers to *the diverse ethnic, racial, religious, or social make-up of a society and the participation of various groups in their traditional culture or special interest within the confines of a common region or jurisdiction.*

Dualism refers to *the coexistence of distinctly different economic systems within the same region or jurisdiction.* Consider the coexistence of traditional agrarian society and modern culture within the same geographic region. In this case, dualism may be viewed as a contrast in development or as a consequence of a resistance to change. Moreover, a dualism exists in both urban and rural settings.

> The coexistence of two broad sectors of employment in both rural and urban areas of the Third World has given rise to the notion of economic dualism and to the application of the terms "informal" and "formal" sector to these different kinds of income-generating activities. (Chant, 1991, p. 185)

Production activities may also be defined as activities which occur in either the formal or informal sector of an economy. The "formal sector" is composed of wage-earning activities and falls under the heading of the so-called **capitalist mode of production**— *a commercial production system that is market- and profit-driven.* Informal sector (non-wage) activities fall under the heading of the so-called **peasant mode of production**—*a traditional production system that is predominantly subsistence oriented.* Note that the peasant mode of production can still be influenced by market forces in the sense that market incentives encourage the production of surplus and exchange.

The concept of "dual economy" is something that applies to virtually all regional production systems to one degree or another. Consider how large-scale, export-oriented agriculture stands alongside small-scale, subsistence farming operations in America's rural farming communities. Economic dualism can also be found in the partitioned and alienating landscapes of urban systems around the globe. As only a fraction of the urban labor force is able to secure gainful employment in the formal sector, many must resort to casual labor and self-employment. Those unable to find work or those that cannot earn a living must rely on the public sector for support. Still others resort to **black market activities**—*the hidden and illicit income-generating activities of the so-called "underground economy"* (Chant, 1991, pp. 184–185).

The attributes of formal and informal sectors are summarized in Table 5.11.

The coexistence of a modern production system and traditional economy is highly visible in the world's least-developed regions. The dense urban landscape of the primate city is a diversion from the prevailing reality of the large number of people who live very traditional lives and are unaffected by changes in the formal sector. In most underdeveloped nations of the SOUTH, the majority of people are far removed from development efforts targeted toward urban settlements. Nonetheless, the expansion of the formal sector has promoted urbanization and growth in urban employment opportunities. These opportunities have encouraged people to migrate from rural to urban areas. The subsequent rural-to-urban migration has created further development problems by expanding the informal sector and underground economies of urban areas. Though sizable portions of the regional labor force will ultimately relocate to urban settlements, only a small percentage will successfully compete for the very limited number of formal sector jobs. Policy responses to allevi-

ate the pressures of soaring urban populations in light of the scarcity of formal employment have been slow in coming. Primate cities have been hit particularly hard by the rural-to-urban migration phenomenon. As a result, government planning efforts have focused on diverting growth and people away from the larger cities. Government has employed strategies which combine "flow-deterring policies" and "decentralization policies."

1. **Flow-deterring policies** *either promote economic opportunities in rural settlements or prohibit people from migrating out of a given region.*
2. **Decentralization policies** *encourage the diversion of migration flows away from major urban areas or primate cities into smaller urban settlements or satellites* (Chant, 1991, pp.183–184).

These two approaches have been pursued with a moderate degree of success in Mexico, Brazil, Egypt, China, and Malaysia. Government efforts to slow rural-to-urban migration have been less successful in the more peripheral nations of the SOUTH; particularly in regions which have shown robust population growth.

Over the past 30 years, the highest urban population growth rates have been observed on the continents of Africa, Asia, and South America (see Table 5.12). Despite this trend, the overall percentage of people classified as urban, or the percentage of the total population residing in cities of at least 750,000 people, is relatively small in both Africa and Asia (less than 15% in 1995). The rural-to-primate city migration pattern does well in explaining why South America has relatively few cities of significant size and why such a large percentage of its population is classified as urban.

MEGACITIES AND WORLD CITIES

Cities are generally defined as population clusters or settlements with at least 5000 people. Atop the world hierarchy of cities are the so-called **megacities**—*built-up urban environments with 5 million or more people.* A list of the world's largest megacities is given in Table 5.13. Data are also presented on the number of telephones per 100 people residing in those areas. This index is useful as a relative measure of modernity.

The spatial distribution of the "major cities" of the world (those with populations greater than or equal to 1 million and less than 5 million) reveals a strong concentration of cities in both the NORTH and in geographic areas with the largest population concentrations (see Figure 5.4). Note, however, that the majority of the world's megacities are located in the underdeveloped regions of the SOUTH. In fact, the ten fastest-growing megacities of the last two decades, Bombay, Dacca, Delhi, Jakarta, Karachi, Lagos, Manila, São Paulo, Seoul, and Teheran,

TABLE 5.11	**Economic Dualism** *Formal and Informal Sectors*
Formal Sector	**Informal Sector**
large-scale operations	small-scale operations
modern landscapes	traditional landscapes
corporate ownership	family/individually run
capital-intensive	labor-intensive
imported technology/inputs	indigenous technology/inputs
protected by labor laws	absence of formal labor laws
difficult entry (skilled labor market)	easy entry (unskilled labor market)
requiring formal training, skills, and knowledge	requiring informal training, skills, and knowledge
capitalist mode of production (commercial activities)	peasant mode of production (subsistence activities)
market- and profit-driven	market-influenced
legitimate and sanctioned activities	legitimate and illicit activities

Source: Originally from David Drakakis-Smith, *The Third World City,* London: Methuen. (1987) Table 5.5, p. 65, adapted from S. Chant, 1991, p. 185, with modifications.

TABLE 5.12	Urban and Rural Population Statistics

Region	Urban Population as a Percentage of Total Population		Average Annual Population Change 1965–1995		Cities of at Least 750,000 Inhabitants	
					Percent of Total Population	Number of Cities
	1965	1995	Urban	Rural	(1995)	(1990)
World	35.5	45.2	2.7	1.3	17.4	376
Africa	20.6	34.7	4.7	2.2	10.9	34
Asia	22.0	34.0	3.5	1.5	13.3	149
N. and C. America	67.4	74.0	1.8	0.7	38.6	66
South America	55.9	78.0	3.3	0.2	36.8	35
Europe	63.8	75.0	1.0	0.8	24.0	56
CWI	52.8	68.1	1.7	0.5	17.9	30
Oceania	68.6	71.0	1.8	1.4	41.3	6

Source: *World Resources 1994–95* (pp. 286–287).

are all located in the SOUTH. Despite the fairly large number of sizeable urban areas in Asia, only 34% of the Asian population is classified as urban (the same percentage as Africa)! Note also that most major cities in the SOUTH, particularly in South America and Asia, are generally oriented toward coastal areas—representing the significance of that all-important transport link in the global economy. It is also a visible reminder of the lasting influence of colonialism (Figure 5.4).

The largest cities of the NORTH and the semiperipheral SOUTH, those cities that are highly connected in the global system, may be labeled as **world cities**—*large, highly developed, and globally prominent metropolises or megacities, which are favored centers of global development and commerce* (Friedmann and Wolff, 1982; Friedmann, 1986). World cities are the hubs of the world economy.

Paul Knox (1995) describes world cities as

nodal points that function as control centers for the interdependent skein of material, financial, and cultural flows which, together, support and sustain globalization (of production). They also provide an interface between the global and the local, containing economic, sociocultural and institutional settings that facilitate the articulation of regional and metropolitan resources and impulses into globalizing processes while, conversely, mediating the impulses of globalization to local political economies. (p. 236)

World cities are an outgrowth of the ultraconsumer-oriented, metro-centric global material culture. They are the ultimate expressions of individualism in the capitalist world economy.

The phenomenon of **global metropolitanism**—*the interconnectiveness and synchronization of globally conscious urban economic agents who share a common, materialistic cultural identity and the increasing interdependence of urban areas in the globalization process*—has dissolved the traditional barriers of space, time, and cultural difference. Global metropolitanism has also led to the diffusion of "Western" ideas on progress, development, and change. The spread of American culture and consumerism have led to the "Americanization" of global culture.

As centers of international business and finance, as well as the largest origins and destinations for investment capital, world cities have benefitted most from the expansion of global markets, the intensification of world trade, and the integration of the world's production systems. Notwithstanding the numerous and large cities in the SOUTH, the "hierarchy" of world cities shows a connective structure that mostly favors the movement and flow of capital amongst the highly connected cities of the NORTH (see Figure 5.5).

Rapid urbanization, high population growth, and a shortage of production and investment capital in the SOUTH continue to aggravate the world poverty situation. Growing numbers of people join the ranks of the world's impoverished each year, and there is no reason to expect a reversal of this trend in the near future. The escalating urban development problem in the SOUTH requires that government leaders find new and innovative

TABLE 5.13	The World's 25 Largest Megacities

Country Megacity	Estimated and Projected Population Size (× 1000)		Annual Population Growth Rate 1980–1990	Telephones per 100 People
	1990	2000		
Argentina				
Buenos Aires	11,448	12,822	1.4%	14
Bangladesh				
Dacca	6578	11,511	7.2%	2
Brazil				
Rio de Janeiro	10,948	12,162	2.2%	8
São Paulo	18,119	22,558	4.1%	16
China				
Beijing	10,867	14,366	1.9%	2
Shanghai	13,447	17,407	1.4%	4
Tianjin	9249	12,508	2.4%	4
Egypt				
Cairo	8633	10,761	2.3%	4
France				
Paris	8720	8803	0.1%	n/a
India				
Bombay	12,223	18,142	4.2%	5
Calcutta	10,741	12,675	1.8%	2
Delhi	8171	11,692	3.9%	5
Indonesia				
Jakarta	9206	13,380	4.4%	3
Iran				
Teheran	9779	14,251	4.6%	n/a
Japan				
Osaka	10,482	10,601	0.5%	42
Tokyo	25,013	27,956	1.4%	44
Korea, Rep.				
Seoul	10,979	12,949	2.9%	22
Mexico				
Mexico City	15,085	16,190	0.8%	6
Nigeria				
Lagos	7742	13,480	5.8%	1
Pakistan				
Karachi	7943	11,895	4.7%	2
Philippines				
Manila	8882	12,582	4.1%	9
Russia				
Moscow	10,446	11,121	0.7%	n/a
United Kingdom				
London	9115	9574	0.5%	n/a
United States				
Los Angeles	11,456	13,151	1.9%	35
New York City	16,056	16,645	0.3%	56

Source: *World Resources 1994–1995* (p.288); *World Almanac 1994* (New York: St. Martin's Press).

"n/a" = not available.

ways in which to promote urban economic growth while deterring both rural-to-urban migration and the expansion of the population base. The high incidence of poverty in the global economy provides evidence that decades of economic growth have not led to a better way of life in all regions.

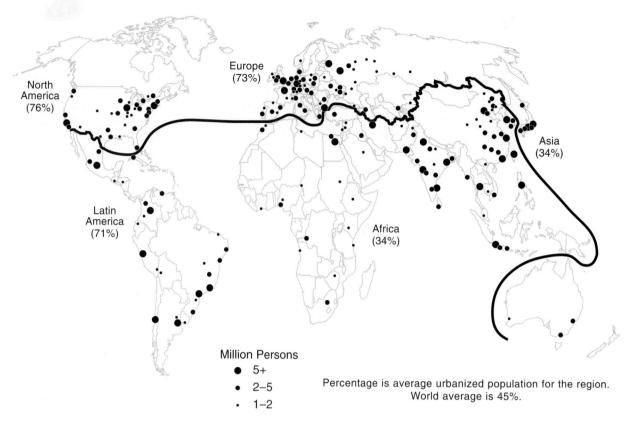

Figure 5.4   Major cities of the world and regional levels of urbanization.

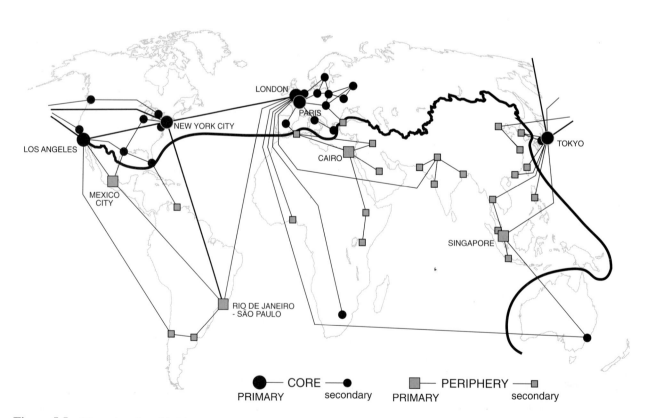

Figure 5.5   Hierarchy of world cities.

THE INCIDENCE OF POVERTY

Although much has been written on development strategies to alleviate poverty in the SOUTH, there is no consensus on a surefire or universal approach to reducing the incidence of poverty. It is recognized, however, that poverty alleviation measures must be region-specific in their design. In most situations, abject poverty is an outcome of **relative scarcity**—*a situation or condition brought about by a shortage of a particular resource at a given location in comparison to other resources or the same resource in other geographic locations.*

The lack of physical and financial resources has placed the underdeveloped world at tremendous comparative disadvantage. In addition, the **absolute scarcity** of resources (that is, *a condition of global scarcity*) is a condition which has also inhibited the growth and development of all regions. It is important to note that limits to growth born out of relative or absolute scarcity do not necessarily imply limits to development (Daly, 1991, p. 243). Scarcity, as determined by the needs and wants of the members of a society, is truly a dynamic concept. Certainly the incidence of poverty as an outcome of relative scarcity is something that can be reduced if development efforts are geared toward satisfying the most basic of human needs and not just controlling population growth in underdeveloped regions.

> Birth control without poverty reform will, at best, reduce the number of poor people but will not eliminate poverty . . . (recognizing) that without population control of both human bodies and their extensions in artifacts, all other social reforms will be canceled by the growing burden of . . . scarcity. (Daly, 1991, p.169)

The increase in world population and increasing global consumption move us closer to test both the limits of our production systems and our level of poverty tolerance. As more people, in a divided world, compete for resources and access to the means by which they may improve their quality of life, it is only a matter of time before a worldwide realization is reached on the necessity to combat the forces which preserve poverty (as poverty promotes inefficiencies). Policy makers must find tangible solutions to relative scarcity problems before true progress can be made.

Strategies to reduce poverty must include

1. "targeting human capital investments in health care and education for the poorest segments of society";
2. economic and legal reforms "to extend land tenure and other rights to the rural poor" to support subsistence endeavors;
3. reforms to "legitimize" and/or formalize the informal sectors of regional economies; and
4. promoting social and political changes that enhance participation in economic matters by increasing access of all members of society to the means of

production, with special provisions for elevating the status of women in the development process *(World Resources 1994–95,* 1994, p. 24).

In an attempt to expand agricultural trade possibilities worldwide, convincing arguments have also been made for the removal of region-based subsidies and trade barriers which protect agricultural sectors. Such measures could significantly increase export income in the SOUTH as part of a global trade-liberalization policy.

> Trade barriers prevent market development and job creation among low-cost producers, perpetuating poverty, subsistence agriculture, and their environmental effects in developing countries . . . Removing such barriers could increase employment, promote economic diversification, and lower consumer costs in industrialized countries. (*World Resources 1994–95,* 1994 p. 24)

Yet such measures, as they expand world agricultural production and trade, may also lower market price levels worldwide. Lower product prices in the global market would depress export income and earnings in peripheral regions and lessen the potential for investment in agricultural production. This outcome would be contrary to the global development goal of increasing the self-reliance of underdeveloped regions and would aggravate the poverty situation.

Poverty has trapped millions worldwide in deplorable living conditions. Despite the growth of exports and emerging markets, the world's poor became poorer and more numerous during the 1980s and 1990s. Economic change has not removed the hardships placed on the world's impoverished people. The underlying cause of poverty is not solely population based. The consequences of production and industrialization have also contributed to the problem.

> Failure of the world community to stem the rising tide of poverty has many roots. Rapid population growth is [only] one, as is the failure of many governments to reform their economic and political systems . . . [Also,] the once separate issues of environment and development are now linked . . . as environmental degradation is driving a growing number of people into poverty. And poverty itself has become an agent of ecological degradation, as desperate people consume resources on which they depend. Rather than a choice between the alleviation of poverty and the reversal of environmental decline, world leaders now face the reality that neither goal is achievable unless the other is pursued as well. (Brown et al., 1991, pp. 24–25).

Rapid population growth may contribute to poverty in many regions. Yet, the lack of personal empowerment and development may also contribute to poverty and population growth.

> Poor people have many children because children are needed to work and to support their elders . . . Having children is one of the few powers the poor can exert over

their own lives, and one of the few hopes [they have] of getting ahead . . . [Moreover,] world fertility surveys indicate anywhere from a third to a half of the babies born in the Third World would not be, if their mothers had access to cheap, reliable family planning and had the personal empowerment to stand up to their husbands and relatives and choose their own family size. Economic development brings lower birthrates because it brings to women the pill, literacy, and self-determination. (Meadows, 1991, p. 61).

Finding solutions to the world's poverty crisis is much more complicated than just controlling excessive population growth or promoting increases in consumption per capita. While the extent and magnitude of poverty varies greatly between regions, it is a phenomenon that is noticeably biased toward the world's peripheral economies. Before examining the geographic incidence of poverty, let us put the crisis in its proper perspective.

Alan Durning (1990) highlights a series of astonishing findings on the subject of poverty. The following eye-opening excerpts are highlighted in his work:

1. By 1990, the world had 157 billionaires, approximately 2 million millionaires, and over 100 million homeless.
2. As water is bottled from natural springs around the world and sold to the prosperous, approximately 2 billion people drink and bathe in water contaminated with deadly parasites and pathogens.
3. While even the less fortunate of people in the developed world may secure access to the most basic of necessities, more than half of humanity lacks sanitary toilets.
4. "In 1988, the world's nations devoted $1 trillion—$200 for each person on the planet—to the means of warfare, but failed to scrape together the $5 per child it would have cost to eradicate the diseases that killed 14 million in that year alone." (Durning, 1990, p. 22)

The economic and demographic realities have led to an all-out crusade to fight the spread of poverty worldwide. While the eradication of poverty may prove difficult in a world economy which is said to be partly responsible for both creating and perpetuating the problem, poverty relief has become a most-worthy development goal.

Poverty may be described as "a condition of life so limited by malnutrition, illiteracy, disease, squalid surroundings, high infant mortality, and low life expectancy as to be beneath any reasonable definition of human decency." (Durning, 1989, p. 24)

In general, **absolute poverty** is characterized by *the lack of sufficient income in cash or kind to meet the most basic biological needs for food, clothing, and shelter.* Note that the cost of providing adequate food, clothing, and shelter varies from country to country. In general, it ranges from approximately $50 to $500 per person per year (in 1995 dollars) depending on variabilities in price levels and access to subsistence resources and the availability of public services.

World Bank estimates place the global poverty rate at about 22.3% in 1980, with approximately 988 million people worldwide classified as living in poverty. In 1989, the poverty rate climbed to 23.4%, for a total of 1.23 billion people. Estimates for the year 1994 fell at around 24%, a figure which translates into approximately 1.35 billion people. A regional breakdown of the poverty population is shown in Table 5.14.

The data on the geographic incidence of poverty reveal four disturbing trends:

1. The percentage of the world's population living in absolute poverty has been rising over time (a by-product of high population growth rates in the poorest regions combined with inadequate economic development policies);
2. Asia has *the greatest number* of people classified as living in absolute poverty, with about 755 million people in 1994 (equivalent to slightly less than 14% of the world's total population);
3. sub-Saharan Africa and Latin America are the two regions of the world with *the greatest share* of people living in poverty (in relative terms), with 64% and 35% of their populations classified as living in absolute poverty, respectively; and
4. The share of sub-Saharan Africa's population classified as impoverished is growing at a fairly rapid rate, whereas the share of Asia's population classified as living in poverty is showing signs of declining (in a relative sense).

SOME FINAL OBSERVATIONS

The positive correlations between the "peripheralness" of a region and the percentage of its population classified as impoverished is something which is not purely coincidental. The economic causes include the relative scarcity of development capital, debt and dependence, and leakage.

Developing nations' debt (of over $1 trillion) caused a reversal of the flow of international capital. Today, poor nations are paying rich ones $50 billion each year in debt and interest payments beyond what they receive in new loans. Capital flight from wealthy people in poor lands may bring the exodus up to $100 billion each year. Trade protectionism in industrial countries results in annual losses on a similar scale, as Third World export prices fall and markets shrink. . . . Poor countries are forced to exploit mineral deposits, forests and fisheries to meet debt obligations. And they are left with few resources to alleviate poverty (Durning, 1989, pp. 25–26).

According to Brown et al. *Vital Signs*, 1993, more money was paid by poor nations to cover their loans than

TABLE 5.14	The Geography of Absolute Poverty			
	Number of People (in millions)		**Share of Total Population (%)**	
Region	**1989**	**1994**[*]	**1989**	**1994**[*]
Asia	675	755	25	23
sub-Saharan Africa	325	340	62	64
Latin America	150	165	35	35
N. Africa and Middle East	75	85	28	28
WORLD	1225	1345	23.4	24.0

Source: Durning (1990), p.139, with additional estimates provided for 1994 based on the trend.

[*] Estimated.

they received in additional funds for development during the mid- and late-1980s. Total debt in the underdeveloped world (that is, the SOUTH) stood at about $1.4 trillion in the mid-1990s and is expected to climb even higher. Based on the current trend, the underdeveloped world is likely to amass a debt approaching $2 trillion by the year 2002.

There is evidence of growing regional disparities in the world economy as real GNP per capita in the 1980s has been dramatically rising in Asia and steadily declining in Latin America and sub-Saharan Africa.

> According to calculations made by the World Bank, the numbers of absolutely poor in Asia (including China and India) may decline over the next ten years. However, in Africa . . . there will be a marked increase in the proportion of the population below the World Bank's poverty line. (*Third World Atlas*, 1994, p. 73)

Hope remains that the development success of the more prominent Asian nations will be duplicated in Africa and Latin America, although it is unlikely that such a miracle will ever occur. There are several reasons for this. First, the preconditions for regional economic development and elevation of economic status are very different than they were 20 years ago. It takes much more than a productive and abundant labor force to sustain economic growth. The social, economic, and demographic trends in the periphery simply do not support the possibility of such a miracle occurring.

Second, the failure of most core economies to eradicate poverty is a sure sign that peripheral regions must contend with this problem for some time to come. Furthermore, just as many of the world's most pressing economic and environmental problems can be traced to individualist culture and uneven development, poverty may be blamed for the acceleration of the economic and ecological devastation of the periphery as "desperate people" overexploit world resources and sacrifice the future to salvage the present. (Brown, 1990, p. 144)

Ecological decline is a self-reinforcing spiral that does little to help break the cycle of poverty. As a result, a high incidence of poverty not only increases the likelihood of economic failure (as economies fail to grow and people become more reliant on external assistance), but it also leads to disastrous changes in structure and performance of an economy as environmental degradation reduces production output and potential. Hence, the consequences of nonsustainable economic activities and the subsequent damage to the environment have aggravated the poverty problem and have created further long-term development problems.

Although there is debate on whether there exists an **optimal population**—*an ideal population size based on resources, technology, and an ascribed quality of life*—for a regional economy or the world itself, many experts contend that the population sizes of many underdeveloped regions have already exceeded the optimum. But even the definition of the optimum can change as human ingenuity is likely to find new and innovative solutions to regional development problems (Simon, 1990). Much damage can be done, however, before real solutions are found and implemented.

Hand in hand with the rise of poverty will be the rise of hunger and malnutrition. In light of population growth trends and limits to economic growth, feeding the world's underdeveloped regions has now become a major development concern. Discussion in the next chapter will focus, therefore, on economic and food security issues.

Poverty in the Core

Leading core economies have not been spared from poverty. Currently, it is estimated that well over 40 million Americans live in absolute poverty. More than two-thirds of America's impoverished live in metropolitan areas. While poverty affects roughly 14% of America's urban population, poverty rates in rural America are even higher. More than 20% of the U.S. population residing in nonmetropolitan areas lives in poverty.

The official poverty line stood at about $12,000 for a family of four in 1990, or $3,000 per family member in 1990. Those most affected by poverty are minorities, female-headed households, children, and the elderly. It is estimated that one out of every five children in America will grow up in poverty.

Growing disparities in income and earnings potential has led many analysts to reexamine the effectiveness of growth and trickle-down economics, as three decades of economic growth have done little to alleviate the poverty situation. Consider the fact that between 1973 and 1992, "the poorest 10% of American families suffered an 11% drop in real income, while the richest 10% enjoyed an

18% increase in real income" (*The Economist,* p. 20). The poverty crisis in the United States has been worsened by recent downturns in industry, and employment losses due to foreign competition, firm relocation and job displacement, and trade deficits. Falling real wages and the loss of jobs in manufacturing-for-export sectors have diminished the recirculation potential of money income. Smaller multiplier effects have meant less trickle down and less chance that the benefits of growth will reach the poorest members of society. This situation has been made worse as wages and earnings in the rapidly expanding tertiary sector have not kept pace with increases in the cost of living.

Poverty contributes to underdevelopment and is responsible for preserving a status quo which magnifies social and economic inequalities. Under siege from poverty, urban areas have become a source of embarrassment for many of the world's richest nations. Flourishing white-collar financial districts and bustling suburban retail corridors of many U.S. cities stand in stark contrast to the abandoned industrial zones, expanding slums, and inner-city blight that have relentlessly survived through three decades of "urban renewal." Chronic unemployment, the lack of sufficient employment opportunities, insufficient access to financial resources, depressed living conditions, declining educational systems, inadequate food and health care, and insufficient levels of human services are conditions which epitomize the culture of poverty in urban settings. Within this culture we find feelings of hopelessness and despair among those caught in the midst of a seemingly inescapable cycle.

Poverty knows no geographic boundaries or limits. Even once-thriving centers of production and prosperity have become home to an expanding welfare population. It is an environment which has created its own set of unique social and economic problems. The exodus of industry and jobs from America's inner cities has led to erosion of the urban tax base. Aging infrastructure and housing, fewer employment opportunities, and shrinking revenues have led to a spiral of decay that seems, for all intents and purposes, irreversible. Nonetheless, urban revitalization efforts continue as cities try to cover up the symptoms of a deeply entrenched economic development problem. Without good jobs, improved housing, and adequate public services, those caught on the "poor side of town" may continue only to dream of a better way of life.

KEY WORDS AND PHRASES

absolute poverty   112
absolute scarcity   111
black market activities
  (underground economy)   107
capitalist mode of production   106
decentralization policies   107
demographics   96
developable land   96
dualism   106

flow-deterring policies   107
forecasts   99
global metropolitanism   108
hinterland   106
megacities   107
optimal population   113
peasant mode of production   106
pluralism   106
population bomb   98

population concentrations   99
population density   99
population doubling time   103
population voids   99
primate city   104
quality of life   96
relative scarcity   111
rule of 70   103
world cities   108

REFERENCES

Birdsall, N. "Population Growth and Poverty in the Developing World." *Population Bulletin* 35 (5). Washington, D.C.: Population Reference Bureau (1980).

Brown, L.R. (ed.). *State of the World 1989.* Washington, D.C.: World Watch Institute (1989).

Brown, L.R. (ed.). *State of the World 1990.* Washington, D.C.: World Watch Institute (1990).

Brown, L.R., C. Flavin, and S. Postel. *Saving the Planet: How To Shape an Environmentally Sustainable Global Economy.* New York: W.W. Norton (1991).

Brown, L.R., C. Flavin, and H. Kane. *Vital Signs.* New York: W.W. Norton (1993).

Brunn, S.D., and J.F. Williams. *Cities of the World: World Regional Development* (2nd edition). New York: Harper & Row (1993).

Chant, S. "National Perspectives on Third World Development. In *Global Change and Challenge: Geography for the 1990's.* R. Bennett and R. Estall (eds.). London: Routledge (1991), pp. 175–196.

Clark, D. *Urban World/Global City.* London: Routledge (1996).

Daly, H. *Steady State Economics.* Washington, D.C.: Island Press (1991).

de Blij, H.J., and P.O. Muller. *Geography: Realms, Regions, and Concepts* (7th edition). New York: J. Wiley & Sons (1994).

Drakakis-Smith, D. *The Third World City.* New York: Methuen (1987).

Durning, A. "Life on the Brink." *World Watch 3* (March–April 1989)

Washington, D.C.: World Watch Institute, pp. 22–30.

The Economist. "For Richer, For Poorer" November 5 (1994), pp. 19–21.

Ehrlich, P.R. *The Population Bomb.* New York: Ballantine Books (1971).

Ehrlich, P.R., and A.H. Ehrlich. *The Population Explosion.* New York: Simon and Schuster (1990).

Fisher, J.S. *Geography and Development: A World Regional Approach.* New York: Macmillan (1992).

Friedmann, J. "The World City Hypothesis," *Development and Change* 17 (1986), pp. 69–83.

Friedmann, J., and G. Wolff. "World City Formation: An Agenda for Research and Action." *International Journal of Urban and Regional Research* 6 (1982), pp. 310–343.

Gilbert, A., and J. Gugler. *Cities, Poverty and Development: Urbanization in the Third World.* New York: Oxford University Press (1992).

Goode's World Atlas (19th edition). Chicago: Rand McNally (1995).

The GREAN Initiative: Global Research on the Environment and Agricultural Nexus for the 21st Century. A Report published by the University of Florida: Office of International Studies (in collaboration with Cornell University), June 1995.

Hartshorn, T.A., and J.W. Alexander. *Economic Geography* (3rd edition). New York: Prentice-Hall (1988).

Knox, P. "World Cities and the Organization of Global Space." In Johnston, R.J., P.J. Taylor, and M.J. Watts (eds.) *Geographies of Global Change: Remapping the World in the Late Twentieth Century.* Oxford, U.K.: Blackwell (1995), pp. 232–247.

Meadows, D.H. *The Global Citizen.* Washington, D.C.: Island Press (1991).

Meadows, D.H., D.L. Meadows, and J. Randers. *Beyond the Limits: Confronting Global Collapse, Envisioning a Sustainable Future.* Post Mills, VT: Chelsea Green (1992).

Seitz, J.L. *Global Issues.* Cambridge, MA: Blackwell (1995).

Short, J.R. (ed.). *Human Settlement.* New York: Oxford University Press (1992).

Simon, J.L. *Population Matters: People, Resources, Environment, and Immigration.* New Brunswick, NJ: Transaction Publishers (1990).

Smith, D.M. *Human Geography: A Welfare Approach.* New York: St. Martin's Press (1977).

Student Atlas of World Politics. Guilford CT: The Dushkin Publishing Group (1994).

Third World Atlas, A. Thomas (ed.) Washington, D.C.: Taylor and Francis (1994).

Weeks, J. *Population: An Introduction to Concepts and Issues* (4th edition). Belmont, CA: Wadsworth (1992).

World Almanac 1994. New York: St. Martin's Press (1994).

World Development Report 1995: Workers in an Integrating World. (published for the World Bank) New York: Oxford University Press (1995).

World Resources 1992–93: Toward Sustainable Development. New York: Oxford University Press (1992).

World Resources 1994–95: People and the Environment. New York: Oxford University Press (1994).

CHAPTER

6

FOOD AND ECONOMIC SECURITY ISSUES

LAND SCARCITY AND DEGRADATION

Population density measures provide important information about the average number of people per unit land area. Population density measures, however, fail to expose the burden a given population places on a regional production system. Population density also tells us very little about the spatial distribution of people or settlements. For example, the population density of Egypt is estimated at slightly more than 150 people per square mile (just slightly above the world average). Yet the spatial distribution of Egypt's population is highly concentrated and in close proximity to water. Consider the fact that 97% of Egypt's population lives in settlements along the rich agricultural landscapes of the Nile River Valley. In reality, the true population density of settlements along the Nile is well over 3200 people per square mile (a density that is roughly 30 times greater than the world average) (Figure 6.1).

There are many factors that influence regional population distribution and density. They include the physical environment, the organization of a region's economic or political system, the availability of land and water resources, and the settlement preferences of human agents. There is, in general, a tendency for population to cluster in regions that have hospitable climates and/or near bodies of water. Locations along river systems and coastal areas offer not only access to water but lower costs of transport. In general, settlement patterns are consistent with the avoidance of regions characterized as having

"extreme" environmental conditions—regions deemed as inhospitable due to excessive heat or moisture, high elevation or mountainous landscapes, arctic zones, remote arid lands, or areas in which transportation costs are prohibitive due to the physical terrain.

Population density is also an inadequate measure of scarcity. Although land scarcity means different things to people in different regions, the more people per land unit, the greater the burden placed on the production system. For the many people crowded in the high-density coastal areas of Asia, limitations imposed by land scarcity (though severe in comparison to what rural agriculturalists may experience) are of little consequence to the people who have learned to adapt to their environment. Effective utilization of space in these regions is essential for people to cope with high-density urban living. The multitude of high-rise apartments and skyscrapers serves as a common backdrop for the flotilla of boats, which provides additional space for families living and working in the nearby waterways. For urban dwellers along the coast, land scarcity may encourage some to work a home garden, perched on a balcony or roof of a crowded tenement. Others may choose to work a land parcel at some distance away from the settlement's fringe to grow food and supplement the family diet, while members of the boat community engage in fishing activities to secure food and earn a living (Figure 6.2).

Even though space is limited in many places around the globe, land scarcity is not a major problem in all high population density areas. For instance, less-peripheral

Figure 6.1 Rural agricultural: Children harvesting crops along the Nile River Valley.

6.1: ©Owen Franken/Stock Boston

Figure 6.2 Land scarcity and large urban populations mean tight quarters in many of Asia's high density coastal areas.

6.2: ©Alan Oddie/Photo Edit

regions may find economic solutions to land scarcity or food shortage problems. Some regions are able to buy (import) food from surplus-producers with money obtained from other economic activities. For example, consider the large number of urban dwellers residing in the affluent high-density settlements of Hong Kong and Singapore. Although the urban residents may not be able to grow enough food to be considered self-sufficient, their wealth and income enables them to buy what they need at the local markets or restaurants. The large majority of peripheral regions in the SOUTH, however, face land scarcities and food-production shortage problems that do not have immediate economic solutions. Low-income levels, deficit-production in agricultural sectors, and competitive disadvantage in nonagricultural sectors

render these regions unable to support their swelling populations. The internal pressures of high population density and high population growth rates have placed an enormous burden on land resources in these areas.

A preferred measure of land scarcity and the impact that a population places on its regional food-production system is "physiological density." **Physiological density** is defined as *the amount of "arable land" per person in a given region*. Note that **arable land** refers to *land suitable for the cultivation of crops* (that is, cropland).

The geography of physiological density tells us a good deal about regional variations in agricultural land scarcity. Physiological density statistics by major geographic regions are given in Table 6.1. Peripheral regions tend to be most severely affected by land scarcity. In particular, Africa, Latin America, and Asia have physiological densities that are critically low and declining over time. The increasing scarcity of arable land in these regions is a global development concern. It is a condition that has been accelerated by the subsistence orientation of the production system and the increasing vulnerability of land to degradation. Cropland per capita statistics continue to decline with increasing populations and the misuse and mismanagement of land.

Physiological density measures have one major shortcoming. They fail to take into account the relative *quality* of a region's arable land. Marked differences in land quality are attributed to local and regional variations in soil conditions, land characteristics, climate, farming practices, and susceptibility to natural hazards. Moreover, declines in physiological densities have been precipitated by **land degradation**—*the loss, damage, and destruction of productive agricultural land due to forces of nature or human practices and the general misuse or mismanagement of land (including the encroachment of urban landscapes on nearby rural areas)*. In many instances, damage to cropland is preventable. Nonetheless, decades of misuse and mismanagement have led to the destruction of millions of acres of farmland worldwide, without any hope of "reclamation" (that is, without any chance of reclaiming the land for its original or traditional use). Areas hardest hit by land degradation are the subsistence farming regions of the SOUTH.

Eight principle causes of land degradation have been identified in the literature (after Postels, 1989, 1990; Brown and Young, 1990; and Grigg, 1993). A summary is provided here.

1. The **overgrazing** (*excessive and nonsustainable grazing*) of rangelands beyond their carrying capacity is a contributing factor to land degradation. Overgrazing promotes erosion, the loss of protective vegetation cover and soil nutrients, and leads to **soil compacting**—the excessive hardening of topsoil from the continual trampling of the soil by livestock.

TABLE 6.1	Physiological Density Measures by Major Geographic Region

Region	Total Hectares (in millions)	Arable Hectares per Person 1990	Arable Hectares per Person 2025*	Rating**
Europe	140	.28	.25	good
N. America	236	.86	.71	good
Oceania	51	1.90	1.24	good
Former USSR	232	.80	.67	fair-to-good
Latin America	179	.40	.24	fair
Africa	185	.28	.15***	poor
Asia (excluding regions of the former USSR)	451	.15	.09***	poor

Sources: Berry, Conkling, and Ray (1993), p. 116; *World Resources 1992–93* (1992); *World Resources 1994–95* (1994); and *Global Outlook 2000,* United Nations (1990).

Note: One hectare is an area equal to approximately 2.471 acres (an acre is 4840 square yards—an area that covers roughly 120' by 120').

* Forecasts based on population projections assuming no further land degradation.

** As based on land availability and quality.

*** Critically low levels.

2. **Overcultivation** (*excessive cultivation without adequate nutrient regeneration*) of cropland beyond a sustainable level is a significant contributor to land degradation. It is known to result in **soil burnout**—*loss of soil fertility and productivity.*

3. **Waterlogging** and **salinization** (*rising groundwater levels and the subsequent increases in salinity of soils in irrigated farming regions*) has increased the amount of toxins in topsoils. This has been brought about by the upward, waterborne movement of natural salts from subsoils as groundwater levels rise. Waterlogging also increases the likelihood of exposure and infiltration of man-made chemicals, which find their way into the underground water supply.

4. The rapid and unnatural transformation of land by humans and environmental change has led to degradation. The processes of **deforestation** (*the loss of natural forest cover and canopy*) and **desertification** (*the intrusion of a desert on marginal farmland at its fringe*) have played major roles in the loss of marginal farmland. Regions of the SOUTH have been especially hard hit.

5. **Soil erosion** (*the loss of topsoil and soil nutrients due to excessive exposure to wind and rain*) continues to reduce productivity of the world's farmland. It is a particularly large problem in underdeveloped regions, where methods of erosion control are not applied.

6. **Cropland mismanagement** (*the inadequate use or handling of farmland*) continues to promote the degradation of land and the loss of productivity. Shorter-than-normal fallow periods (the time cropland is idled), intensified cultivation and production cycles, soil compaction from machinery, water drainage problems, and nonsustainable practices have taken a toll on land in commercial farming systems.

7. **Urban sprawl** (*the outward advancement of the urban fringe*) continues to absorb productive rural farmland for urban land uses.

8. Damage due to **external agents** (*air pollution, acid rain, and increased ultraviolet exposure*) and **natural hazards** (*physical phenomenon such as storms, typhoons, floods, tidal waves, etc.*) have contributed greatly to the loss of topsoil and lower overall productivity of agricultural land. Environmentally sensitive farming areas near high-polluting industrial zones in core and semiperipheral economies have been negatively impacted by pollution. Storm-prone tropical lowlands in Asia and Latin America have also experienced considerable amounts of damage to farmland over time.

Estimates of the annual loss in grain output as a result of land degradation run as high as 20 million metric tons per year, with soil erosion responsible for the largest production losses. Roughly 17% of the world's vegetated land has been degraded to one extent or another. Estimates of human-induced soil destruction by selected geographic regions are shown in Table 6.2. About 3% of the earth's vegetated land has experienced strong-to-extreme degradation. Asia, Africa, and Europe have each exhibited significant amounts of land degradation. At least 20% of the arable land in these regions has been severely affected by erosion or physical and external chemical agents. Africa accounts for 42% of the world's most severely degraded land, Asia is a close second with 36%.

TABLE 6.2	The Geography of Soil Destruction (1945–1990)

Region	Total Degraded Area TDA (millions of hectares)		Degradation as % of Vegetated Land (P)	Erosion	Physical and External Agents (millions of hectares)
World	**1964.4**	**(2.6)**	**17%**	**1642.0**	**322.4**
L-M	1659.5		6	1392.1	267.4
S-E	304.9		11	249.9	55.0
Asia	**748.0**	**(2.9)**	**20%**	**662.8**	**85.2**
L-M	638.8		17	573.7	65.0
S-E	109.2		3	89.1	10.2
Africa	**494.2**	**(5.7)**	**22%**	**413.9**	**80.3**
L-M	365.4		17	302.5	62.9
S-E	128.8		5	111.4	17.4
S. America	**243.4**	**(1.4)**	**14%**	**165.1**	**78.3**
L-M	218.3		13	152.9	65.3
S-E	25.1		1	12.2	13.0
Europe	**218.9**	**(1.5)**	**23%**	**156.7**	**62.2**
L-M	205.0		22	143.8	61.2
S-E	13.9		1	12.9	1.0
N. and C. America	**158.1**	**(1.4)**	**8%**	**145.3**	**12.8**
L-M	131.4		7	120.2	11.2
S-E	26.7		1	25.1	1.6

Source: Adapted from *Student Atlas of World Politics* (1994, p. 91) with data obtained from United Nations Environmental Program; International Soil Reference and Information Center, and *World Resources 1992–93*. (Figure 6.3).

Figures for Oceania omitted.

Categories:

L-M: Light-to-Moderate degradation.

S-E: Strong-to-Extreme degradation.

Values for a degradation-severity index, as defined by the weighted ratio: (S-E/TDA) × P are shown in parentheses.

Note that "vegetated land" includes all natural vegetated areas and all agricultural land.

The overuse of cropland is an outcome of many factors, including increasing global population pressures and demand, and the intensification of production in agricultural systems. Yet overutilization of the world's agricultural land is a problem that is much more than just a demand-driven phenomenon. The dwindling supply of agricultural land is now a growing concern. Significant land losses have come about due to urban sprawl and the development of rural areas. The loss of farmland in surplus-producing developed economies is a serious global development problem. As critical shortages of reserve agricultural land now exist in most of the world's underdeveloped regions, the remaining supply of arable land in developed regions must be treated as a **world resource**—*something that is of utility to the global community.* The relatively low and slow-rising land prices in the rural areas of the developed world surely do not reflect the growing land scarcity problem, which is global in scope.

Although Latin America and Africa can support urban and industrial expansion without a significant loss of arable land, many experts contend that Asia does not have land reserves to spare. During the 1980s, China has spawned more than 200 new cities, with over 100 million peasants relocating from rural to urban areas. Prolifera-

tion of the urban landscape in China and elsewhere in Asia has claimed well over 1 million hectares of arable land. In this decade alone, China has converted more than 400,000 hectares of cropland into urban land uses, land enough to feed about 10 million people. Encroachment of the urban landscape poses a formidable threat to surrounding agricultural areas in cities such as Jakarta, Bangkok, and Dacca. Most troubling is the fact that expansion of the urban fringe often claims the very best agricultural land, given that urban settlements have been founded near rich and fertile agricultural land (Gardner, 1996, pp. 80–81).

The loss of farmland is a problem that is not limited to the world's poorest regions. Consider the recent land losses in the state of Florida, one of the premier agricultural areas in the United States. While Florida boasts over $7 billion a year agricultural output in the production of citrus, sugarcane, and winter vegetables, economic and population growth have combined to increase the rate of urbanization. During the late 1980s and early 1990s, Florida lost farmland at the rate of about 150,000 acres a year (a figure equivalent to roughly an acre every five minutes during that period). Much of the loss of farmland in the Sunshine State was attributed to urban population

Figure 6.3 Farmland scarcity and land degradation have become global development concerns. Dani people tilling sweet potato plot in Baliem Valley, Irian Jaya, Indonesia, and a man tending to his crops in the Dogon village of Tougoume, Mali; regions where farmland is becoming increasingly scarce.

6.3a: ©Michele Burgess/Stock Boston; 6.3b: ©Michael A. Dwyer/Stock Boston

growth and sprawl. Note that Florida was home to nine of the twelve fastest-growing metropolitan areas in the United States from 1970 to 1990.

The loss of farmland throughout the United States to urban sprawl and alternative land uses could spell impending doom to agricultural exports at a time when overseas demand and production deficits in the periphery are soaring. The United Nations estimates that total farm output must increase fourfold over the next 100 years if the world is to adequately feed a world population of 11 billion. This presents a formidable challenge given the enormity of the losses of commercial farmland in surplus-producing core economies. As rural land in the developed world remains grossly undervalued, global food security initiatives and the preservation and maintenance of the world's food production system remain in jeopardy. Attempts to recover rural farmland from urban land uses have already begun in China and elsewhere on the Asian continent.

While it is unnerving to think that North America, a region hailed as the "foodbasket" of the world, is losing an estimated 2 million acres of farmland each year to urban residential and commercial uses, much optimism remains as technology has allowed agriculturalists the luxury of producing more per unit land than ever before. Despite the impact of technology in food production systems, there have been noticeable slowdowns in the growth of food production per capita over the last decade. Efficiency and technology-related gains in commercial agricultural systems have made the recent production trend seem less critical than it actual is.

Fuelwood scarcity and the loss of residual vegetation have presented additional problems in many underdeveloped regions (particularly in Africa). Dried animal dung, a good source of natural fertilization, is used as a supplemental heating and cooking fuel as wood becomes increasingly scarce. The use of animal excrement for purposes other than fertilization does little to help sustain soil fertility. As increasing numbers of people compete for increasingly scarce fuelwood, they will be forced to travel great distances (usually on foot) from established settlements to collect or harvest this resource. As the distance they travel increases, so too does the area affected by this activity. The loss of trees, vegetation, and soil cover will only serve to speed up land degradation.

WORLD GRAIN TRADE

The economies of the SOUTH are becoming increasingly dependent upon food production systems in the NORTH (see Table 6.3 and Figure 6.4). This increased dependence has come at a time when surplus and export limits are being reached. Consider that in 1992, the United States exported 40% less grain to overseas markets than it did in 1984, due to production shortfalls. If recent trends in grain trade continue, deficit-producing nations in Africa and Asia may not find enough outside suppliers to meet their growing demand for grain. This situation has been aggravated by cropland and production losses from environmental degradation and the proliferation of urban areas (Figure 6.4).

In addition to the increasing dependence of peripheral regions on core economies, world grain harvest area per capita is on the decline (see Table 6.4). By the year

TABLE 6.3	Changing Pattern of World Grain Trade, 1950–1995 (grain trade in millions of metric tons)					
	Average					
Region	**1950**	**1960**	**1970**	**1980**	**1988**	**1990–1995***
N. America	+23	+39	+56	+131	+119	+140
W. Europe	−22	−23	+30	−16	+22	+35
Oceania	+3	+6	+12	+19	+14	+24
E. Europe and the USSR	0	0	0	−40	−27	−44
Latin America	+1	0	+4	−9	−11	−15
Africa	0	−2	−5	−15	−28	−39
Asia	−6	−17	−37	−63	−89	−105

Source: L.R. Brown, "The Changing World Food Prospect: The 90's and Beyond," *World Watch Paper 85,* Washington, D.C.: World Watch Institute (1988); and D. Grigg, *The World Food Problem,* Oxford, U.K.: Blackwell (1993), p. 240.

Note that "grain" refers to "cereals"—a plant whose seeds and/or residuals are suitable for food and other products (for example, wheat, rice, corn, oats).

* Estimate of average based on 1970–1988 trend.

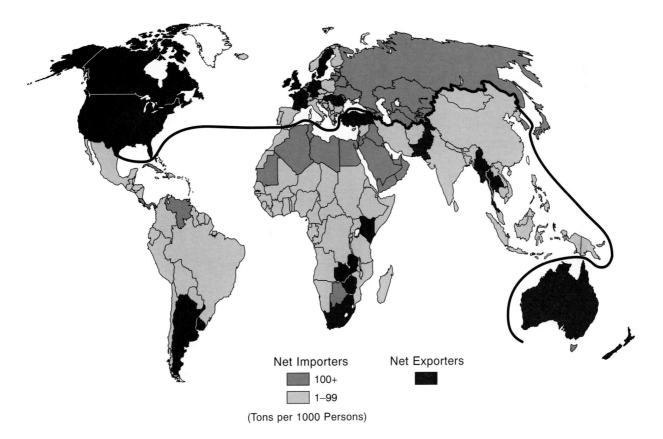

Figure 6.4 World grain trade.

2000, total harvested grain area per person will drop to a level that is about 50% of the average value observed in the 1950s (Brown et al., 1993). The growth in world irrigated farmland per capita has also showed signs of slow- ing down. This is due to a combination of several factors including water scarcity problems, competition between urban and agricultural systems for freshwater, and lim- ited possibilities for the further expansion of the world's

TABLE 6.4	World Grain Harvest Area (1950–1990)	
Year	**Total Harvested (million hectares)**	**Hectares per Capita**
1950	593	.23
1955	646	.23
1960	651	.21
1965	657	.20
1970	666	.19
1975	711	.17
1980	724	.16
1985	717	.15
1990	695	.13
2000	709*	.12*

Source: Brown et al., *Vital Signs 1993*, p. 41.

Note: A hectare is a metric unit of area equal to 10,000 square meters.

* Forecast.

marginal farmland. Note, however, that these trends are not so disturbing when one considers the fact that yields (output) per hectare have increased dramatically since the 1960s.

MALTHUSIAN THEORY

Thomas Robert Malthus (1766–1843) produced several controversial essays on the implications of world population growth. His contributions represent the very foundations of the "limits to growth" school. In his works entitled "An Essay on the Principle of Population" (first published in 1798) and "The Power of Population" (1803), Malthus argued that the world faced an imminent crisis as food production would not keep up with population growth (Menard and Moen, 1987, pp. 97–103). Originally cast as a global doomsday prophecy and dismissed by leading scholars of the last several decades, the writings of Malthus have once again sparked the interest of the academic community.

Malthus made several startling predictions on the future of the world in which we live. He contended that population growth, if left unchecked, would result in war, civil unrest, hunger, disease, and poverty, conditions born out of increasing food scarcity. Malthus concluded that the reproductive power of the human population is indefinitely greater than the ability of the earth to produce subsistence. Arguing that the earth possessed a finite capacity to support human life, he concluded that the earth's carrying capacity would be exceeded as human population increased exponentially and as food production systems would not be able to keep pace. Eventually some segments of the human population would suffer as total production fell below that which is required to subsist. He

eluded to regional variations in food shortages and human suffering, stating that

> population, when unchecked, increases (geometrically). Subsistence increases only (arithmetically). By that law of our nature which makes food necessary to the life of man, the effects of these two unequal powers must be kept equal. This implies a strong and constantly operating check on population from the difficulty of subsistence. This difficulty must fall somewhere and must necessarily be severely felt by a large portion of mankind. (as reprinted in Menard and Moen, 1987, p. 97)

Malthus goes on to write how the natural laws of population growth posed the ultimate threat to human development.

> It is undoubtedly a most disheartening reflection that the great obstacle in the way to any extraordinary improvement in society is of a nature that we can never hope to overcome. The perpetual tendency in the race of man to increase beyond the means of subsistence is one of the general laws of animated nature which we can have no reason to expect will change. (p. 103)

At the time Malthus was writing his essay on population, it was well known that world population was growing "geometrically." A geometric progression is a sequence of numbers in which each individual number is greater than its predecessor by a constant ratio. In the case of population growth, the numbers double each time period assuming a constant ratio of 2. It was also known that the world food supply was growing "arithmetically" or at a linear rate. As growth of the world population exceeded the growth of the world's food supply, regional and worldwide shortages of food would eventually lead to panic, chaos, and conflict. Limitations to food production were said to be inherently tied to land resource scarcity and the "law of diminishing marginal returns," as less and less additional output would be produced per increased production inputs. As world population increased beyond the maximum number that could be adequately supported and sustained by the existing food production system (that is, as the carrying capacity of the earth was exceeded), it would result in mass starvation and worldwide disorder.

Consider the following number sequences shown in Table 6.5 (a hypothetical situation contrasting geometrically increasing population and arithmetically increasing food output). Note that the population/food output ratio increases dramatically over the range of values considered. The greater this ratio, the greater the burden placed upon the land and related food production systems. Malthus argued that, eventually, this ratio would exceed a critical threshold and cause production to fall below a subsistence level when viewed on a per capita basis.

In retrospect, the doom and gloom prophecy of Thomas Malthus has not come to pass. On the contrary, world food production systems have boasted outputs that

TABLE 6.5	Malthusian Growth Rates: A Hypothetical Example

Progression	Type of Growth
Population: 1, 2, 4, 8, 16, 32, 64, 128, 256 . . .	geometric
Food Output: 1, 2, 3, 4, 5, 6, 7, 8, 9 . . .	arithmetic
Population/Food Output Ratio: 1, 1, 1.3, 2.0, 3.2, 5.3, 9.1, 16.0, 28.4 . . .	explosive

have well exceeded the growth of world population since the time of Malthus. The more recent and explosive growth of global food production has been associated with technological innovations introduced since the outset of the so-called "green revolution."

THE GREEN REVOLUTION

The extraordinary increases in agricultural output worldwide have been attributed to the **green revolution**—*a period (roughly 1960s to the present) marked by major biogenetic and technological advancements and breakthroughs in crop production and the rapid diffusion of new production technology, inputs, and information throughout agricultural regions of the world.* The green revolution brought about the use of miracle seeds, fast-growing varieties of high-yield and genetically engineered hybrid crops (that were disease and drought resistant), chemical fertilizers, pesticides and herbicides, and advanced soil analysis technology. These innovations have helped farmers to expand the output of regional food production systems by as much as 500% when compared to prerevolution levels.

Despite the enormous increases in food output associated with the use of hybrid crops and green revolution technology, production ceilings are now being reached in most agricultural regions and sectors. Diminishing marginal returns have set in to provide a sobering reminder of the limitations of technology and techno-driven change. Production output per capita has leveled off as world population continues to increase and as people are living longer than ever before due to advancements in medicine.

The use of hybrid crops has not been without controversy. Industry analysts have been concerned over the loss of many original or natural crop strains. The world's current food production systems have been limited to the production of a handful of crop varieties. The spread of disease in any of these systems could virtually wipe out any number of the preferred hybrids. This represents a very real threat to global food security. In response, many publicly and privately funded agencies have emerged to

combat the loss of natural strains. Government-funded research centers and seed banks have been established to intensify efforts to preserve crop/seed diversity should one of the few and popular varieties of crops be wiped out due to an unforeseen event.

The green revolution was hailed by many as a phenomenon that would dramatically reduce the incidence of (or eradicate) hunger and poverty in the world economy. In retrospect, the green revolution failed to achieve this most basic development objective. Critics of the green revolution have identified five major reasons why it failed to bring about progress in the human condition in many peripheral regions.

First, the green revolution was a "class-biased" revolution. Agriculturalists in many underdeveloped regions simply could not afford to use the latest production inputs or methods. Benefits accrued only to those regions and agriculturalists with sufficient land and capital assets to afford the innovations. The cost of hybrid seeds and chemical fertilizer has prohibited the large majority of farmers in the periphery from applying green revolution technology.

Second, farmers in underdeveloped regions were not entirely keen on becoming increasingly dependent on core economies for production inputs such as fertilizer and seed. Cultural resistance has also played a major role in slowing the diffusion of production technologies from NORTH to SOUTH as farmers in peripheral economies continue to be reluctant to change. Some were suspicious of the green revolution in general and the underlying motives behind technology transfers from the core to the periphery. As traditional crop varieties and established production methods have worked well for generations, agriculturalists in the SOUTH felt little need to increase risk using unfamiliar technology. In short, subsistence farmers were hesitant to replace the old crops and methods with new crops and methods that had no proven track record.

Third, the green revolution brought about the consolidation of land holdings by landowners and the subsequent displacement of peasant farmers. Wealthy landowners began to withdraw their land holdings from the rental and sharecropping markets and drive peasant farmers off their land. The displacement of peasant farmers and sharecroppers aggravated the poverty and hunger situation. Landless peasants were no longer able to live and farm the very land they needed to subsist.

Fourth, local and regional food shortages emerged as the emphasis switched from the production of subsistence crops to the production of "cash crops" for export. Landowners and government sought to exploit the benefits of producing crops for foreign markets as their attention turned toward agricultural exports. Hard currency obtained from the sale of cash crops abroad was partly used to fuel regional economic growth and finance development projects. Agricultural exports led to critical shortages of food in many regions and an increased reliance on outside sources to meet internal food needs. Landowners in the tropical low-latitude areas of Latin America, Asia, and Africa began to intensify their efforts to produce coffee, tea, spices, bananas, rubber, or other commodities for export. These products were earmarked for consumers in the NORTH. Utilization of cropland to produce exportable commodities, rather than subsistence crops for local and domestic consumption, created regional shortages of staple food crops and grain.

Last, many of the benefits of green revolution technology have gone largely unnoticed as the population explosion of the SOUTH and the rapid diffusion of medical technology paralleled increasing agricultural output. World population growth over the past four decades has been unprecedented as increased access to medicine and health care translates into less mortality, more people, and longer life spans. Increasing agricultural output has been easily absorbed by the growing demand for food.

The green revolution did, nonetheless, allow more land to be brought into production worldwide. The frontiers of the world's production regions were expanded with the introduction of faster-growing and drought-resistant crops. Land that was once deemed as marginal or unusable, particularly land in cold and drought-prone areas, became highly productive cropland. Reduction in the required length of growing seasons for corn and other cereals, and genetic engineering of crop types for increased resistance to harsh environmental conditions or disease, had profound effects on the geography of agriculture.

The green revolution was responsible for increasing crop yield and production efficiency. It was instrumental in supporting a vast array of improvements in planting, harvesting, transport, and crop storage technologies. In addition, the green revolution brought about the emergence of specialized and commercial agricultural zones that produced a level of surplus that could more than adequately feed the world's hungry. Even as the world's agricultural revolution is ongoing, the greatest benefits of the green revolution may already be behind us.

POPULATION AND ECONOMIC SECURITY ISSUES

Today, food security issues continue to be debated. With a rapidly growing population in the SOUTH and visible slowdowns in the growth of food production worldwide, several schools of thought have surfaced on dealing with development problems in relation to overpopulation and resource scarcity in the periphery.

Neo-Malthusians, *modern-day followers of Malthusian doctrine, argue that underdevelopment, poverty, and hunger are direct by-products of unbridled population growth.* Neo-Malthusians warn that the current global population growth situation will not be reversed without intervention to reduce birth rates. Hence, neo-Malthusians are staunch supporters of family planning efforts and the distribution and use of contraceptives. Neo-Malthusians contend that poverty and food-crisis situations may be averted by social, economic, and demographic reform. The regulation of births to limit future population growth is essential before higher development objectives can be met.

There is one major obstacle to the interventionalist approach prescribed by neo-Malthusians—cultural resistance. Many people and cultures are strictly opposed to the use of contraceptives on religious and moral grounds. Adherents of several religions, including Roman Catholic, Fundamental Protestants, Muslims, and Hindus, cite that religious convictions prevent them from using birth control.

Cornucopians, by contrast, are those who entertain a more optimistic look at the future. **Cornucopians** *argue that intervention is not necessary, because the world does not face impending development problems which are population based for the simple fact that we are far from exhausting our human and physical resources.* In other words, cornucopians believe that the earth possesses an indefinitely large carrying capacity that we are in no danger of exceeding anytime in the near future. Cornucopians also believe that humankind is not currently confronting any technological or production limitations. They contend that scientific breakthroughs will lead the way toward redefining scarcity in all economic regions and endeavors. Cornucopians argue that technology will lead us to a virtually limitless source of cheap energy and the continued expansion of production capacities in all economic sectors (including agriculture). Firm believers in the adage that "necessity is the mother of invention," cornucopians insist that the human spirit cannot be broken in matters of development and that progress will be achieved at any cost. They cite history itself as proof of how human beings continue to meet challenges, overcome adversity, and redefine limitations.

The insistence of the cornucopians in technology as a panacea for development problems is far from true when one considers the simple fact that many underdeveloped regions will simply be unable to afford to use many of the innovations that are likely to come about. As technological change will be championed by the large and prosperous economies of the NORTH, peripheral economies will continue to be dependent upon the core.

Marxists attribute the dependence of the periphery on the core as an outcome of global capitalism. They are convinced that problems of overpopulation can be eliminated if the very system responsible for creating the conditions of uneven development is abolished. By raising the overall standard of living and shrinking the gap between rich and poor (through the global redistribution of wealth and capital for starters) the labor-value of children would be reduced worldwide. This would lower the incentive for families to have a large number of children. Although some argue, as Malthus did, that affluence could have a positive effect on family size (that is, the more prosperous a family would become, the more it is likely to expand), such a pattern has not been readily observed in more-developed regions.

Despite these observations, culture plays a big part in the preference toward large families in many areas of the SOUTH. In Africa and the Middle East, for instance, many leaders oppose lower birth rates for political and national security reasons. Government officials, for instance, encourage large families to increase the supply of young men who could potentially serve in the military. In addition, various cultures take pride in large family size, as it provides feelings of prosperity and security.

MALTHUS REVISITED

The growth of world food production has steadily outpaced population growth from 1950 to the early 1990s. As a result, output per capita dramatically increased over this period. In recent years, however, production slowdowns and ceilings have become a reality in many regions. Consider the data in Table 6.6. Although world grain output continues on an upward trend, output per capita has stagnated. Note that increases in world grain production have occurred despite recent declines in harvested acreage, a legacy of the green revolution. Declining production per capita statistics have been observed in the United States, Central America, and Africa (see Table 6.7). Further reductions in the agricultural export potential of North America will mean greater pressure on other regions to satisfy the growing demand for grain and other agricultural commodities in deficit-producing regions of the SOUTH. The more productive agricultural zones of Europe, Oceania, South America, Asia, and the former Soviet Union (particularly Russia and Ukraine) will most likely be called upon to increase output.

Shortly after the turn of the century, grain output per capita is expected to decline to levels experienced during the 1960s and 1970s. This is not to say that the world is in

TABLE 6.6	World Grain Output	
Year	Grain Output (in millions of metric tons)	Kilograms per Capita
1950	631	246
1960	847	278
1970	1096	295
1980	1447	323
1985	1664	341
1990	1780	335
1991	1696	314
1992	1745	318
2000	1790*	315*

Source: Brown et al., *Vital Signs 1993*.
* Forecast.

danger of not being able to sustain its people, but concerns have arisen over the ability of the world's production systems to provide for 11 billion people by the year 2100. Whether or not the demand for food will be met is highly dependent upon (a) the severity of land and soil degradation and associated losses in crop output due to misuse and mismanagement of agricultural systems; (b) the extent to which diminishing marginal returns to additional production inputs will slow future increases in output in the world's more productive grain-growing regions; and (c) how rapidly population growth rates decline in the SOUTH. Given the impending population explosion and the wide array of development problems found in many regions of the SOUTH, it is easy to see why Malthusian-based concerns have resurfaced.

The doom and gloom visions of the neo-Malthusians, however, are not globally oriented. They see the problem as "regional," mostly pertaining to high-populated nations of the SOUTH. In the worst-case scenario, the food-crisis situation will most likely mean feast in the core, famine in the periphery.

Increasing tensions between neighboring nations, and the scores of ethnic, political, and religious conflicts throughout the SOUTH have increased the likelihood of a crisis emerging. Regional food shortages are more apt to be brought about by disruptions in food distribution rather than production shortfalls. The use of food as a political tool or weapon in regional conflicts will become increasingly more commonplace, as factions attempt to exercise authority and leverage over people in need. Development agendas, therefore, must include provisions for increasing the self-reliance and self-sufficiency of people in regions most susceptible to manipulation.

The average growth rate of world grain output has slowed to a mere .5% a year during the 1990s. As the world's population growth rate is more than three times that amount, feeding the world's people will become

TABLE 6.7	Indexed Agricultural and Food Output per Capita by Selected Geographic Region

Region	Index of Agricultural Production per Capita (1979–1981 = 100)		Index of Food Production per Capita (1979–1981 = 100)	
	1980–1982	1990–1992	1980–1982	1990–1992
World	**101**	**104**	**101**	**105**
Africa	**99**	**93**	**99**	**94**
Egypt	100	108	101	117
Ethiopia	99	85	99	85
Nigeria	99	122	99	122
Somalia	101	54	101	54
Sudan	98	80	98	81
Zaire	100	92	100	92
Zimbabwe	100	79	103	71
Asia	**102**	**121**	**102**	**121**
Bangladesh	99	95	99	96
China	104	139	102	137
India	101	121	101	122
Indonesia	102	132	103	135
Pakistan	101	117	101	112
N. and C. America	**100**	**96**	**101**	**96**
United States	101	97	101	97
Mexico	101	98	101	100
S. America	**101**	**106**	**102**	**108**
Argentina	100	98	100	97
Brazil	102	107	103	111
Chile	101	115	101	116
Commonwealth (former USSR)	**100**	**99**	**99**	**101**
Europe	**101**	**104**	**101**	**104**

Source: *World Resources 1994–95* (pp. 292–293).

more and more of a challenge over time. Nonetheless, there is a fair amount of optimism on the ability of technology and agricultural science to overcome production obstacles both now and in the future (Gardner, 1996).

In general, agricultural development and "intensification" of agricultural systems are viewed as population related. Ester Boserup (1965, 1981) argued that *agricultural change is driven by population growth*—forming the basis of the so-called **Boserup thesis.** In other words, agricultural intensification is viewed as dependent upon population growth. As a region's population increases, it forces agricultural change, intensification of the use of agricultural "factors" (land, labor, etc.), and the increasing use and development of technology. Hence, population growth can actually lead to increasing food output over time.

The Boserup thesis is contrary to Malthusian theory. Recall that **Malthusian theory** suggests that *production limitations in agriculture provide the ultimate check to population growth.* In the case of many peripheral or

semiperipheral economies, however, it is difficult to ascertain whether population growth is a response to, or a cause of, agricultural development and intensification. This has led to a resurgence in the interest to explore region-specific relationships between population growth and agricultural production.

FERTILIZER USE

According to statistics from the World Watch Institute, world fertilizer use has been on the decline (see Table 6.8). From 1950 to 1985, world fertilizer use increased by approximately 7% per year, from 14 to over 130 million tons. From 1985 to 1990, fertilizer use expanded by less than 2% annually. The decline in the use of fertilizer has paralleled slowdowns in world grain production. Worldwide, the average annual growth rate of grain output from 1984 to 1990 was a mere 1%—one-third of the average annual growth rate from 1950 to 1984. Slowdowns in

TABLE 6.8	World Fertilizer Use	
Year	Total Fertilizer Use (in millions of tons)	Kilograms per Capita
1950	14	5.5
1960	27	8.9
1970	66	17.8
1980	112	25.1
1985	131	27.0
1989	146	28.0
1991	137	25.4
1992	131	23.9
2000	135*	23.0*

Source: Brown et al., *Vital Signs 1993.*
* Forecast.

Figure 6.5 The widespread use of chemical fertilizers has become an essential ingredient to modern-day farming operations. Here a tractor is plowing and tilling a field in Canadaigua, New York; preparing it for planting and fertilization.
6.5: ©Fred R. Palmer/Stock Boston

production and the use of fertilizer have been attributed to diminishing marginal returns (Figure 6.5).

During the 1960s, one additional ton of fertilizer in the U.S. Cornbelt boosted output by as much as 20 tons on average. Today, one additional ton of fertilizer will boost output by only a few tons at best. Diminishing marginal returns to fertilizer use combined with cropland scarcity and land degradation have raised concerns over the limitations of the world's agricultural systems. Some limits-to-growth theorists see production ceilings and declining food output per capita as just the tip of the iceberg. While it is true that the world's largest food-producing regions are using about the same amount or less fertilizer than a decade ago and that crop output is becoming insensitive to additional doses of fertilizer, there are several other reasons behind the overall decline in fertilizer use.

First, economic reforms in many underdeveloped nations have begun to eliminate government subsidies for fertilizer to reduce dependency on developed economies—the major suppliers of chemical-based fertilizers. Without subsidies, agriculturalists have demanded less fertilizer. Second, the inability of many underdeveloped nations to afford increasing amounts of fertilizer has also lessened worldwide demand. Third, the fall of the Soviet Union and the abandonment of state-set prices in favor of world market prices has made fertilizer more costly to farmers in areas formerly under Soviet command economy. Fourth, the lack of incentives to expand output when grain prices are low has caused farmers in commercial systems to limit the use of additional production inputs. Fifth, precision farming methods, including soil analyses and local fertilizer-crop-condition matching, have reduced waste in commercial agricultural systems. Recent declines in the demand for fertilizer may also be attributed to the wise and efficient use of production inputs.

FOOD SECURITY: LINES OF DEFENSE

Of paramount importance to world food security is the amount of grain held in storage at any given time by grain brokers, industry, and governments. These so-called **carryover stocks**—*grain inventories held in storage*—can act as a first line of defense in the event of a worldwide production shortfall. Recently, however, carryover stocks have been reduced to dangerously low levels (see data in Table 6.9) (Figure 6.6).

In 1990, as is the case today, the world held just over two month's worth of carryover stocks. This is substantially less than the ideal level of grain inventory necessary to allow for the uninterrupted consumption of grain in the event of a global production shortfall; a figure that stands at about 200 consumption days. The 200-day mark gives producers adequate time to deploy a second line of defense—recultivation. As a third line of defense, the world's idled cropland (much of which is in the United States) could be brought into production. Note that in the early 1990s, some 5 million hectares of cropland were idled in the United States as part of a government-led land management program. Note, however, that the use of this cropland is viewed as a weak line of defense given that the amount of idle land that is readily available for production is less than 1% of the world's arable land.

As a fourth line of defense, considerations may be made to alter the end use of grain output. For example, much of the grain that is used to feed livestock could be used for human consumption should a world food crisis arise. The mix of animals supported by a given grain

TABLE 6.9	World Carryover Stocks	
Year	Carryover Stock (millions of metric tons)	Consumption in Days
1963	190	81
1970	228	76
1979	328	84
1987	465	104
1990	301	65

Source: Brown et al., *Vital Signs 1993.*

TABLE 6.10	Corn: Some Facts on the Average Annual Usage in the United States

Production: Approximately 8 billion bushels per year.

Use	Percent of Output
Seed	.2%
Food (whole kernel and processed)	1.4%
Starch (food thickeners and industrial uses)	1.8%*
Alcohol (ethanol and bourbon)	3.7%
Sweeteners (corn syrup)	5.8%
Exports (esp. Russia, Japan)	16.8%
Carryover Stocks	25.6%
Animal feed	44.7%**

Source: *National Geographic* 183, (6) (June 1993), "Corn, the Golden Grain," pp. 92–117.

* Including new and exciting uses and end-products such as biodegradable packing materials (eco foam), golf tees, and eating utensils.

** Note: 10 lbs of corn yield 1 lb of beef; whereas 3 lbs of corn yield 1 lb of chicken.

Figure 6.6 A large grain storage facility in America's grain belt. Grain inventories have become a necessary line of defense in the food security battle.

6.6: ©Spencer Grant/Photo Edit

supply could also be altered. For example, the production of beef utilizes considerably more grain per unit output than does poultry.

From 1988 to 1993, the amount of grain fed to livestock on an annual basis was in excess of 600 tons, over 30% of the world harvest. Put another way, more than one out of every three tons of grain is used to feed and fatten animals to produce meat and milk for human consumption. The production of feed grain is especially high in the farming systems of North America and Europe. Note that very little corn output in the United States, for example, is actually used for human consumption (see Table 6.10). Currently, about 45% of U.S. corn output is used to feed livestock, mostly in the major food- and milk-producing regions of the U.S. Corn- and Dairybelts. Covering large segments of the Midwest (including the states of Iowa, Indiana, Illinois, Ohio, Minnesota, and Wisconsin), the Cornbelt and Dairybelt regions produce more than one-third of the world's corn output. Given these numbers, it is easy to see why there is no immediate food crisis as far as humans are concerned. Yet, there are questions about the stability of feed-crop systems in a world economy where the overall growth in agricultural output is slowing down. As feed grain prices will inevitably rise over time, a smaller percentage of grain output might be allocated to

livestock. This, of course, depends upon profit margins in the livestock industry and the extent to which rising prices will reduce the demand for meat and dairy products.

PRICE AND PURCHASING POWER

The world possesses an enormous potential to feed its growing numbers, yet the forces of supply and demand are what govern the price and distribution of food. In many instances, market forces may prohibit food from reaching those who need it the most. Price changes in world grain markets and subsequent variations in purchasing power are of great concern to consumers in peripheral regions, particularly those with limited income or bartering power.

Consider that in the United States, a loaf of bread priced at $1.50 contains roughly 6 cents worth of wheat input. The other 96% of the cost to consumers is related to the value added from processing and transport, distribution and delivery, wholesale and retail marketing, as well as advertising and packaging. A doubling of the price of wheat will raise the price of bread to $1.56 per loaf. This represents a 4% increase in the overall cost to consumers. Such an increase would have very little effect on the overall purchasing power of consumers in the market.

Large numbers of people, however, in peripheral nations are severely affected by rising price levels. In regions where food is sold or bartered in an unprocessed form, the direct purchasing of grain and other food at the local market means that consumers are extremely vulnerable to price fluctuations. For example, an increase in world grain prices directly translates into a proportional decrease in buying power to those purchasing grain at the market. For those already spending large portions of their income on food, price increases can be devastating. Radical price hikes can literally force many to the brink of starvation. Should price levels of agricultural commodities double, the purchasing power of consumers would be literally cut in half.

Note that the market price levels of agricultural products are extremely volatile. Consider the data on quoted world grain prices for the month of July in the years 1987 and 1988 as shown in Table 6.11. Note the dramatic changes in price of grain over the period in question. Understanding the potential impact of price changes on the world's poor, leaders around the globe have come to acknowledge the importance of promoting price stability in agricultural markets. A global development itinerary must also include food security initiatives to help ensure that the need for food is met despite market conditions, production constraints, and problems in food distribution.

TABLE 6.11	Price Changes for Wheat, Rice, and Corn Over a One-Year Period

	Prices (in U.S. dollars $)*		
Grain	1987	1988	Percent Change
Wheat (per bu.)	2.85	4.07	+43%
Rice (per ton)	212	305	+44%
Corn (per bu.)	1.94	3.22	+66%

Source: International Monetary Fund Financial Statistics 1989 (Washington, D.C.: IMF).

* Quoted price per bu. or per ton (month of July).

TABLE 6.12	World Fisheries Output

Year	Total Catch (in millions of tons)	Output per Capita (kilograms)
1950	22	8.6
1960	38	12.5
1970	66	17.5
1980	72	16.2
1985	86	17.7
1988	99	19.4
1989	100	19.2
1990	97	18.3
1991	97	18.0
1992	97	17.8
1993	99	17.7
2005*	99*	16.8*

Source: Brown et al., *Vital Signs 1993*.

* Forecast.

WORLD FISHERIES

Food security has also meant both the expansion of existing industries, which supply food and protein, and the development of alternative food sources. This includes the reexamination of world fisheries as a supplemental food source. Research findings indicate that most of the world's largest fisheries are already operating at or near their production capacity. Furthermore, there is evidence that several of the more productive fisheries are functioning at nonsustainable levels. It is no surprise, therefore, that total world fisheries output has leveled off and that fisheries output per capita is also on the decline (see Table 6.12).

Note that from 1950 to 1989, the world's food supply was expanded considerably as over 2 million tons of fisheries output were added each year (on average). World fisheries output peaked by the late 1980s. Since then, total output in the industry has stagnated. Experts in the field agree that there is little or no room for any further expansion.

The stagnation of world fisheries output has been attributed to at least five factors: (1) overharvesting of the world's fisheries and the utilization of regional fisheries beyond sustainable levels; (2) damage to coastal areas, wetlands, mangrove swamps, and river systems from pollution and the infiltration of chemical agents into the water supply; (3) an increase in the incidence of algae blooms and the subsequent depletion of water-oxygen levels; (4) stratospheric ozone damage and increasing ultraviolet radiation exposure (a phenomenon that negatively affects sea life by causing decreases in the growth of "phytoplankton"—the lowest link in the aquatic food chain); and (5) increased competition among nations for scarce ocean resources in key fishing regions. Hardest hit by declines in fish population were the northwest, west-central, and southeast Atlantic fisheries, and the east-central Pacific region. These fisheries have been identified as operating at or above their capacity.

Approximately 200 million people worldwide rely on the $80 billion fishing industry for their livelihood. The number of commercial fishermen who have been forced out of the industry is staggering. Many more will lose their jobs as fish stocks become depleted and as degradation of the world's oceans and river systems continues.

Further expansion does exist however for **aquaculture** (also known as "fish farming")—*the growing of shellfish, lobsters, and other edible aquatic species of fish in controlled or man-made environments.* Note, however, that aquaculture accounts for a relatively small portion of fish output worldwide despite its increasing importance in commercial production systems (Figure 6.7).

Note that ocean-based (saltwater) fisheries dominate world production, accounting for roughly 85% of the

Figure 6.7 Commercial fish farming operations have become increasingly important as a supplemental food source. Pictured here is a saltwater salmon pen in Lubec, Maine.

6.7: ©Mark C. Burnett/Stock Boston

TABLE 6.13	World Fisheries Statistics: *Production and Consumption*

Fish Production (Average for period 1990–1992)

Region	Percent of World Output
China	12.6
Former USSR	10.7
Japan	10.6
Peru	7.1
Chile	6.1
United States	5.9
Africa	5.0
India	4.0
Indonesia	3.2
Thailand	2.9
South Korea	2.8
Philippines	2.3

Leading Fish-Consuming Nations (1992)

Rank*	Region	Per Capita Consumption (in kilograms per year)
1	China	12.2
2	Japan	66.6
3	United States	20.4
4	Russia	25.2
5	India	4.1
6	Indonesia	15.6

Source: Parfit, M. *National Geographic* 188 (5) (November 1995), p. 12; and *Goode's World Atlas* (19th edition), p. 40.

* Rank as defined in terms of total fish consumed, in millions of metric tons.

world's fish output in 1990. Inland or freshwater fisheries account for the remaining 15% of output. Fish farming yields from either freshwater or marine sources stood at an average of more than 12 million tons per year in the early 1990s—roughly 10% of world output. Note that the global aquaculture yield is expected to surpass the 20-million-ton-per-year mark by the year 2000. While aquaculture may not be the answer to the world's food problems, it must be viewed as an industry that will play an important role as a supplemental food source in the world economy.

Regional variations in the annual per capita consumption of fish for the world's leading fish-consuming regions are highlighted in Table 6.13. Note that with the exception of producers in the southeast Pacific region, total fish consumption is positively correlated with population size.

Capacity restrictions of world fisheries will most severely impact people in underdeveloped regions. As world population increases faster than world fisheries output, regional production deficits will become more pronounced over time. As price levels rise to reflect increasing scarcity, consumers in peripheral regions are most likely to suffer the consequences. Serious consideration, therefore, must be given to global initiatives aimed at preserving and sustaining world fisheries output. Laws governing restrictions on total catch must be strictly enforced to lessen the likelihood of overharvesting. In addition, pollution and damage control of coastal marine environments must be ongoing to prevent any further degradation of this most valuable world resource.

FOOD EXPORTS

There is growing evidence of a possible food crisis in the SOUTH, as peripheral regions continue to become more dependent upon core economies to meet their increasing demand for food. Peripheral regions have also become less able to utilize agricultural exports to support economic growth and modernization efforts. In short, the loss of agricultural exports has hampered global development efforts.

Table 6.14 shows the share of world agricultural exports for selected time periods and regions. The shrinking export potential of peripheral nations has also meant a loss in export diversity and less export income. The agricultural export base of the SOUTH has been mostly limited to a handful of commodities—coffee, tea, cocoa, bananas, and sugar.

Increased reliance on imported foodstuffs has placed many of the most peripheral nations, specifically those without a globally competitive manufacturing base at a great disadvantage.

> The period since the end of the Second World War has seen a great boom in agricultural trade and marked alterations in its direction. Since the early 1960's there has been a great increase in food imports by the developing countries. This has been due first to rapid

| TABLE 6.14 | Share of World Agricultural Exports by Region |

	Average			
Region	1961–1963 (%)	1972–1974 (%)	1980 (%)	1988 (%)
North America	19.7	23.1	21.4	20.7
Western Europe	17.6	26.1	36.2	42.5
Oceania	8.2	7.1	5.4	4.2
Eastern Europe and former Soviet Union	7.9	6.0	4.2	3.8
Africa	9.1	6.5	4.5	2.8
East Asia	12.5	8.3	8.5	9.9
Latin America	17.1	15.4	13.8	9.5
West Asia	3.7	3.2	2.2	1.6

Source: Adapted from D. Grigg (1993), p. 253; from FAO, "The State of Food and Agriculture" 1975, Rome, 1976, p. 68; FAO "Trade Yearbook" 1981, vol. 39, Rome, 1982; and FAO "Trade Yearbook" 1988, vol. 42, Rome, 1990.

population growth and the inability of some countries, particularly in Africa, to maintain per capita food supplies from home production, and second to the imports of wheat, livestock products and livestock feeds into the richer developing countries. The latter has been possible only where there have been oil or manufacturing exports. The poorer developing countries, often dependent upon one or two crops for much of their exports, have found it increasingly difficult to pay for food imports. (Grigg, 1993, p. 255)

Despite the growth of exports and markets over the last several decades, peripheral nations have found themselves increasingly more in debt (see Table 6.15). In fact, the external debt of developing nations is growing faster than the value of world exports. The opportunity costs of paying off that debt is high, as it will bring about a critical shortage of investment capital in the SOUTH. Increasing debt and capital flight have forced many developing nations to squander precious natural resources or specialize in one or a few agricultural exports. Much of the developing world continues to receive only a small share of the benefits of expanding world trade, a situation that has only exacerbated the world poverty problem. The benefits of export-oriented development have been negated by increasing debt and the increasing dependence of the periphery on the core. Many fear that this situation will ultimately result in defaulting of development loans throughout the SOUTH.

> The indebted LDCs are attempting to improve their economic conditions by renegotiating their debts, requesting new forms of aid, and examining non-traditional trade agreements. They are seeking new, large loans on a more concessional and less conditional basis.

| TABLE 6.15 | World Exports and Debt for Selected Years |

Year	Value of World Exports	External Debt of All Developing Nations
	(in billions $1990)	
1970	1326	—
1980	2328	639
1990	3331	1355
2000	4230*	2450*

Source: World Watch Institute 1993; and *Vital Signs 1992* (p. 68–69).
* Forecast.

They also want debt forgiveness or at least more favorable terms for their extensive external debts. In addition, they want improved access to foreign markets for their exports; better terms of trade, including higher and more stable prices for their exportable raw materials; and major transfers of technology. To achieve these objectives, they are asking for fundamental restructuring of existing international trade and the world monetary system. (Furlong, 1989, p. 137)

Inextricable links have been established between poverty, underdevelopment, and hunger. The rising incidence of hunger and malnutrition continues to undermine human development efforts in the periphery. The global food problem has now become a top priority in development circles. The focus has turned to strategies for improving food distribution and on finding ways in which to increase protein and caloric intake. During the 1980s, the Food and Agricultural Organization (FAO) of the United Nations estimated that the number of people suffering from undernourishment ranged from 300 to 700 million people. By the year 2000, it is expected that more than 1 billion people worldwide will suffer from inadequate nutrition. Nations on the continent of Africa continue to have the highest incidence of hunger-related problems. Today, well over 40% of Africa's population is classified as "malnourished." Hunger and malnutrition must be viewed as unacceptable conditions in a world economy where market forces continue to promote the efficient use of human and physical resources. Not only does inadequate food intake and protein deficiency lead to premature death and a higher incidence of disease, but it promotes lower overall productivity levels of labor.

Bradley and Carter (1989) discuss four central elements of the glaring world food crisis as it relates to matters of uneven development. First, the world food crisis must be viewed as something that has a long history. Its roots can be traced to the origins of the capitalist world economy. The emphasis on export-oriented production (as motivated by profit) and the subsequent shifts in wealth and power from the expropriation of resources to an emergent core at the expense of the periphery are largely responsible. The unfolding of this global melodrama has inevitably led to the deterioration of pre-existing rural agrarian economies and a reduced ability of indigenous cultures to support their growing numbers. The profit potential of agricultural exports has led to displacement of subsistence farmers from the land as the emphasis turned to the production of cash crops for export. Second, the subjugation of the periphery (and peasant mode of production) to the core (and the capitalist mode of production) in virtually all matters of economy, has prohibited the periphery from directing its future. Many regions of the SOUTH do not earn enough export income to allow them to purchase some of the very agricultural commodities they have been constrained to produce.

Third, the interference by foreign interests in the economic matters of the periphery (under the guise of foreign aid, development loans, and trade agreements) has maintained the status quo. It is difficult to imagine a situation where the main beneficiaries of global trade —- nations and corporate interests of the NORTH—are likely to advance global economic development initiatives that would lead to the demise of the system responsible for initially elevating their status. Radical geographers argue that monetary and technocratic solutions to economic development woes, or strategies aimed at increasing global integration of the periphery, will always fall short of the desired objective. They see the capitalist world economy as a system that tends to maintain the status quo. Playing by the rules of capitalism only serves to reenergize the very system that perpetuates regional inequalities. All in all, the capitalist mode of production is viewed as a self-reinforcing phenomenon that will ensure the continuation of the processes responsible for yielding the core-periphery distinction.

Fourth, concerns have mounted over food distribution problems in peripheral regions. Questions of production and carrying capacity aside, there has been a greater focus on solving the problem of how to get food to those who need it most. Administrative inadequacy, civil disruption, military conflict, graft and corruption, inadequate transportation infrastructure, and the use of food as a tool of manipulation, have presented formidable obstacles to the movement of food in the SOUTH. While green revolution technology and free trade play vital roles in averting a worldwide production shortfall, widespread hunger or famine cannot be overcome without finding real solutions to global food-distribution problems.

With inadequate food distribution in the periphery, regional production deficits have become an even greater problem. While science and technology may help the world's farmers avert large-scale food crises, many argue that the speed and depth of technological progress have not been sufficient. According to World Bank estimates, output

in world food production systems will have to double over the next 30 years to keep pace with population growth. As there is so little room for expansion of the world's production zones, attention has turned to both the scientific community and the private sector to find new and innovative ways to increase world food supply.

Efforts to promote **biotechnology**—*applied plant and biological science for the advancement of agriculture and medicine*—have been hampered by uncertainty over intellectual property rights, patents, product-use compensation, and the degree to which government will regulate the industry. Radical factions are worried that commercial interests will exploit agriculturalists or regions most in need of a production gain. Furthermore, product licensing agreements between the private and public sectors (that is, industry and government) must be made before biotechnology applications become more readily available. Many fear the long-term implications of biotechnology in light of the failures of the green revolution, citing not only the cost prohibitiveness of new technology but the unforseen economic effects as well.

> Widespread acceptance of a limited selection of bio-engineered seeds or laboratory-produced clones may lead to a loss of genetic diversity among farm crops as local strains are replaced, much in the same way that modern hybrid seed has displaced many traditional local varieties. The movement toward genetic uniformity could mean greater susceptibility to large-scale pest or disease attack or other environmental disruptions, a problem that has long plagued monoculture farming. Furthermore, dependence on outside sources for clones and specialized seed could erode the self-reliance of farmers and raise costs, as high-tech planting materials tend to be more expensive than cuttings or seeds supplied locally. (*World Resources 1994–95,* 1994, p. 125)

As the green revolution failed to significantly reduce the incidence of poverty and hunger in the underdeveloped world, so too may the biotechnology revolution fail . . . but for a different reason. Biotechnology applications may be limited if sufficient economic incentives do not exist for developing commercially viable products.

> The real threat is not that the Earth will run out of land, topsoil or water but that nations will fail to pursue the economic, trade and research policies that can increase the production of food, limit environmental damage and ensure that resources reach the people who need them. (Budiansky, 1994, p. 60)

Nonetheless, advancements in biotechnology may result in redefining the nature and scope of the world's food crisis without attacking the underlying causes. While technology has allowed us to expand the earth's carrying capacity, we must not assume that technology will solve food allocation problems.

THE MYTHS OF HUNGER AND DEPRIVATION

In the seminal work entitled *World Hunger: Twelve Myths* (1986), Frances Moore Lappe and Joseph Collins provide compelling discussions on the nature of the world food crisis. The **twelve myths of hunger,** as they define them, are summarized in the following text:

Myth #1: There's simply not enough food to go around, a situation which inevitably leads to some going hungry.

This statement is far from being true when one considers that the world's food production systems currently produce enough grain and meat output to supply each person on the planet with well over 3000 calories per day and an adequate amount of protein. In short, "abundance, not scarcity, best describes the supply of food in the world today" (Lappe and Collins, 1986, p. 9). Although there are regional variations in food output per capita and evidence of recent production slowdowns, the growth of agricultural output at the global level consistently outpaced population growth from 1950 to the early 1990s (Grigg, 1993). Subsequently, the world's expanded surplus production potential has yet to be tested despite unprecedented population growth.

Myth #2: Natural hazards and events which are beyond human control are the leading cause of starvation and famine.

Notwithstanding the periodic disruptions which arise in relation to region-specific shortages in the supply of food caused by natural disasters such as floods, storms, and droughts, most famines are man-made. Critical shortages of food are most likely to arise as a result of

1. production shortages due to the cash-crop (export) orientation of food production systems;
2. severe land degradation in subsistence areas due to nonsustainable agricultural practices;
3. the ineffectiveness of technology as a result of cultural barriers, cost, or the remoteness of regions in which the peasant mode of production is dominant;
4. disruptions in food production and distribution as a result of political unrest, conflict, war, terrorism, and/or misguided government policy;
5. the widespread presence of poverty in the SOUTH and the general failure of markets in regions which lack the ability to buy food, despite demand; and
6. the food-dependency relationships established between the NORTH and SOUTH. (Bradley and Carter, 1989; Wennergren, 1989)

As food is now a commodity, the forces which govern its production are those which relate to buying and selling at a profit. If the poor . . . cannot afford to purchase it, then production and distribution will fail to adjust to their real needs, even though there may be widespread malnutrition and starvation. Thus we observe a paradox whereby a global production system has the capacity to feed the world, but does not do so because people are too poor. (Bradley and Carter, 1989, p. 120)

Myth #3: Too many people, too few resources, and a growing population are testing the limits of our finite capacity to support our increasing numbers.

While slowing the population engine of the SOUTH is desirable to help lessen the burden placed on regional food production systems, there is no direct evidence of a connection between hunger and population density or between hunger and a lack of resources. Consider that China, the world's most populated nation, has had very little problems feeding its people, despite having roughly one-half the physiological density of India. Note also that there is evidence that increasing agricultural productivity in developing nations has lagged behind productivity increases in the developed world, implying that considerable "slack" may exist in the agricultural systems of the periphery. Once this slack is taken up, it is likely that further increases in food output will be realized (Grigg, 1993).

Certainly population growth is a concern, yet to bring the world's population into balance with the natural environment, governments have no choice but to address the issue of maldistribution and unequal access to basic **survival resources**—*land, jobs, food, clothing, education, health care, and personal security* (Lappe and Collins, 1986, p. 32).

Governments must simultaneously focus their attention to the underlying structures which support excessive population growth and inequality. Development policies that seek to attack high birth rates without attacking the causes of poverty and hunger will most certainly fail. Simple solutions do not exist, as deprivation and hunger are outcomes of the complex interrelationships between human agents and market forces in economic, geographic, social, and political space. In short, deprivation may be viewed as an outcome of imbalances brought about by institutional and social relations (Findlay, 1993).

From this perspective, so-called "population crises" may arise because of the powerlessness of a group in political space, as may be the case with deprivation . . . in a situation where the bounding of food resources results from specific exploitative relations enforced relative to particular production relations. (Findlay, 1993, p. 169)

Myth #4: Pressure to produce increasing amounts of food is destroying the very systems and resources that are necessary to produce that food.

Yes, there are growing concerns over environmental degradation and subsequent losses in food output, yet the resilience of the earth and the adaptive qualities of its people must not be ignored. Increasingly, governments and other organizations are cooperating to halt the destruction of agricultural land in their push toward promoting "sustainable" forms of agriculture, which include land management and erosion controls and alternatives to hazardous agents such as chemical pesticides. Efforts such as these, along with strategies to reduce dependence on outside suppliers, are likely to slow the devastation. Protection and conservation of land-based resources will both minimize future losses and allow some of the already ravaged lands to recover with time. Interventionalist policies have already proven successful in Latin America, South and Southeast Asia.

The push toward **green development**—*environmentally sensitized development*—has helped establish global development priorities. Green development efforts are part of a global conservation strategy. The notion is based on the recognition that many of the world's precious resources lie partly or wholly beyond jurisdictional boundaries. It is essential, therefore, that we gain an understanding of ecological processes and interdependencies if we are to effectively outline strategies for the management and maintenance of land, air, and water resources, habitat and ecosystems. In addition we must preserve genetic and species diversity, and promote the sustainable utilization of cropland, firewood, and other land-based resources (Adams, 1990).

Plucknett and Smith (1989) see international collaboration in agricultural research and international extension programs as key to expanding global food production. Cooperative efforts in the exchange of **germplasm**—*genetic material*—have provided researchers with access to wild species and cultivars that have boosted agricultural productivity levels.

The use of biological controls such as integrated pest management and weed management has also enhanced productivity. For example, harmful insects may be held in check by releasing predators, rotating crops, and dispersing species-specific hormones into the air to disrupt breeding. These types of interactions are likely to bring about further increases in productivity as technology, discovery, and interaction continue to redefine the limits of agricultural systems. While such efforts are costly, the total benefits are great indeed.

> Successful biological control requires a constant search for pests and predators of problem plants or insects. . . . In more recent years, as cooperative activities have increased, we have come to recognize our dependency in other areas as well—or at least, the great advantages to be gained from collaboration in agricultural research, of all kinds, by all parties to the collaboration. . . . If the past is prologue to the future, it is clear our interest lies in open lines of communication, in sharing our expertise, and in financial investment to promote international research on an ever wider scale. (Plucknett and Smith, 1989, pp. 126–127)

Scientists are, however, in a race against **genetic erosion** —*the loss of naturally diverse germplasm.* Maintaining the diversity of genetic resources has become a priority among those concerned with vulnerability of modern-day production systems. Surplus-producing regions continue to rely on just a few crop strains, a situation that could have disasterous consequences should a single pest or disease wipe out an entire crop in a given production year. History has taught us some valuable lessons about the fragility of food production regimes and the rapidity of pest and disease diffusion.

> In the U.S., for instance, billions of rows of essentially identical corn are planted each year, making the entire crop vulnerable . . . United States farmers learned the hard way in 1970, when an unexpected epidemic of corn leaf blight wounded the pride of the world's most agriculturally advanced nations. A virulent new strain of fungus appeared in south Florida that winter and raced north like a killer flu. Since each ear of corn was a copy of every other, there was no margin of safety. The fungus destroyed half the crop from Florida to Texas. Nationwide losses amounted to 15 percent, at a cost of perhaps one billion dollars. Such disasters are nothing new. Throughout history the sowing of uniform crops has led to a harvest of tragedy.
>
> The collapse of classic Maya civilization around A.D. 900, some anthropologists speculate, resulted from farmers' planting a mere handful of maize varieties, which were destroyed by a virus. Ireland's infamous potato famine of 1845 started with a fungus accidentally introduced from Mexico. That scourge, spreading through millions of genetically similar spuds, left the Irish without their main food source, and nearly a million people starved to death. A few decades later a fungus wiped out the homogeneous coffee plantations of Ceylon, transforming that island into one of the world's major tea producers. And as recently as 1984 a bacterial disease struck Florida, forcing 135 nurseries to destroy 18 million citrus trees and seedlings. (Rhoades, 1991, pp. 84–85)

Governments and industry have responded to the challenge and are now attempting to manage the world's genetic resources as part of a global agricultural development effort. Organizations such as the Consultative Group of International Agricultural Research (CGIAR), the International Agricultural Research Centers (IARCs), the International Board of Plant Genetic Resources, the International Potato Center, the International Rice Research Institute, the United States Department of Agriculture, the International Center for Tropical Agriculture, along with other government agencies and seed companies, are making concerted efforts to collect, store, evaluate, and maintain germplasm from around the world.

Explorations to discover fresh genetic material are underway. These efforts have been challenged, however, by clear-cutting of virgin land, destruction of natural vegetation and habitat, and degradation of tropical areas (which are home to roughly one-half of the world's plant and animal species). Human intervention in natural systems has accelerated the rate of genetic erosion. It is estimated that by the year 2050, over one-quarter of the earth's 250,000 plant species may be lost or extinct if environmental degradation continues at its current rate. In response, seeds of wild and modern varieties of crops are being kept in storage in research stations and laboratories around the globe. This is all part of a strategy to protect human civilization from the risks of **monoculture**—*food production systems which rely on a single or predominant crop type.* Yet, there are realizations that science can do little without supportive development policies. As Plucknett and Smith (1982, p. 219) lament, "agricultural research alone is not a panacea for the food problems of the developing world. Social, economic, and ecological issues must also be tackled."

The resiliency of any food production system hinges greatly upon the ability of researchers to continuously supply new crop varieties through plant breeding. Hence, the success of global agricultural development efforts is highly dependent upon access to genetic material. As such, development policies must help protect those geographic regions that act as reservoirs of plant and genetic resources. Protection and proper management of these resources will require a cooperative and coordinated effort on the part of the scientific community, the public sector, private land holders, and the indigenous people living in degradation-prone areas. It is paramount that these groups work together to promote human-environment relations, which help conserve biodiversity while meeting the growing demand for food.

It is essential to form a better understanding of the impacts of **bioinvasion**—*the unwanted introduction of a foreign organism or species into a regional food production system.* Avoiding the ecological devastation of a bioinvasion is becoming a major preoccupation with scientists researching the disruptive influence of "invaders" and the cascading effects on an ecosystem. Successful invaders can dramatically alter the natural balance and harmony amongst resident plant and animal communities. As a result, many ecosystems lose diversity as invading species establish themselves, outcompete other species common to that area, and transform the ecology of the affected area. New limits are then placed on old systems, leading to greater uncertainty over the sustainability of those systems. Loss of biodiversity renders these areas more susceptible to degradation and natural hazards. Whether we are talking about damage to cotton crops in Asia from boll worms or from an insect-transmitted virus, a new form of potato blight in North America, or poisonous algae "blooms" in association with "red tide" (the discoloration and poisoning of a body of water due to the proliferation of brown and red marine algae known as dinoflagellates and the subsequent release of algae toxins) and its devastating impact on fish populations off Florida's Gulf Coast, invasions are known to radically

alter the delicate balance established by nature. This is not to say that all invasions have detrimental effects on the environment or that all invasions should be controlled.

The red tide phenomenon has had devastating effects on the fish populations off the coast of Australia (transmitted to this region by Japanese fishing vessels). It is responsible for huge losses to the Australian fishing industry. The occurrence of red tide, however, has been economically beneficial to Florida's fishing industry. After a red tide incident, several high-value marine species (for example, shrimp) tend to rapidly multiply in and along the Gulf Coast, allowing marine harvests to reach record levels. Moreover, these species tend to become even more abundant with each recurring red tide event.

There are, nonetheless, relatively few success stories in comparison to the many problems caused by invading organisms. Slowing the rates of invasion in cases where there are few or no recognizable net benefits has recently become an important consideration in food and environmental resource management. Furthermore, identifying the pathways of invasion by understanding the human and economic processes (for example, patterns of trade and migration), which promote the transmission of invaders, is key to outlining regional and global bioinvasion control strategies.

Myth #5: While the green revolution did not end world hunger, it did cause tremendous harm to regional production systems.

In fact, many food production regions have yet to fully realize the benefits of the green revolution. Many scholars argue that we are in the midst of yet another revolution in agriculture due to the multitude of advancements in biotechnology and agronomy. It has been well established that food security issues must be addressed in the policy arena, especially in situations where technology or markets fail to find solutions to the problems. Consider how the green revolution led to food shortages in regions where the production emphasis turned toward cash crops or feed grain for livestock. As agents on the economic landscape responded to technologically guided market forces, vital grainland that was once used to grow crops for local human consumption was displaced at an unacceptable rate. In many regions of the SOUTH, the demand for animal feed has been growing considerably faster than the demand for human food over the past few decades. The switch to feed grains or cash crops for export is a trend that can only be reversed by intervention.

In many cases, the application of technology boils down to a compatibility issue. Agricultural development policies must be viewed in a much wider context, and must be compatible to macroeconomic reforms which seek to reduce trade imbalances, diversify export and import opportunities, and promote efficiency and free-market activity.

Improved technology has underpinned much of the agricultural growth in both the developed and the developing countries. Some countries have been at a disadvantage because the available technology was less suited for their ecologies and farming systems. R&D for agriculture will be essential for agricultural development in the 1990's, but more emphasis should be placed on identifying and promoting more efficient ecologies and farming systems. They offer high pay-offs not only in terms of potential increases in global production, but also in terms of rural income generation, poverty alleviation, and improved nutrition. (*Global Outlook 2000*, p. 133–134)

Plucknett and Smith (1993) see the green revolution as a success story yet untold. They highlight many accomplishments including: higher yields from genetic engineering, discovery of new crop varieties, improved breeding techniques and a deeper understanding of the nature and historical relationships between crops, farmers and farming practices, and the various pests and pathogens which affect regional food production systems. The green revolution has helped to advance genetic-resource conservation by underscoring the value of **gene banks**—*storage and research facilities oriented toward the preservation of germplasm.* The green revolution also assisted in the intensification of food production in geographic areas with once limited production potential. It is important to note that widespread use of green revolution technology helped to spare much of the world's marginal farmlands. As output of productive farmland increased dramatically, marginal farmland was made less susceptible to misuse and degradation. Production gains at the global level actually slowed the proliferation and spillover of nonsustainable agriculture to many unspoiled areas. By sparing virgin land from the plow and the axe, the green revolution helped preserve land and important genetic resources for future generations. Nonetheless, it is not clear that the preservation of these lands and the impact of new production technology will bring us closer to finding a nonpolitical solution to hunger.

Myth #6: Land reforms that support the peasant mode of production and subsistence farming would actually lower food output and hurt the very people they intend to help.

As most wealthy landowners in the agricultural regions of the SOUTH utilize only a small percentage of their land for food production, typically less than one-quarter of the total acreage, consolidation of land holdings for commercial purposes represents an inefficient use of land resources. Individual peasant farmers could easily cultivate portions of the land and achieve higher total output on a per land unit basis. This is true, as they tend to work the land more "intensively" than commercial farmers cultivating large tracts. Hence, the notion that big operations are better or more efficient than smaller

operations is open for debate when it comes to agricultural systems in the developing world. This debate reminds us that issues of land reform need to be examined on a region-specific basis.

Myth #7: Market forces alone can bring an end to world hunger.

Such thinking would be all well and good if every regional economy relied solely upon the market and money as a way in which to guide the production, distribution, and consumption of food. As established earlier, regions vary in terms of the degree to which market forces affect production and exchange of agricultural output. Many regions produce for use value and not exchange value. Variations in the distribution of income and wealth add further complications. Market processes may work best for core economies, but do not necessarily work well in semiperipheral and peripheral regions, where market signals sometimes fail to account for the subsistence value of agricultural commodities and/or where questions arise as to the purchasing or bartering power of consumers. The far-too-rigid notion of a "government/free-market tradeoff" only constrains the choice of agricultural development policy options. There is always a role for government (no matter how small or even if just as a fail-safe), for markets are known to falter. This is particularly true in regions where markets are "blind to the social and resource costs of the production engine it is supposed to drive." (Lappe and Collins, 1986, p. 79)

Myth #8: Free trade is the answer to the problem of ending world hunger.

While the global elimination of all trade barriers is a worthy goal (in theory), it may actually serve to reinforce production specialization and comparative advantage. Nations would export what they produced best and import what they cannot produce (competitively speaking). Since most peripheral nations are unable to compete with core economies in high-value-added production, they would be limited to produce and export extractive resources (agricultural products, precious and strategic minerals, and oil). Global free trade would bias the agricultural systems of the SOUTH toward the production of cash crops for export. We have already established that this has only made matters worse in the periphery.

Myth #9: The poor and hungry are too weak to revolt.

The world's poorest regions survive with very few resources, a condition that requires tremendous efforts on the part of individuals in these systems. Depicting the poor and hungry as helpless and weak, or as people "conditioned into a state of passivity" is not realistic (Lappe and Collins, 1986, p. 95). In actuality, the impoverished masses work very hard at trying to cope and survive. They travel great distances to work, put in long hours, and expend much energy to work the land, hunt and trap, or gather food from nature. Hence, if the poor were truly passive, they would not be able to survive in such adverse economic conditions. Their ability to overcome adversity and the forces which oppress them is testimony to their strength. Many at the bottom of the world's social hierarchy appear to be uninterested in change; this is not to say that they do not understand their situation or are unaware of the obstacles that stand in the way of progress. As roughly one-fifth of the world is severely afflicted by hunger and poverty, the poor and hungry must be viewed as a formidable force in their own right.

Myth #10: More aid from the United States and other core economies is necessary to halt world hunger.

Hunger may be viewed as a direct result of "antidemocratic political and economic structures that trap people in poverty" (Lappe and Collins, 1986, p. 103). The governments of core economies have already practiced interventionalist measures in countries where the incidence of hunger is high. Financial and food aid have done little to reverse the mechanisms which cause and perpetuate hunger and depravation. More times than not, aid has been a tool of foreign policy and manipulation. The following generalizations may be made about U.S. involvement in aiding underdeveloped economies:

1. Economic assistance has been highly concentrated in a handful of countries. Since the early 1980s, Egypt and Israel have received over half of all U.S. economic assistance dollars; Russia and Mexico have also fared well as recipients of U.S. foreign aid.
2. The distribution of food aid, when viewed on a per capita basis, has not been related to need.

   Most food aid is used to bolster politically allied governments. During the war in Indochina in the early 1970s, for example, the U.S. government allies there received nearly 20 times more food aid than the five African countries then suffering famine. In 1985, the United States shipped nearly four times more food aid per capita to the countries of Central America than it did to the entire famine ravished region of sub-Saharan Africa. (p. 106)

3. Food aid has been characterized as a direct disincentive to production, as it is likely to deter the expansion of food production systems in regions receiving aid. This has led governments and people to rely on cheap and plentiful external food sources. Moreover, it has allowed them to postpone confronting the problems inherent to their own region and food production systems.
4. Much economic and food aid has been concentrated in regions where the governments are fundamentally opposed to economic reforms on the behalf of the poor (as long as they pledge their political alliance to the United States). For example,

substantial funds and food aid have been dispensed to El Salvador, Pakistan, Bangladesh, and the Philippines, nations that have shown little interest in land reforms which benefit subsistence farmers or landless peasants.

5. Economic assistance, food aid, or food exports can be cut-off as a form of punishment for noncompliance in economic and political matters. Jimmy Carter's grain embargo against the former Soviet Union (after their invasion of Afghanistan) or Ronald Reagan's dramatic reductions in food and economic aid to Nicaragua (under Sandinista control) are two fine examples of the politics of food.
6. Military aid from the core, in the form of foreign assistance, has helped to arm governments and protect those in power from the poor, hungry, and disillusioned. When corrupt governments and businessmen remain in power and control, backed by a strong military presence, foreign aid does little to support positive social, political, and/or economic change.
7. Development assistance fails to reach and help the poor and hungry in many regions due to the high cost of government bureaucracy, graft, and corruption. A disproportionately large share of the benefits of foreign transfers earmarked for development projects seems to go to those who need it the least—wealthy businessmen and landowners. More times than not, development infrastructure is put into place to enhance the economic performance of industries controlled by wealthy individuals in society.

It is true that the United States and other donor nations have had a fair amount of success in offering food aid in emergency situations around the globe. Yet, it is critical that relief efforts be viewed as temporary solutions to an ongoing problem. Continued development assistance must work on defeating the processes which cause crises to emerge. Increasing the innate production capacities of peripheral regions is of great importance, but it offers only a partial solution. Attacking the underlying causes of poverty is equally important (Figure 6.8).

Myth #11: Core economies of the NORTH benefit from hunger and poverty in the SOUTH.

The argument goes like this: The world's poor and hungry are willing to work for less money and subsequently produce a wide variety of products at a low cost. Core economies as the world's major markets, therefore, have much to lose in a world without poverty and hunger, as this would reduce the overall standard of living in the core due to increasing price levels. Yet, the fate of people in all regions is truly interwoven. Low wages and low-cost products from peripheral areas reflect polarization,

Figure 6.8 Food aid to avert famine. Nations of the developing world have become increasingly reliant on outside suppliers of grain.
6.8a: ©Mark Richards/Photo Edit; 6.8b: ©Michael A. Dwyer/Stock Boston,

inequalities, and exploitation. In a global economy, low wages in one corner of the world are an indication of the "powerlessness of workers everywhere." While it is appealing for regions to promote their own self-interest, it is something that is usually done at the expense of the poor and powerless. In essence, all of humanity falls victim to poverty and hunger in a system that shows little compassion for the poor.

Investing in people is key to reversing this unwanted economic trend and providing a better way of life for all. Educating and up-skilling the world's labor force will not only help empower labor, but increase their productivity and expand their creative potential. Nations such as Japan, Taiwan, South Korea, Thailand, China, and Malaysia have understood the importance of investing in "human capital." Surely, economic progress in Asia and elsewhere has improved the quality of life for people around the world.

Myth #12: Eradication of hunger means elimination of freedoms (that is, there is a "food-versus-freedom" trade-off).

To many people, government involvement in any matter ultimately translates into less personal freedoms. Redistribution of wealth, as a solution to abolishing hunger and poverty, is viewed as contrary to the "right to unlimited accumulation of wealth"—the hallmark of modern materialistic culture.

Some see a sudden end to poverty and hunger as incompatible with the goal of global economic prosperity. This is based on the belief that poverty is an outcome of the accumulation of wealth, a situation that has inevitably led to economic inequality and unequal power and leverage of various members or groups in world society. Redistribution schemes go against the right to secure wealth. On moral grounds, society does have an obligation to see that all individuals are at least allowed to subsist and exercise the right to pursue a better way of life for themselves and their families.

Feeding the hungry and educating the poor is a sound long-term investment. Such an investment is likely to bring about efficiency gains, economic growth, sociopolitical stability, and new forms of regional economic cooperation. Providing the means to basic survival resources will eventually allow the needy to provide for themselves and make contributions to their local economies. The elevation of each member of society furthers the advancement of society at large. It is a philosophy that is rooted in the Jeffersonian ideal that the economic security of all citizens will ensure the preservation of liberties. Elimination of hunger and poverty will most assuredly increase both the productivity of human capital and the general quality of life.

While the evolution of world economy has led to efficiency gains in production, globalized production has led to a loss of control of the destiny of regions and industries. As the world economy becomes increasingly interdependent, so too have the world's food production systems become increasingly interwoven. According to Whatmore (1993), the globalization of the food production network has led to unevenness, instabilities, and a series of three overlapping crises: the crises of production, regulation, and legitimation.

The **crisis of production** *centers on problems of surplus production and indebtedness of agricultural producers as they face escalating input costs, production ceilings, and declining product prices.* Declining profits and income in regional food production systems have threatened the commercial longevity of these systems. The **crisis of regulation** *is a direct result of the growing political and institutional tensions and conflicts of interests in the agro-policy arena* as policy makers attempt to regulate food trade, on the one hand, and stabilize regional and national farm incomes, on the other. The **crisis of legitimation** *centers on the increasing politicization of "agro-industry" and the mounting concerns about "the consequences of industrialized agriculture."* This includes concerns over the nature of food security, food safety, and the uncertain future of farmed environment at national and local levels (Whatmore, 1993, p. 48).

Some argue that the intensity of each crisis may be dampened by government intervention. The crisis of legitimation is of special concern for environmentalists. International environmental management has now become a big part of the liberal global development agenda. For government intervention to be effective, however, programs must be set up to both provide incentives to producers to minimize the consequences of industrialized agriculture and reward those producers that are least damaging or nondestructive. Even the crisis of legitimation could be rationalized as a temporary inconvenience, as new markets develop for environmental quality and new partnerships emerge between business and government. Certainly tougher international environmental standards and laws (when properly enforced) would not only change the face of international competition in agricultural markets, but reduce the degradation of farmland. Many **moderates**—*those residing somewhere in between the conservative and liberal viewpoints in matters of political economy*—argue that the role of government should be limited to (a) establishing environmentally related development priorities; and (b) the translation of those priorities through market channels. Mandating global food quality standards (in the growing, harvesting, transportation, processing, and handling of food) and levying taxes against land degradation are examples of how governments may promote food security and sustainable agriculture.

Marxists contend that the crises of production, regulation, and legitimation will never be eliminated as surplus-producing regions of the world continue to produce agricultural commodities for exchange value. In this light, the globalization of agricultural production in the capitalist world economy is responsible for the origination and continuation of the crises.

From a universal standpoint, governments must effectively meet the challenge of creating region-specific, agro-development agendas to stave off the ill-effects of the crises of production, regulation, and legitimation. This can be accomplished by tailoring agro-development policies to the needs of a region and its people, placing people first and economic growth objectives second. Differentiating regional production systems and coming to grips with the complex array of attributes that define the processes which effect production economics is fundamental to practical policy formation. Policy makers and practitioners must build new contextual frameworks by which to classify or portray regional economies and identify the factors responsible for economic change. Hence, it is necessary to dispense with the far-too-rigid notion that economies are of a dualistic nature. Stepping outside the limitations of current development theory to more fully explore the social, political, economic, and cultural realities of food production and security will require a more thorough examination of the human geographic aspects of the problem.

Logan (1995) maintains that a multidimensional continuum of production economies exists. Within the continuum between pure subsistence economy and pure commercial economy, regions are viewed as having their own unique mix of attributes, interactions, limitations, problems, and challenges. Instead of thinking in terms of either the traditional economy, which relies heavily on barter as a dominant system of exchange, or the "Westernized" version of economy, which is dominated by production in response to market signals and exchange value, it is essential that regional-based agro-development policies be formulated in ways that are sensitized to the uniqueness of regions and their physical, economic, and cultural conditions. Agro-development policy, therefore, must be tailor-made to fit each region.

Consequently, a greater understanding of how the decision-making processes and the behavior of production agents are influenced by local conditions is essential. Policy makers must be in tune with the perspectives of agriculturalists as influenced by (a) attitudes toward optimization and survival; (b) economic orientation and setting (traditional versus modern, and informal versus formal); (c) geographic location with respect to markets; (d) physical limitations; (e) knowledge and experience; (f) prevailing mechanisms of exchange (for example, pure barter, semibarter, and/or currency); and (g) local or region-specific development goals and obstacles.

STABILIZATION AND STRUCTURAL ADJUSTMENT: IN THEORY AND PRACTICE

Two general development strategies have been implemented to reverse the effects of nonsustainable macroeconomic trends: namely, "stabilization" and "structural adjustment." **Stabilization** *involves efforts aimed at stabilizing a regional economy or enhancing its performance through macroeconomic policy and economic reforms.* This includes the implementation of policies to control inflation, minimize national unemployment, and correct a debtor nation's immediate balance-of-payment problem. Stabilization of the value of a nation's currency is a top priority. Economic stability reduces the likelihood of a nation defaulting on its current development loans.

Structural adjustment *is the implementation of strategies that support long-term regional growth and development goals and economic change through market forces.* This includes measures to promote the internal restructuring of a region's economy through the privatization of industry and the acquisition of development loans and foreign investment. The objective is to increase the competitiveness and status of a region in the global economy, thereby enabling that region to both "develop" and find long-term debt relief.

Nations that face increasing poverty are often forced to try and secure additional loans. As a condition of attaining this money, heavily indebted countries have often had to make "structural adjustments" to their economies; eliminating subsidies, removing tariffs, and privatizing government-owned enterprises. These reforms aim to help them grow out of indebtedness by removing glaring economic inefficiencies. (Pearce et al., 1995, p. 54)

Target regions typically include nations in which currencies are overvalued or artificially supported, regions in which industrialization efforts were organized around inward-directed, import-substitution approaches (necessitating the use of tariffs and market barriers), and economies which maintain a higher-than-average ratio of government expenditure to economic output. Traditional "top-down" structural adjustment policies of the International Monetary Fund and the World Bank generally encourage trade liberalization, market-based pricing and exchange of commodities and currency, budget-balancing initiatives, tax reform, minimized government intervention, and measures to reduce trade deficits by reducing the domestic demand for imports. The intensity by which structural adjustment policies are applied is directly related to the degree of peripheralness. As such, sub-Saharan African nations have been the largest recipients of development loans.

While theoretically sound, structural adjustment policies have failed to promote development in many peripheral regions. The dismal performance of structurally adjusting economies may be attributable to the short-term imbalances and upheaval caused by economic restructuring. The structural adjustment philosophy has proved somewhat successful for nations like India and China, as both continue to push toward self-reliance in many economic sectors; nations like Russia have been less than satisfied with the economic gains of structural adjustment.

One commonly held misconception, is that structural adjustment brings about a new set of priorities that are oriented toward meeting economic growth objectives at the expense of the environment. In other words, there is the belief that all other objectives, including social and environmental ones, must be subordinate to growth and debt repayment. Opponents of structural adjustment maintain that adjustment policies not only fail to meet most designated macroeconomic objectives, but restructuring is harmful to both the poorest members of society and the most fragile portions of the environment (*World Resources 1994–95*, p. 228). While debate continues over the effectiveness of development loans and the growth versus environmental quality trade-off, many policy makers have refocused their attention on what they see as an out-of-control spiral of debt and degradation in the SOUTH. Consider the fact that total external debt of underdeveloped nations, with the exception of middle-income economies, is becoming an

increasing percentage of nations' exports and GNP (see Table 6.16).

While it is difficult to deny that some development efforts lead to human economic hardship, there is little evidence to support the claim that development loans are directly connected to ecological turmoil. On the contrary, there is much evidence to support the argument that regional development loans and subsequent structural adjustment of regional economies are, in many cases, beneficial to the environment. Arrangements between international lending institutions and target regions have brought about many positive changes including (a) the reduction or abandonment of destructive farming practices; (b) the lessening of market controls and curtailing of subsidies on agricultural commodity prices (a situation that has limited export-oriented production and slowed deforestation and soil erosion); (c) the reduction of subsidies and use of environmentally harmful chemicals in agro-systems; (d) limiting the use and expansion of government-supported irrigation techniques (which has helped reduce waterlogging and salinization); and (e) forcing governments to clarify their positions on land-ownership and, in many instances, allowing them to institute much needed agricultural land reforms (Pearce et al., 1995).

Despite the lackluster performances of many stabilized or structurally adjusted economies, proponents of these policies have conceded that the poor would have probably been much worse off without these development efforts. Moreover, the experiences of the past several decades have offered much in the way of learning. In retrospect, the inadequacies of development assistance have been linked to the failure of institutions to include social and environmental costs in the assessment of the net benefits of a development project, as well as a general lack of

TABLE 6.16	Total External Debt Statistics			
	Net % value of external debt as % of			
	Exports		GNP	
Selected Region/Group	**1980**	**1993**	**1980**	**1993**
Low-income economies	95.3	170.9	13.7	37.1
Low-income economies (excluding China and India)	106.9	269.1	27.0	75.6
Middle-income economies	148.6	128.7	36.1	30.8
sub-Saharan Africa	96.8	151.4	26.8	47.4
East Asia and Pacific	89.5	91.0	16.7	28.5
South Asia	162.7	206.2	17.3	31.1
Latin America and Caribbean	202.8	227.6	35.1	34.0

Source: *World Development Report 1991* (pp. 250–251); *World Development Report 1995* (pp. 206–207).

communication and insufficient consultation with the people most affected by the regional economic change (*World Resources 1994–95*, p. 228). In light of these pitfalls, future development policies must take on a more balanced approach—promoting growth and self-reliance while preserving environmental assets. In terms of agro-development, structural adjustment policies must be designed in ways that make agriculturalists more aware of the adverse effects of nonsustainable practices, and directly accountable for actions which degrade the environment.

A comprehensive approach to reducing environmental degradation must include (1) reforms which promote "individual ownership" of agricultural land to allow the poor to sustain themselves; (2) an end to subsidies that promote environmentally insensitive behavior and practices, excessive habitat destruction, and/or loss of biodiversity; and (3) a well-enforced system of degradation and pollution control that would monetarily punish those unwilling to comply to imposed "standards" of maintaining environmental quality in production.

Fifty years of structural adjustment in the SOUTH has brought about mixed results. Nonetheless, analysts from around the world are beginning to realize the importance of advancing both environmentally friendly development policy and reforms that "cushion" the poor as regional economies deal with the growing pains of adjustment (French, 1994). Land tenure issues continue to remain at the very forefront of the human development priority list.

> The need for some degree of land reform and the expediting of secure land titles has been recognized for a long time in parts of Brazil and other areas of Latin America. Confusion over land titling has fuelled deforestation, since one of the most visible ways to place a stamp of ownership on the land is to clear it, even if the opened space is not used productively. . . . [Yet,] land reform could exacerbate social problems by reducing agricultural production and dismantling managerial expertise for certain agricultural enterprises. (Smith et al., 1995, pp. 57–58)

Securing land-ownership titles is essential to the process. Land ownership will allow individual agriculturalists to obtain loans or credit and more responsibly manage the land. Hence, development efforts must concentrate on the needs of people at the "local" *and* "regional" levels. Though stabilization and structural adjustment policies which focus on both economic and environmental objectives may work on a region-specific basis, the challenge of how to coordinate these development efforts on a broader geographic scale remains.

> Equally important to ensuring long-term sustainability is reform of the international trade and financial systems that still favor industrialized nations. The debt of many Southern countries makes investment in human development and environmental protection difficult, if not impossible. Exacerbating this situation is the tendency of Southern countries to draw on their natural resource base in an effort to expand exports and thereby finance external debt. Some countries, in fact, are paying more to service their debt than they receive in new resources. . . . The combination of high debt and inability to expand exports in environmentally sound ways is a major impediment to sustainable development in the SOUTH. (World Resources 1994–95, p. 232)

Attention has also turned toward safeguarding the environment. No longer can economic development come at the expense of the environment. "Making polluters pay" has become a battle cry of environmentalists as they attempt to influence development policies that would steer local and regional economies and the environment toward greater "sustainability" (Postels, 1991). Opponents of these types of measures argue that imposing a heavier tax burden on individuals and corporations, by taxing income and value added, would discourage production (and thereby limit growth-related opportunities) and provide disincentives for investment and savings. Nonetheless, taxing waste and pollution, and "costing" for environmental damage remains as a viable policy option.

Efficiency gains associated with improving environmental conditions are likely to improve future economic conditions and lead to unforeseen opportunities. With evidence of a global economic slowdown (see Table 6.17), issues of uneven development and environmental degradation have become more relevant with time. It all boils down to a question of finding ways to satisfy the basic needs of people without further disrupting or destroying an economy's support system (Brown, 1993, p. 26).

CONCLUDING REMARKS

Two-hundred years of relatively unbridled growth and industrialization (with little or no concern for the environment) has reduced the world economy's current growth

TABLE 6.17	World Economic Growth for Selected Periods	
	Average Annual Growth (%)	
Decade/Period	**World Economy**	**Output per Capita**
1950–1960	4.9	3.1
1960–1970	5.2	3.2
1970–1980	3.4	1.6
1980–1990	2.9	1.1
1990–1993	0.9	–0.8
1990–2000	1.0*	–0.9*

Source: Brown, L.R., 1993, p. 24.
* Forecast.

potential. The cumulative effects of environmental degradation are now imposing real obstacles to development. As output per capita declines across most economic sectors, we must now face up to the fact that modest limitations to growth do exist. We must also acknowledge that the food and economic security objectives of rich and poor nations are not separate issues. Environmental concerns have taken center stage in the policy arena as leaders from the NORTH and SOUTH support responsible economic growth and development.

KEY WORDS AND PHRASES

aquaculture   129
arable land   117
bioinvasion   135
biotechnology   132
Boserup thesis   126
carryover stocks   127
cornucopians   124
crisis of legitimation   138
crisis of production   138
crisis of regulation   138
cropland mismanagement   118
deforestation   118
desertification   118

external agents   118
gene banks   136
genetic erosion   134
germplasm   134
green development   134
green revolution   123
land degradation   117
Malthusian theory   126
Marxists   125
moderates   139
monoculture   135
natural hazards   118
neo-Malthusians   124

overcultivation   118
overgrazing   117
physiological density   117
soil burnout   118
soil compacting   117
soil erosion   118
stabilization   139
structural adjustment   139
survival resources   133
twelve myths of hunger   133
urban sprawl   118
waterlogging and salinization   118
world resource   119

REFERENCES

Adams, W.M. *Green Development: Environment and Sustainability in the Third World.* New York: Routledge (1990).

Berry, B.L.J., E.C. Conkling, and D.M. Ray. *The Global Economy.* Englewood Cliffs, NJ: Prentice Hall (1993).

Boserup, E. *The Conditions of Agricultural Growth: The Economics of Agrarian Change under Population Pressure.* Chicago: Aldine (1965).

Boserup, E. *Population and Technology.* New York: Blackwell (1981).

Bradley, P.N., and S.E. Carter. "Food Production and Distribution—and Hunger." *In A World in Crisis,* R.J. Johnston, and P.J. Taylor (eds.). Oxford, UK: Basil Blackwell (1989), pp. 101–124.

Bright, C. "Understanding the Threat of Bioinvasion." In *State of the World 1996.* Washington, D.C.: World Watch Institute (1996), pp. 95–113.

Brough, H.B. "A New Lay of the Land." *World Watch* 4 (January-February, 1991), pp. 12–19.

Brown, L.R. (ed.). *State of the World 1989.* Washington, D.C.: World Watch Institute (1989).

Brown, L.R. (ed.). *State of the World 1990.* Washington, D.C.: World Watch Institute (1990).

Brown, L.R. (ed.). *State of the World 1992.* Washington, D.C.: World Watch Institute (1991).

Brown, L.R. "A Decade of Discontinuity." *World Watch* 6 (July-August, 1993), pp. 19–26.

Brown, L.R. (ed.). *State of the World 1996.* Washington, D.C.: World Watch Institute (1996).

Brown, L.R., C. Flavin, and H. Kane. *Vital Signs 1992* (published for World Watch Institute). New York: W.W. Norton (1992).

Brown, L.R., C. Flavin, and H. Kane. *Vital Signs 1993* (published for World Watch Institute). New York: W.W. Norton (1993).

Brown, L.R., C. Flavin, and S. Postel. *Saving the Planet: How to Shape an Environmentally Sustainable Global Economy.* New York: W.W. Norton (1991).

Brown, L.R., and J.E. Young. "Feeding the World in the Nineties." In *State of the World 1990,* L.R. Brown (ed.). Washington, D.C.: World Watch Institute (1990), pp. 59–78.

Budiansky, S. "10 Billion for Dinner, Please." *U.S. News and World Report* (September 12, 1994) pp. 57–62.

Burbach, R., and P. Flynn. *Agribusiness in the Americas.* New York: Monthly Review Press (1980).

Business Week. "Inequality: How the Gap between Rich and Poor Hurts the Economy." (August 15, 1994), pp. 78–83.

Cairncross, F. *Costing the Earth: The Challenge for Governments, The Opportunities for Business.* Boston, MA: Harvard Business School Press (1992).

Cernea, M.M. (ed.). *Putting People First: Sociological Variables in Rural Development.* New York: Oxford University Press (1985).

Cole, J. *Development and Underdevelopment.* London: Methuen (1987).

The Economist. "For Richer, For Poorer." (November 5, 1994), pp. 19–21.

Findlay, A. "Population Crises: The Malthusian Specter?" In *Geographies of Global Change,* R.J. Johnston et al. (eds.). Oxford, UK: Blackwell (1993), pp. 152–174.

French, H.F. "The World Bank: Now Fifty, But How Fit?" *World Watch* 7 (July-August, 1994), pp. 10–18.

Furlong, W.L. "Hunger, Poverty, and Political Instability." In *The United States and World Poverty,* E.B. Wennergren, D.L. Plucknett, N.J.H. Smith, W.L. Furlong, and J.H. Josji (eds.). Washington, D.C.: Seven Locks Press (1989), pp. 129–151.

Gainesville Sun (AP wire story) entitled "Florida Losing Farms Faster Than in Any Other State." Sunday edition (June 28, 1992), p. 7B.

Gardner, G. "Preserving Agricultural Resources." In *State of the World 1996,* L.R. Brown (ed.). Washington, D.C.: World Watch Institute (1996), Chapter 5, pp. 78–94.

Goode's World Atlas. (19th edition). Chicago: Rand McNally (1995).

Grigg, D. *The World Food Problem* (2nd edition). Oxford: Blackwell (1993).

Hardin, G. "The Tragedy of the Commons." Reprinted in *Perspectives on Population: An Introduction to Concepts and Issues,* S.W. Menard, and E.W. Moen (eds.). New York: Oxford University Press (1987), pp. 106–112.

Hayami, Y., and V.W. Ruttan. *Agricultural Development: An International Perspective.* Baltimore, MD: John Hopkins University Press (1985).

Head, C.M., and R.B. Marcus. *The Face of Florida* (2nd edition). Dubuque, IA: Kendall/Hunt Publishing Co. (1987).

Jackson, R.H., and L.E. Hudman. *World Regional Geography: Issues for Today.* New York: J. Wiley & Sons (1990), Chapters 3 and 4.

Johnston, R.J., P.J. Taylor, and M.J. Watts. *Geographies of Global Change: Remapping the World in the Late Twentieth Century.* Oxford, U.K.: Blackwell (1995).

Kane, H. "Growing Fish in Fields." *World Watch* 6 (September-October, 1993), pp. 20–27.

Lappe, F.M., and J. Collins. *World Hunger: Twelve Myths.* New York: Grove Weidenfeld (1986).

Lappe, F.M., and J. Collins. "Food First." In *Perspectives on Population: An Introduction to Concepts and Issues.* S.W. Menard, and E.W. Moen (eds.). New York: Oxford University Press (1987), pp. 113–119.

Logan, B.I. "The Traditional System and Structural Transformation in sub-Saharan Africa." *Growth and Change* 26, (1995), pp. 495–523.

Malthus, T.R. 1798. "An Essay on the Principle of Population." Reprinted in *Perspectives on Population: An Introduction to Concepts and Issues.* S.W. Menard, and E.W. Moen (eds.). New York: Oxford University Press (1987), pp. 97–103.

Meadows, D.H. *The Global Citizen.* Washington, D.C: Island Press (1991).

Meadows, D.H., Meadows, D.L., and J. Randers. *Beyond the Limits: Confronting Global Collapse, Envisioning a Sustainable Future.* Post Mills, VT: Chelsea Green (1992).

Menard, S.W., and E.W. Moen (eds.). *Perspectives on Population: An Introduction to Concepts and Issues.* New York: Oxford University Press (1987).

Menken, J. (ed.). *World Population and U.S. Policy: The Choices Ahead.* New York: W.W. Norton (1986).

Mitchell, B. *Geography and Resource Analysis.* London, UK: Longman (1989).

Mosley, P. *Overseas Aid: Its Defence and Reform.* Brighton, U.K.: Wheatsheaf Books (1987).

Murdoch, W.W. *The Poverty of Nations.* Baltimore, MD: John Hopkins (1980).

Parfit, M. "Diminishing Return: Exploiting the Ocean's Bounty." *National Geographic* 188 (5), Washington, D.C.: National Geographic Society. (November 1995), pp. 2–37.

Pearce, D., N. Adger, D. Maddison, and D. Moran. "Debt and the Environment." *Scientific American* 272 (June 1995), pp. 52–57.

Pearse, A. *Seeds of Plenty, Seeds of Want: Social and Economic Implications of the Green Revolution.* New York: Oxford University Press (1980).

Plucknett, D.L., and N.J.H. Smith. "Agricultural Research and Third World Food Production." *Science* 217 (July 1982), pp. 215–220.

Plucknett, D.J., and N.J.H. Smith. "Benefits of International Collaboration in Agricultural Research," In *The United States and World Poverty,* E.B. Wennergren, D.L. Plucknett, N.J.H. Smith, W.L. Furlong, and J.H. Josji (eds.). Washington, D.C.: Seven Locks Press (1989), pp. 113–128.

Plucknett, D.L., and N.J.H. Smith. "Green Revolution Remains in Force." *Forum* 8 (1993), pp. 65–69.

Postels, S. "Halting Land Degradation." In *State of the World 1989,* L.R. Brown (ed.). Chapter 2 (1989), pp. 21–40.

Postels, S., "Saving Water for Agriculture," in *State of the World 1990,* L.R. Brown (ed.). Chapter 3 (1990), pp. 39–58.

Postels, S. "Accounting for Nature." *World Watch* 4 (March-April 1991), p. 28–33.

Redclift, M. *Sustainable Development: Exploring the Contradictions.* London: Methuen (1987).

Rhoades, R.E. "The World's Food Supply at Risk." *National Geographic* 179 (April 1991), pp. 74–105.

Riddell, R.C. *Foreign Aid Reconsidered.* Baltimore, MD: John Hopkins University Press (1987).

Rubenstein, J.M. *An Introduction to Human Geography* (3rd edition). New York: Macmillan (1992).

Ryan, W. *Blaming the Victim.* New York: Vintage Books (1976).

Sen, A. *Poverty and Famines: An Essay on Entitlement and Deprivation.* Oxford: Clarendon Press (1981).

Smith, N.J.H., E.A.S. Serrao, P.T. Alvim, and I.C. Falesi. *Amazonia: Resiliency and Dynamism of the Land and Its People.* New York: United Nations University Press (1995).

Smith, N.J.H., T.J. Fik, P.T. Alvim, I.C. Falesi, and E.A.S. Serrao. "Agroforestry Developments and Potential in the Brazilian Amazon." *Land Degradation and Rehabilitation* 6, (1995), pp. 251–263.

Streeten, P. *What Price Food? Agricultural Pricing Policies in Developing Countries.* New York: St. Martin's Press (1987).

Student Atlas of World Politics. Guilford CT: The Dushkin Publishing Group (1994).

Tarrant, J. (ed.). *Farming and Food.* New York: Oxford University Press (1991).

Third World Atlas. A. Thomas (ed.). Washington, D.C.: Taylor and Francis (1994).

United Nations. *Global Outlook 2000.* Washington, D.C.: United Nations Publications (1990).

United Nations. *New Compact for Cooperation.* New York: United Nations Publications (1991).

Wennergren, E.B. "Hunger, Poverty, and Constraints on Development." In *The United States and World Poverty.* E.B. Wennergren, D.L. Plucknett, N.J.H. Smith, W.L. Furlong, and J.H. Josji (eds.). Washington, D.C.: Seven Locks Press (1989), pp. 1–32.

Whatmore, S. "From Farming to Agribusiness: The Global Agro-Food System." In *Geographies of Global Change,* R.J. Johnston et al. (eds.). Oxford, UK: Blackwell (1993), pp. 36–49.

World Atlas for Students (New Revised Edition). Maplewood, NJ: Hammond (1993).

World Bank. *Adjustment Lending: An Evaluation of Ten Years of Experience.* Washington, D.C.: World Bank (1989).

World Development Report 1990: Poverty. (published for the World Bank) New York: Oxford University Press (1990).

World Development Report 1991: The Challenge of Development. (published for the World Bank) New York: Oxford University Press (1991).

World Development Report 1995: Workers in an Integrating World. (published for the World Bank) New York: Oxford University Press (1995).

World Resources 1992–93: Toward Sustainable Development. New York: Oxford University Press (1992).

World Resources 1994–95: People and the Environment. New York: Oxford University Press (1994).

DEMOGRAPHIC TRANSITIONS AND MOVEMENT

OVERVIEW

Regional trends in population change throughout the SOUTH have been interesting and encouraging. To appreciate the nature of these trends, it is essential to gain an understanding of the components responsible for regional population change. Specifically, the population size of any geographic region is related to three factors: fertility, mortality, and migration.

Fertility is *the rate at which births are taking place in a region over a given period of time;* **mortality** is *the rate at which deaths occur in a region over a given period of time.* The nations of the SOUTH have earned a reputation as being regions with both high fertility and mortality rates. Together, fertility and mortality define a region's **natural increase**—*that portion of a region's population change that is solely attributed to births and deaths.* Hence, the natural increase of a region's population is an outcome of internal forces.

Migration is *the physical movement and relocation of people from one region to another.* It is a factor that has become increasingly important as an agent of regional population change. Migration is influenced by forces which are both internal and external to a region. In other words, the movement of people to a given destination is influenced by what is going on inside and outside their region of origin. People engaged in the migration process are generally referred to as "migrants."

While there is much debate over whether there is a "global population crisis," many demographers have concluded that population crises do exist on a regional level.

These crises are a direct result of the burden placed upon regional and global production systems (as illustrated in Figure 7.1). Most population-related development problems are restricted to regions of the SOUTH, outcomes of a so-called **demographic crisis.** A demographic crisis *occurs when the population growth situation threatens the subsistence ability of people in a given region.* This is particularly common in regions where natural increase is high or excessive. A "crisis" arises when the rate of regional population growth exceeds the rate of growth in food availability (from domestic production and imports). A region might be described as "overpopulated" should the demographic crisis be long lasting. This may result in widespread hunger or famine should food supplies become inadequate (Figures 7.1 and 7.2).

Regional Population Change

Regional population change is attributable to both natural increase (births less deaths) and **net migration**—*the number of people entering a given region less the number of people leaving that same region.* As such, a region's population at a given time period t is equal to its base population in t–1 (the previous time period) plus the population change due to natural increase and net migration. Hence, the **population change equation** may be written as follows:

Regional Population at t	=	Base Population at t–1	+	Natural Increase from t–1 to t	+	Net Migration from t–1 to t

145

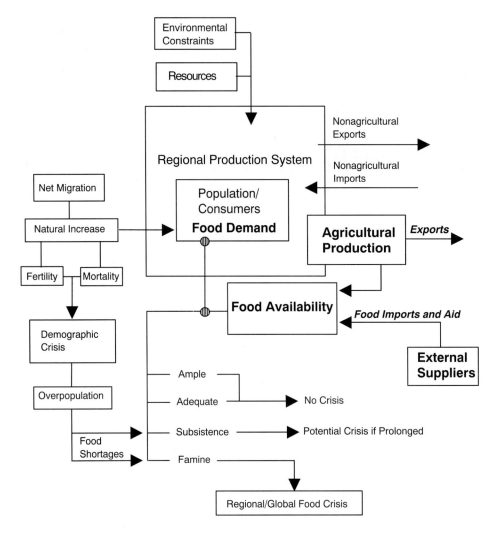

Figure 7.1 Demographic and global population crises.

More compactly, the population change equation may be expressed as

$$P(t) = P(t-1) + NI(t-1,t) + NM(t-1,t) \qquad (1)$$

where P = population of region in question; NI = natural increase; NM = net migration; and (t–1,t) refers to the time period t–1 to t (usually one year).

Typically, the largest share of regional population growth is due to natural increase. The **rate of natural increase** (RNI)—*the rate at which a region's population is growing due to fertility and mortality*—is defined in terms of a region's "crude birth rate" and "crude death rate"; and may be expressed as follows:

$$RNI = (\text{crude birth rate} - \text{crude death rate}) / 10 \qquad (2)$$

Crude birth rate (CBR) is *the total number of births in a region (in a given year) per 1000 people.* **Crude death rate** (CDR) is *the total number of deaths in a region (in a given year) per 1000 people.* As shown in equation (2), the rate of natural increase is defined as

CBR less CDR divided by 10 (a factor which allows RNI to be expressed in percentage terms).

Note that net migration (both legal and illegal) is becoming increasingly important over time as a contributor to regional population change. Over the past several decades, core economies of the NORTH have absorbed large numbers of migrants from the SOUTH. Many regions of the SOUTH have also absorbed large numbers of people from neighboring regions or nations hit hard by famine or civil unrest.

The Geography of Natural Increase

Regional comparisons of average estimated crude birth rates and crude death rates for selected periods are highlighted in Tables 7.1 and 7.2, respectively. Note the contrast in rates by geographic region and implied development status. Low crude death rates demonstrate the impact of the diffusion of medical technology to the SOUTH. Although crude birth rates remain high through-

Figure 7.2 Overpopulation, war, economic crisis, and food scarcity leave disturbing and unbearable impressions on the landscape, driving many to the brink of starvation. Here Afghanistan refugees await the arrival of food.

7.2: ©Mark Richards/Photo Edit

TABLE 7.1	Crude Birth Rates for Selected Periods and Regions	

	Average Estimated Crude Birth Rates[*]	
Region	1970–1975	1990–1995
Africa	46.8	43.0
Asia	34.9	26.3
South America	33.0	24.2
North and Central America	29.3	22.8
Oceania	23.9	19.3
North America	19.0	14.5
former USSR	18.1	16.5
Europe	15.7	12.7
World	**31.5**	**26.0**

	Crude Birth Rate[**]	
Region or Group	1970	1993
Low-income economies	39	28
Low-income economies (excluding China and India)	45	40
Low- and middle-income economies	36	27
sub-Saharan Africa	48	44
Middle East and N. Africa	45	33
East Asia and Pacific	35	21
South Asia	41	31
Europe and Central Asia	20	16
Latin America and Caribbean	36	26
Severely indebted countries	33	26
High-income economies	17	13

Source: *World Resources 1994–95*, pp. 270–271; and *World Development Report 1995*, pp. 212–213.

Note: [*] Figures rounded to the nearest .1 percent; [**] rounded to the nearest percent.

TABLE 7.2	Crude Death Rates for Selected Periods and Regions	

	Average Estimated Crude Death Rates	
Region	1970–1975	1990–1995
Africa	19	14
Asia	12	8
South America	10	7
North and Central America	10	7
Oceania	10	8
North America	9	9
former USSR	9	10
Europe	10	11
World	**12**	**9**

	Crude Death Rate	
Region or Group	1970	1995
Low-income economies	14	10
Low-income economies (excluding China and India)	19	13
Low- and middle-income economies	13	9
Sub-Saharan Africa	21	15
Middle East and N. Africa	16	7
East Asia and Pacific	10	8
South Asia	17	10
Europe and Central Asia	9	11
Latin America and Caribbean	10	7
Severely indebted countries	11	8
High-income economies	10	9

Source: *World Resources 1994–95*, pp. 272–273; and *World Development Report 1995*, pp. 212–213.

Note: Figures rounded to the nearest percent.

out the less-developed regions of the SOUTH, crude death rates have been dramatically reduced in virtually all regions of the world.

The rates of natural increase for selected geographic regions and time periods are shown in Table 7.3. Note that virtually all geographic regions have experienced a decline in the rate of natural increase over the last two decades, a trend that is expected to continue as world population climbs toward the 11 billion mark by the year 2100. The average annual population growth rates of low-income economies (excluding China and India) have been an exception to the rule. Increasing rates of natural increase have been observed in sub-Saharan Africa over the last two decades. This has raised concerns over the possibility of a growing demographic crisis in the region. Note that a small yet significant increase in population growth has also been observed in South Asia. Recent demographic evidence suggests, however, that the rates of natural increase for nations in Africa and South Asia are now on the decline.

Recall from Chapter 5 that the average annual rate of global population growth during the 1990s is approximately 1.6%. This figure is substantially lower than 1.95% observed over the period 1965–1980 (as shown in Table 7.4). As the global rate of natural increase contin-ues to decline, world population growth is expected to slow and world population is expected to stabilize in roughly 100 years. As the global rate of natural increase continues its descent to zero, the world will eventually experience "zero population growth."

Zero population growth (ZPG) *occurs when the crude birth rate equals the crude death rate (on average) over an indefinite period of time.* In other words, if the average crude birth rate equals the average crude death rate over an extended period of time, it follows that the average rate of natural increase will be equal to zero. It is highly probable, given observed demographic trends, that world population will approach a zero population growth scenario around the year 2100.

Recent demographic trends and forecasts support the argument that world population will eventually stabilize. Nonetheless, it will be difficult to verify this outcome given that accurate counts or estimates of the number of people residing in many of the world's regions are simply not possible. Many peripheral regions lack the human and economic resources to carry out a **census** (*a complete enumeration of the number of people residing in a region at a given point in time*) or monitor population growth. By tracking trends in crude birth and death rates and migration, however, demographers are able to produce reasonable "guesstimate" of regional and world population.

LIFE EXPECTANCY AND TOTAL FERTILITY RATES

Estimates of "life expectancy" from birth are commonly used to gauge the development status of a region.

TABLE 7.3	Rate of Natural Increase for Selected Geographic Regions and Periods

	Average Annual Population Change (Rate of Natural Increase %)		
Region	**1980–1985**	**1990–95**	**2000–2005**
Africa	2.91	2.93	2.70
Asia	1.91	1.78	1.39
N. and C. America	1.68	1.65	1.29
South America	2.15	1.67	1.35
Europe	0.31	0.27	0.27
Former USSR	0.89	0.51	0.65
Oceania	1.51	1.51	1.37
Selected Nations			
China	1.44	1.42	0.78
India	2.14	1.91	1.65
Indonesia	2.06	1.78	1.28
Nigeria	3.20	3.13	2.95
Ethiopia	2.12	3.05	2.84
Zaire	3.18	3.17	2.98
Bangladesh	2.68	2.41	2.18
Pakistan	3.31	2.67	2.54
Mexico	2.40	2.06	1.55
Brazil	2.23	1.59	1.22
Thailand	1.83	1.27	0.92
Philippines	2.58	2.07	1.70
Vietnam	3.43	3.47	3.24

Region or Group	**1970–1980**	**1980–1993**	**1993–2000**
Low-income economies	2.1	2.0	1.8
Low-income economies (excluding China and India)	2.5	2.5	2.9
Low- and middle-income economies	2.1	1.9	1.7
Sub-Saharan Africa	2.7	2.9	2.9
Middle East and N. Africa	2.9	3.0	2.5
East Asia and Pacific	1.9	1.5	1.2
South Asia	2.3	2.1	2.2
Europe and Central Asia	1.0	0.8	0.4
Latin America and Caribbean	2.4	2.0	1.6
Severely indebted countries	2.1	1.9	1.7
High-income economies	1.8	1.7	1.5

Source: *World Resources 1994–95*, pp. 268–269; and *World Development Report 1995*, pp. 210–211.

TABLE 7.4	Average Rates of Global Natural Increase for Selected Time Periods

Period	**Global Rate of Natural Increase (%)**
1965–1980	1.95
1980–1985	1.76
1990–1995	1.68
1993–2000	1.50[*]
2000–2015	1.30[*]
2015–2030	1.15[*]
.	.
.	.
.	.
2090–2110 (zero population growth)	approx. .05[*]

Source: *World Resources 1994–95*, p. 268; *World Development Report 1990*, pp. 228–229; *World Development Report 1995*, pp. 210–211.
[*] Forecast.

Life expectancy *describes the average number of years a person in a given region could expect to live once born, based on current trends.* Life expectancy statistics for se-

TABLE 7.5	Life Expectancies at Birth for Selected Regions and Nations (1993)

Region or Nation	Life Expectancy (LE) or Average LE in Years (in 1993)
The World (average)	**66**
SOUTH	62
Somalia	46
Ethiopia	48
Zambia	48
Nigeria	51
Sub-Saharan Africa	52
Zimbabwe	53
Bangladesh	56
Kenya	58
South Asia	60
Bolivia	60
India	61
Pakistan	62
South Africa	63
Indonesia	63
Russia	65
Middle East and N. Africa	66
Brazil	67
Nicaragua	67
El Salvador	67
Philippines	67
East Asia and Pacific Islands	68
Latin America and Caribbean	69
China	69
Saudi Arabia	70
Malaysia	71
Mexico	71
Chile	74
Developed World (NORTH)	74
N. America	77
Europe	75
United States	76
Canada	78
Sweden	78
Switzerland	78
Australia	78
Italy	78
Japan	80

Source: *World Development Report 1995*, pp. 162–163.

lected geographic regions are given in Table 7.5 and Figure 7.3. Today, life expectancies range from the upper 40s to the low 80s depending on geographic location. Note that there is a high positive correlation between a region's life expectancy and its economic development status (Figure 7.3).

Demographers also monitor a region's "total fertility rate," which can be used to predict trends in the crude birth rate. The **total fertility rate** (TFR) is defined as *the average number of children that a woman is expected to give birth to throughout her "childbearing years."* For statistical purposes, a woman's childbearing years are thought of as ranging from age 15 through 45. Total fertility rates currently range from approximately 1.3 to 6.2 (as shown in Table 7.6). The highest TFRs are observed in sub-Saharan Africa, West Asia, and the Middle East. Note that there is an inverse relationship between TFR and a region's economic development status. (Figure 7.4)

Total fertility rates can be a misleading indicator of population growth. Note that many infants born in a given region may never live to see the end of their first year of life. As such, it is useful to observe a region's "infant mortality rate." The **infant mortality rate** (IMR) is defined as *the number of infant deaths in a given year, per 1000 live births.* Infant mortality rates for selected regions are shown in Table 7.7.

Since the mid-1960s, infant mortality rates have declined dramatically worldwide. Note, however, that IMRs in the SOUTH remain high in comparison to the IMRs in the NORTH. Note also that IMR statistics are highly consistent with trends in observed life expectancies and fertility. The higher the life expectancy, the lower the IMR. High IMRs may also explain high fertility rates in peripheral regions. Note that it is not uncommon to observe high rates of fertility in regions attempting to expand the size of their military or labor force, or in regions with a disproportionately large share of "elderly" (age 65 and over) population. It is generally thought that a large and youthful labor force will ensure that the needs of an aging and elderly population will be met.

High fertility rates may also be viewed as a natural response to adverse reproductive conditions. Women in peripheral regions have a relatively large number of children in order to hedge against the high rates of infant and childhood mortality. Total fertility rates are also high to counteract the effects of inadequate health and medical care. Women throughout the SOUTH continue to face a real danger of dying during pregnancy or childbirth when medical complications arise. Subsequently, the "maternal mortality rate" is much higher in the SOUTH than in the NORTH. The **maternal mortality rate** (MMR) is defined as *the number of deaths from pregnancy or childbirth-related causes per 100,000 live births (over a given time period).* A comparison of MMRs for selected nations is given in Table 7.8.

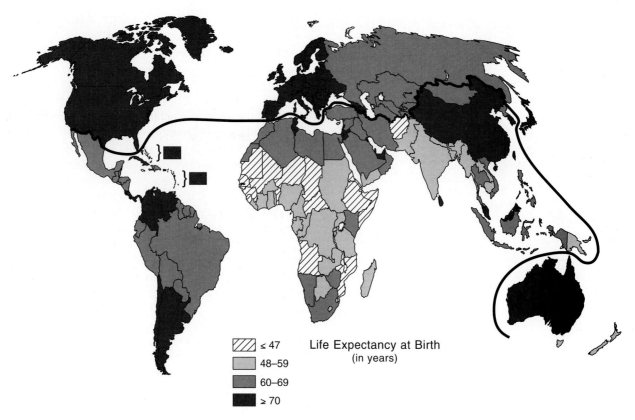

Figure 7.3 Life expectancy at birth (in years).

POPULATION PYRAMIDS

Total fertility and mortality rates affect the structure of a region's "demographic profile." A **demographic profile** *highlights the general demographic characteristics of a region as defined in terms of population size, age, gender, and life expectancy.* One commonly used visual representation of a region's demographic profile is the "population pyramid." A **population pyramid** *is a graphical representation (usually a bar graph) of a region's population age-composition and structure, highlighting the proportion of males and females by age group.* It provides a visual summary of a region's recent demographic history and reveals much information about future population growth and development status.

Population pyramids illustrate the relative distribution of population by age, sex (gender), and **cohort**—*a group of people that share a common demographic experience as determined by their age.* More specifically, population pyramids show the breakdown of a region's population by age cohorts, denoting the number or percentage of males or females in age ranges 0–4, 5–9, 10–14, 15–19, 20–24, 25–29, etc. (see Figure 7.5).

A region's population growth rate is directly proportional to the size of the **base** of a population pyramid—*the bottom segment of a population pyramid, defining the extent of a region's population that is associated with the*

first several age cohorts (typically 0–4 and 5–9). The rate at which a base may expand is related to the following components: crude birth rate, total fertility rate, infant mortality rate, and the number of women in childbearing years (ages 15–45). The top of a population pyramid is typically narrow, tapering off quickly and coming to an abrupt end roughly 10 to 15 years beyond the average life expectancy in that region (given that some people will exceed the average life expectancy).

Five types of population pyramids have been identified, each with their own unique features and implications. Let us discuss each in turn.

1. Rapid-Growth Population Pyramid

The **rapid-growth (RG) population pyramid** is a cone-shaped pyramid. It possesses a wide and expanding base (over time), as illustrated by Figure 7.5. The shape and features of the RG pyramid are an indication of a population that is rapidly growing as a result of high fertility rates. The logic behind this is simple. As the base of the population pyramid widens through time, more and more people (and women) filter through the childbearing years. Hence, more and more children are likely to be born over time if fertility rates are high. In general, the wider the base of the RG pyramid, the more explosive is the expected population growth (Figure 7.5).

TABLE 7.6	Total Fertility Rates for Selected Regions and Years

Region or Group	Total Fertility Rates			
	1965	1988	1993	2000*
Low-income economies	6.3	4.0	3.6	3.3
Low-income economies (excluding China and India)	6.4	5.6	5.5	4.9
Sub-Saharan Africa	6.6	6.7	6.2	5.6
Ethiopia	5.8	7.5	6.9	6.2
Nigeria	6.9	6.6	6.4	5.7
Zambia	6.6	6.7	5.9	5.2
Middle East and N. Africa	6.9	5.3	4.7	4.1
Egypt	6.8	4.5	3.8	3.7
Saudi Arabia	7.3	7.1	6.3	5.7
South Asia	6.3	4.5	4.0	3.6
East Asia and Pacific	6.2	2.7	2.3	2.2
Bangladesh	6.8	6.3	4.3	4.0
Pakistan	7.0	6.6	7.0	6.1
Malaysia	6.3	3.7	3.5	3.0
Indonesia	5.5	3.4	2.8	2.6
Thailand	6.3	2.5	2.1	2.1
India	6.2	4.2	3.7	3.2
China	6.4	2.4	2.0	1.9
Middle-income economies	5.6	3.8	3.0	2.7
Argentina	3.1	2.9	2.7	2.5
Brazil	5.6	3.4	2.8	2.5
Mexico	6.7	3.5	3.1	2.6
Russia	2.7	2.3	1.4	1.7
Chile	4.8	2.7	2.5	2.4
High-income economies	2.8	1.8	1.7	1.8
United Kingdom	2.9	1.8	1.8	1.8
Sweden	2.4	2.0	2.1	2.1
United States	2.9	1.9	2.0	2.1
Canada	3.1	1.7	1.9	2.0
Japan	2.0	1.7	1.5	1.4
Austria	2.7	1.5	1.5	1.6
Italy	2.7	1.3	1.3	1.3
Germany	2.5	1.7	1.3	1.3

Source: *World Development Report 1990,* pp. 230–231; and *World Development Report 1995,* pp. 212–213.

* Forecast.

Figure 7.4 A woman surrounded by many children. Total fertility rates remain high in many developing regions as a matter of convenience, preference, and pride.

7.4: ©Jim Whitmer/Stock Boston

RG population pyramids are most commonly observed in low-income or peripheral economies. Hence, the vast majority of nations in the SOUTH have population pyramids of the RG variety. Currently, the RG pyramid is most prevalent in sub-Saharan Africa, but is also readily observed in the less-developed nations of Latin America, Asia, and the Middle East. The list of nations with RG population pyramids includes India, Bangladesh, Indonesia, Afghanistan, Pakistan, Saudi Arabia, Iran, Iraq, Yemen, Oman, Bolivia, Paraguay, Guatemala, Honduras, Nicaragua, El Salvador, Mexico, Nigeria, Zaire, Zimbabwe, Tanzania, Ethiopia, Uganda, Kenya, Sudan, and Somalia, among others. In general, the less economically developed a region is, the greater the likelihood that it has an RG population pyramid.

2. Declining-Growth Population Pyramid

The **declining-growth (DG) population pyramid** is best described as an indented cone-shaped structure whose features include a narrow or narrowing base and a slightly bulging lower-middle section. The DG population pyramid is illustrated in Figure 7.6. Note that the shape of this pyramid suggests that a region's population growth is showing signs of declining over time due to dramatic reductions in fertility. An eventual stabilization of the region's population is possible if the size of the base stabilizes over several generations (Figure 7.6).

TABLE 7.7	Infant Mortality Rates for Selected Regions and Years

	Infant Mortality Rates	
Region or Group	1970	1993
Low-income economies	108	64
Low-income economies (excluding China and India)	135	89
Sub-Saharan Africa	132	92
Ethiopia	159	117
Nigeria	114	83
Middle East and N. Africa	136	52
Egypt	158	64
Saudi Arabia	119	28
South Asia	138	84
East Asia and Pacific	77	36
Bangladesh	140	106
Pakistan	142	88
Indonesia	118	56
Thailand	73	35
Malaysia	45	13
India	137	80
China	69	30
Middle-income economies	74	39
Argentina	52	24
Brazil	95	57
Mexico	72	35
Russia	29	21
Chile	77	16
High-income economies	19	7
United Kingdom	18	7
United States	20	9
Italy	29	8
Canada	18	7
Sweden	11	5
Japan	14	4
Germany	22	6

Source: *World Development Report 1995, pp.* 214–215.

TABLE 7.8	Maternal Mortality Rates for Selected Nations

Nation	Maternal Mortality Rate (1980–1990)
Kenya	170
Nigeria	800
Sudan	550
South Africa	83
Zaire	800
India	460
China	95
Malaysia	59
Indonesia	450
Mexico	110
Chile	67
Uruguay	36
South Korea	26
Japan	11
Singapore	10
United States	8
United Kingdom	8
Canada	5
Sweden	5
Germany	5
Italy	4
Israel	3
Australia	3

Source: World Resources 1994–95, pp. 272–273.

This type of pyramid is associated with regions undergoing marked declines in the crude birth and total fertility rates. Examples of the DG pyramid include China (from about 1970 to the mid-1990s) and the United States (from the early 1970s to the mid-1980s). Recently, the base of China's population pyramid is showing signs of a modest expansion, though it is not expected that the base will widen beyond its middle-cohort bulge.

Although the base of the DG pyramid is narrow or narrowing, it does not mean that total population has stabilized. It simply means that the region's population growth rate is in the process of declining, suggesting stabilization of population in approximately 50 to 60 years (should the base remain narrow and relatively constant over time). Before stabilization occurs, however, total population will continue to grow at a fairly rapid pace as the population associated with the "bulging middle" filter

their way up and past the childbearing years. The population growth rate for a region with a DG population pyramid will be substantially lower than the population growth rate in regions with the RG population pyramid given the fact that the base of the DG pyramid is narrow relative to its midsection.

3. Slow-Growth Population Pyramid

The **slow-growth (SG) population pyramid** may be represented as a slim conic- or beehive-shaped pyramid. It possesses a fairly narrow and slow-expanding base when compared to the RG pyramid (compare Figures 7.5 and 7.7). The structure of the SG population pyramid suggests that a region is undergoing only slow or modest population growth. The SG population pyramid has a narrow base, which gradually tapers off as one moves toward the top of the pyramid (Figure 7.7).

Generally associated with regions sustaining modest fertility booms, SG population pyramids are commonly observed in high-income and high-middle-income economies. Historically, these types of population pyramids have been observed in many European nations (including Denmark during the 1970s, Spain and Finland in

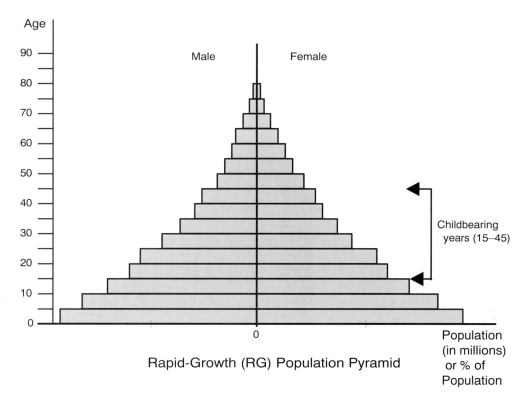

Figure 7.5 The rapid-growth population pyramid.

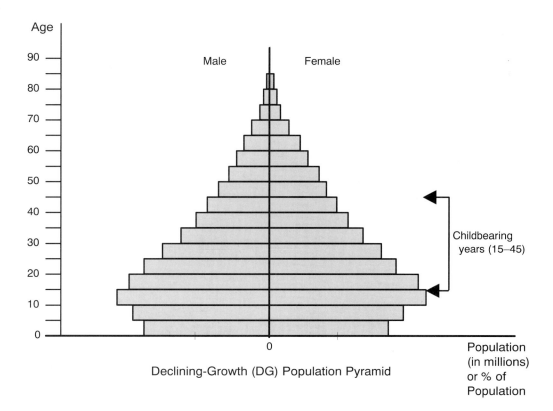

Figure 7.6 The declining-growth population pyramid.

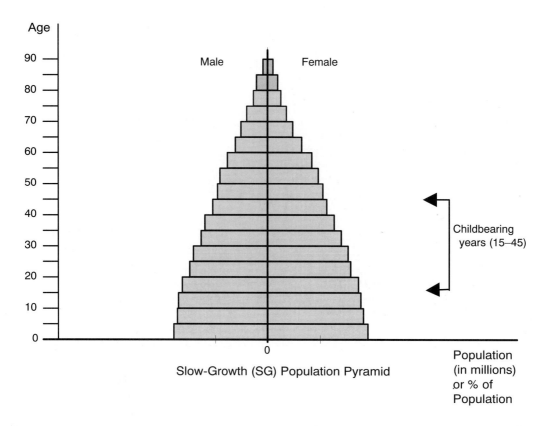

Figure 7.7 The slow-growth population pyramid.

the 1980s, Austria and Germany in the late 1990s), as well as in the more prominent Latin American nations over the last several decades (for example, Brazil and Chile). The SG population pyramid has also been observed in the United States during the 1960s, in association with the so-called "baby boom" and the marked rise in total fertility rates shortly after World War II.

4. Zero-Growth Population Pyramid

The **zero-growth (ZG) population pyramid** is best described as a cylindrical or tube-shaped pyramid. Its structure and shape suggest that population growth has stabilized or that a "zero population growth" (ZPG) status has been achieved (see Figure 7.8). Note that the base and the middle of the pyramid are of approximately the same width, implying that people filtering up through the childbearing years are simply replicating themselves. Generally associated with nations of the highly industrialized world, the ZG population pyramid represents a theoretical profile that ultimately satisfies population management objectives of most regions. Should regional and global population growth trends continue, it is likely that all nations of the world could achieve this profile around the year 2100. Many regions are already in the process of attaining this goal. Examples of ZG population pyramids include the United States by the year 2045, and many of

the more prosperous European nations (for example, Sweden, Denmark, Germany, Austria, and Italy) within the next few years (Figure 7.8).

It is important to note that the ZG population pyramid signifies that the region in question is proceeding at or near its so-called "replacement rate." The **replacement rate** (RR) is defined as *a total fertility rate of 2.1*. Achieving this rate implies that 2.1 children will be born to a typical male and female partnership (on average) during the course of their lifetime. It is important to note that the replacement rate is above 2.0, as a typical male and female partnership (through reproduction) must compensate for losses associated with cohort-specific mortality rates up until the end of the childbearing years. In other words, they must reproduce more than themselves to hedge against cohort attrition due to infant and childhood mortality. In short, the ".1" represents compensation for the number of children and young adults who will not live to see the age of 45.

5. Negative-Growth Population Pyramid

The **negative-growth (NG) population pyramid** is a "mummy-shaped" pyramid that boasts a narrow base, and a bulging upper-middle area (see Figure 7.9). These features suggest that a region is experiencing "negative" population growth. This type of pyramid has been

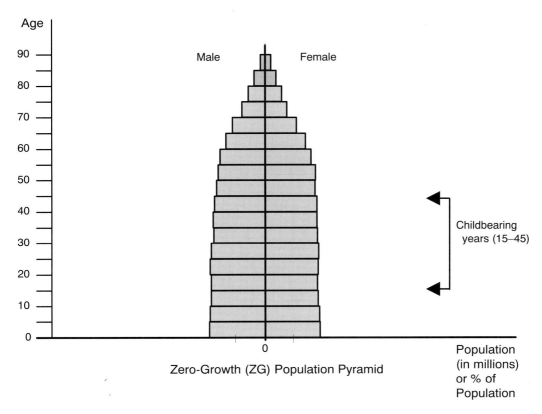

Figure 7.8 The zero-growth population pyramid.

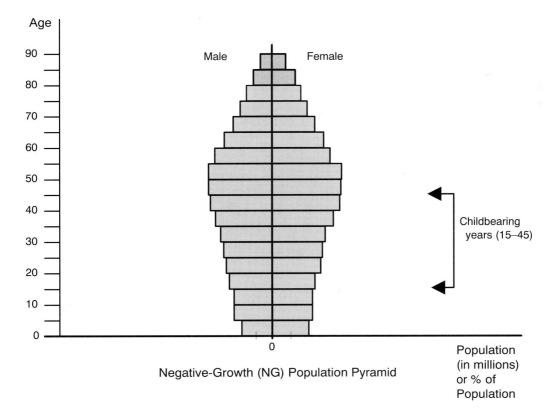

Figure 7.9 The negative-growth population pyramid.

observed in developed regions where the total fertility rate falls below the replacement rate (that is, falls below 2.1) for an extended period of time. NG population pyramids have been observed in industrialized nations that have undergone a sustained decline in fertility. Note that the NG pyramid is also likely in regions where the impact of positive net migration does not compensate for losses associated with low fertility rates. Population pyramids of this variety have been observed in Germany and Austria (1970s through early 1990s), as well as in Belgium and Hungary. (Figure 7.9)

Note that fertility rates less than the replacement level of 2.1 do not necessarily translate into negative population growth rates. For example, the U.S. population continued to grow from throughout the 1970s, 1980s, and 1990s despite the fact the total fertility rate dropped below the replacement rate from 1972 to the early 1990s. The reason for this is fairly simple: the United States experienced the "baby boom" (1946–1967) just prior to the fertility rate dropping below the replacement level. During the height of the baby boom, fertility rates exceeded 3.0 (with a peak fertility rate of 3.7 in 1957). The baby boom was responsible for producing a fairly wide base for the U.S. population pyramid (see Figure 7.10a). The wide base later turned into a mid-pyramid bulge during the 1970s. Although the base of the U.S. population pyramid began to shrink as fertility rates fell below 2.1 (see Figure 7.10b), population growth continued at a steady pace as a result of the large portion of the population in childbearing years (Figures 7.10a, b).

The post-baby boom decline in fertility was attributed to eight major factors:

1. the widespread use of contraceptives;
2. the rise of legalized abortions;
3. changing social attitudes, which favored smaller family sizes;
4. increasing costs of raising a family (especially during the high inflationary period of the mid-1970s);
5. the general rise in the age at which young adults married, in comparison to the 1950s;
6. more women working outside of the home (as part of the trend toward dual-income families);
7. greater social acceptance of "childless couples" and alternative family structures (including the emergence of the dual-income/no-children households); and
8. delayed reproduction (related to individuals' preferences to marry later in life as they pursued careers, and women postponing their first pregnancy until or after age 30—a factor that physically restricted many women from having numerous children (Miller, 1995, p. 133).

Although the "baby boomers" continue to filter through the childbearing-age cohorts, fertility rates have remained relatively low—averaging approximately 1.9 from 1972 to 1997. As a result, the base of the U.S. population pyramid has stabilized (despite the post-1970s population increases supported by positive net migration). Sociologists have labeled the 1968 to the early 1990s as the "baby bust," a period characterized by declining fertility rates due to changing attitudes, behavior, and socioeconomic conditions.

Note that the average total fertility rate in the United States from 1990 to 1995 rebounded to the replacement level (see Table 7.6). The U.S. population pyramid is expected to continue its transition toward a ZG population pyramid as the last of the baby boomers pass beyond the age-bearing cohorts (see Figure 7.10c). This will undoubtedly bring about stability in the **cohort-dependency ratio** (*the number of persons 65 or older per 100 persons ages 18 to 64*) by the mid-21st century. Cohort-dependency ratios for the United States for selected years are shown in Table 7.9. Note that recent demographic trends in the United States are likely to bring about many social and economic changes as the aging baby boomers use their political influence to place higher tax burdens on the baby-bust generation to supplement increasing costs of health care and a social security system that is now in jeopardy (Figure 7.10c).

THE GRAYING OF JAPAN

Aging populations and increasing cohort-dependency ratios are common to the more mature industrialized economies of the NORTH. Forecasts for Japan show that demographic trends will lead to negative population growth sometime in the middle of the next century (a legacy of declining fertility). Liberal population containment policies and laws such as legalized abortion and government-sponsored family planning allowed Japan to cut its fertility rates and contain its population growth from 1949 to 1956. Fertility rates continued their decline right into the 1990s (where in 1993, Japan's total fertility rate dropped to a record low of 1.5). With further declines in total fertility expected, Japan's 1998 population of approximately 130 million is likely to increase by only a few million people by the year 2025 (Figure 7.11).

As less people filter up through the childbearing years, Japan's population is likely to shrink well below the 96 million mark shortly before the end of the next century should the total fertility rate remain well below the replacement rate. As the average life expectancy at birth in Japan is the highest in the world at 80 years of age, the "graying of Japan" will pose increasing pressures on its health and pension systems and shrinking labor force. The money and taxes needed to support Japan's growing elderly population will not only place an enormous financial burden on their economic system, but will discourage future economic growth as the propensity to

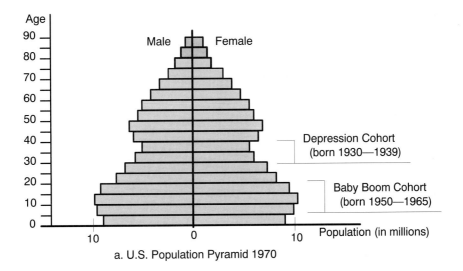

a. U.S. Population Pyramid 1970

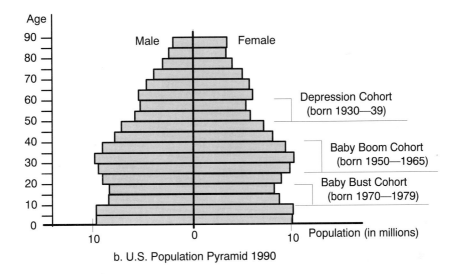

b. U.S. Population Pyramid 1990

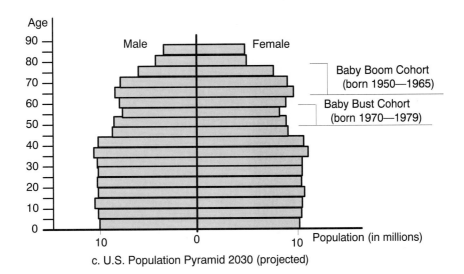

c. U.S. Population Pyramid 2030 (projected)

Figure 7.10 U.S. population pyramids for selected years.

Source: Population Reference Bureau and U.S. Census Bureau.

invest and save will decrease over time, as private and public funds are directed toward social programs.

WORLD LEADERS IN POPULATION GROWTH

The largest population increases are expected in places that have rapid-growth RG population pyramids. The rate of population increase is directly proportional to the size of the base of the pyramid. Expansion potential is high for those nations with the widest or fastest growing bases. The top 20 contributors to world population growth over three decades are listed in Table 7.10. Note that these nations share a common locational attribute: all are located in the SOUTH. Despite this trend, some analysts argue that the seemingly minimal population growth in the NORTH poses a much greater threat to global development. This argument is based on the fact that the average person born in the NORTH is likely to consume a greater amount of world resources than the average person born in the SOUTH.

Of the top-twenty contributors to world population, the nation of India will add more people to the planet than China over the next three decades. The base of India's population pyramid is wide and rapidly expanding, while the base of China's population pyramid has recently stabilized. These differences are, in many ways, indications of variability in the success of population containment policies in these nations.

In light of its impending population growth, India now faces its greatest development challenge. But challenges are nothing new for a region that contains one-sixth of the world's population in an area that is roughly one-third the size of the United States. Democratic rule in India has been slow to bring about positive social and economic changes in a heterogeneous nation that boasts more

TABLE 7.9	Cohort-Dependency Ratios for the United States

Year	Cohort-Dependency Ratio
1960	16.8
1970	17.6
1980	18.6
1985	19.3
1990	19.5
2000*	21.1
2020*	28.7
2030*	37.0
2040*	37.9
2050*	38.0

Source: U.S. Bureau of the Census.
* Forecast.

Figure 7.11 A rapidly aging population in Japan may bring about impending labor shortages and financial difficulties in forthcoming years. Here an old couple enjoys an animated discussion at a temple in Kyoto.

7.ll: ©Sean Sprague/Stock Boston

TABLE 7.10	Top 20 Contributors to World Population, 1990–2025

	Nation	Projected Increase in Millions (1990–2025)
1	India	592.2
2	China	357.1
3	Nigeria	188.3
4	Pakistan	144.4
5	Bangladesh	119.4
6	Brazil	95.4
7	Indonesia	82.7
8	Ethiopia	65.6
9	Iran	65.5
10	Zaire	63.5
11	Mexico	61.5
12	Tanzania	57.5
13	Kenya	52.5
14	Vietnam	50.8
15	Philippines	49.0
16	Egypt	39.9
17	Uganda	36.8
18	Sudan	34.4
19	Turkey	34.0
20	South Africa	28.0

Source: Bergman and McKnight (1993), p. 176.

than twenty political parties and over a thousand different languages and dialects. After fifty years of independence from Great Britain, India still struggles to find solutions to its internal problems while resolving territorial disputes with China and administering to an ongoing conflict with neighboring Pakistan. Although classified as semiperipheral, the benefits of India's ongoing economic reforms have been slow to diffuse across the landscape. While India continues its transition from a state-planned to a free-market economy, the divisions between the old and the new are becoming increasingly obvious (Figure 7.12).

The ill effects of overpopulation and poverty continue to bring misery to the multitudes residing in the highly congested and increasingly polluted inland cities like Delhi, while economic development along the coast continues to attract foreign investors to a new and improved India. The modern skyline of the thriving city of Mumbai (formerly Bombay), one of Asia's busiest and most prominent ports, stands in stark contrast to the blight and decay of the overcrowded streets of the impoverished Calcutta—a one-time Marxian stronghold. As India's population grows by more than 2% per year, it must be resourceful in its attempt to expand both agricultural productivity and urban employment opportunities.

India continues to add an additional 18 million people to the planet each year despite the fact that it was the first nation in the world to institute a national family-planning program (back in 1951). Coerced sterilization of tens of millions of males has done little to stop India's population tidal wave. Population growth in India now threatens to negate the benefits of 50 years of economic progress.

China's success in combating explosive population growth is largely due to economic and social reforms. These reforms helped to dramatically lower fertility rates during the 1970s and 1980s. In contrast to the efforts put forth by China's political leadership during the Great Leap Forward (during the late 1950s), which encouraged population growth to expand China's labor force, recent changes have led to the support of policies which promote population growth reduction as a means to economic development. Industrialization policies were implemented to foster an atmosphere of "self-reliance" in all economic sectors. Economic reforms created a wealth of investment and employment opportunities in China's coastal cities and provinces, where export-led growth was nothing less than explosive throughout the 1970s and 1980s. Economic change has given individuals a chance to secure a better standard of living, thereby providing an even greater impetus for holding population growth in check.

Social reforms have largely focused on population containment and fertility reduction. The late-1960s theme of "later, sparser, and fewer" (in reference to advocating marriage at a later age, less children, and at least a five-year waiting period between births) was later replaced with a more rigid standard.

The precipitous drop in China's fertility rate during the 1970s was attributable to three factors:

1. a nationwide campaign to encourage people to marry later in life (men at the age of 25 or later, and women at the age of 30 or later);
2. the introduction of contraceptives, which were widely made available to all adults; and
3. the enactment of the controversial "one-child" policy, including a wide array of economic incentives that potentially benefitted couples who conformed to the state-imposed one child per family goal. Families complying with the one-child policy where given priorities in the allocation of scarce housing, food, and health care, as well as preferential treatment in the workplace (for example, job promotions and career advancement).

Figure 7.12 Two cities in India: the modern city of Mumbai—formerly Bombay—and the old and deterioratiiong city of Calcutta. Economic reform and development in India has brought prosperity to cities like Mumbai, yet poverty and despair prevail in depressed urban environments like Calcutta.

7.12a: ©Jeff Greenberg/Photo Edit; 7.12b: ©Anna E. Zuckerman/Photo Edit

Conforming households would also be given increased access to educational opportunities, social programs, and community services (Figure 7.13).

The one-child policy was not without shortcomings. Despite its success in terms of lowering fertility rates, China's one-child policy created many disturbing social problems including (a) the sharp rise in abortions during the 1980s; (b) the unnecessary killing of infants born female (given the cultural preference of males in Chinese society, as males are viewed as a more productive source of labor for industry and best able to provide for aging parents); and (c) an age-cohort manipulation that has increased the likelihood of a forthcoming labor shortage as China's population ages.

Note that the one-child policy and the male gender preference have combined to produce yet another problem. A shortage of females is likely to disrupt the reproductive mechanisms of future generations. The impending shortages of females will allow only about one-in-three adult males to reproduce. This is likely to produce further slowdowns in China's population growth over the next 20 to 30 years.

Note that China's one-child policy carries economic disincentives for those who were unwilling to conform to the rule. These came mostly in the form of non-preferential treatment in the workplace and punishment by shame. Enforcement of China's population containment policies, however, are noticeably different between regions and classes. In urban areas, for example, multiple births per household were strictly discouraged, whereas in rural areas a more relaxed implementation of the rule was adopted in accordance with local population density conditions. In many cases, couples in rural areas were allowed to have a second child to combat the problems of gender preferences. Moreover, those most able to afford a larger family were able to do so without penalty. Families with more than two children can be observed, yet this is typically a privilege reserved for the more affluent members of Chinese society.

DEMOGRAPHIC TRANSITION

Many demographers contend that decreasing population growth rates are, in part, due to economic development. As economies transform or become functional in the global production network, individual members of society in affected regions begin to realize the economic benefits of population containment. Gradually, this realization would lead to a smaller average family size. Historical evidence suggests that economic development and modernization are accompanied by parallel declines in the total fertility rate and the rate of natural increase. Experts are convinced that economic transformation in the SOUTH will lead to sweeping demographic changes. In

Figure 7.13 A one-child family in China.
7.13: ©Owen Franken/Stock Boston

short, they see the nations of the SOUTH as proceeding through a "demographic transition."

The characteristics and dynamics of regional population growth are said to be linked to a region's economic development status. It is posited that regional economic transformation will lead to certain sequential changes in the demographic profile of a region as its development status improves. The accompanying shifts in crude birth and death rates are thought to follow a logical transition or sequence. We may define the **demographic transition** as *a four-stage process of demographic change that is expected to occur as regions develop and modernize.* Note that the world economy is also thought of as proceeding through a demographic transition (as will be discussed shortly).

The four stages of the demographic transition are illustrated in Figure 7.14. Let us now discuss each of these stages in turn.

Stage 1: The High-Stationary Stage
The high-stationary stage is characterized by high crude birth rates and high and fluctuating crude death rates. In

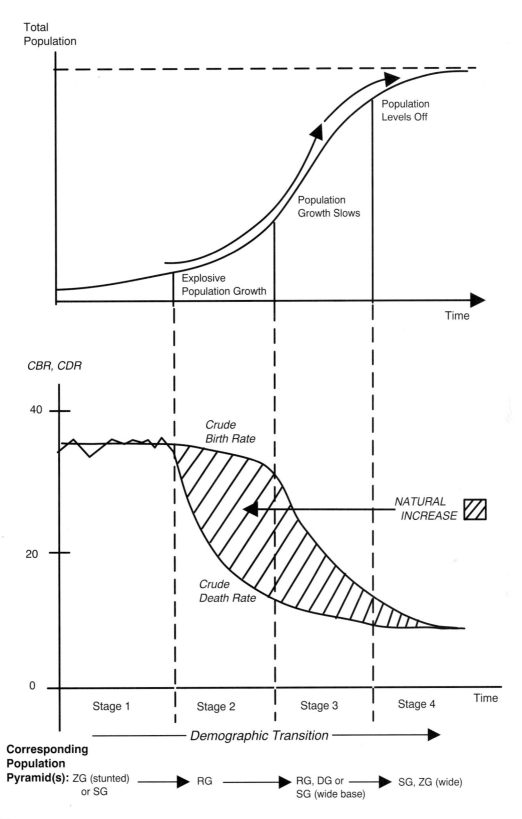

Figure 7.14 Demographic transition.

this stage, the high rates of mortality are attributable to a general lack of access to modern medical technology. Crude death rates fluctuate wildly as disease, epidemics, hunger, malnutrition, famine, or other crisis situations occur sporadically on the landscape, claiming large numbers of lives over short periods of time. Shortages of food and medicine are common and there is an increased susceptibility to natural hazards. In general, the high-stationary stage is characteristic of extremely peripheral economies.

Most of humanity prior to the late 1700s would be characterized as being in stage one of the demographic transition. Yet even today, we find examples of regions in the high-stationary stage. This is particularly true in remote areas of the world, where crude birth rates remain to offset high infant and maternal mortality rates. Stage one is also characterized by a preference toward large families to support labor-intensive food gathering and production efforts. Note, however, that high crude birth and death rates and short life expectancies keep overall population and population density low and stable. Historically speaking, preindustrial Europe is a good example of a region in stage-one of its demographic transition. The "hunting and gathering societies" of the tropical rainforest areas of the Amazon and Congo basins are modern-day examples of regions that are still in stage one.

Stage 2: The Early-Expanding Stage

The early-expanding stage is characterized by rapidly declining crude death rates and high and stable crude birth rates. The decline in mortality is attributed to several factors, including increases in permanent food supply, intensification of agricultural production, the diffusion and use of technology, availability of food and water, improvements in sanitation systems, and increased access to medicine to control the spread of disease.

Crude birth and total fertility rates, nonetheless, remain high throughout this stage. As a result, a stage-two region's population begins to undergo explosive growth. The early-expanding stage is consistent with the presence of a cone-shaped RG population pyramid (that is, a pyramid with a wide and expanding base). Thus, stage two describes a scenario that is common to most of the underdeveloped world. Historically, this stage occurred in Europe and North America just prior to and up until the height of the industrial revolution. Stage two was observed in China from 1940 through the late 1970s. Today, the majority of the nations of the SOUTH would be categorized as falling into stage two of the demographic transition. This would include most of the highly populated and rapidly growing nations in Asia, sub-Saharan Africa, the Middle East, Latin America, and Caribbean. Stage two is typically associated with peripheral and semi-peripheral economies in which the rates of natural increase range from roughly 2.1% to 3.5% (or higher).

Stage 3: The Late-Expanding Stage

The late-expanding stage is characterized by a precipitous decline in crude birth rates and the stabilization or leveling out of the crude death rate. The latter is due to diminishing marginal returns in the use of medical technology. The drop in the total fertility and crude birth rates are usually associated with increased access to health care, education, and birth control. As infant and child mortality rates decline, so does the necessity to have many children as a hedge against mortality. As development and modernization bring about urbanization and the diffusion of technology, large families (while an economic asset in a labor-intensive agricultural system) become a liability in an urban-industrial society, where individuals' aspirations take on greater economic dimensions. Reductions in family size are achieved through a growing awareness of family-planning alternatives and social responsibility. The economic incentives associated with smaller family size increase the likelihood of voluntary compliance.

Note that in stage three of the demographic transition, crude birth rates remain above crude death rates. Subsequently, the late-expanding stage is associated with a population pyramid whose base continues to enlarge over time. Note, however, that in the latter portions of stage three, the expansion of the base slows dramatically. By the end of stage three, the base of the population pyramid will show signs of shrinkage or stabilization. Although total population continues to increase throughout stage three, it grows at a declining rate. Thus, stage three is best represented by the SG and DG population pyramids.

Examples of stage-three nations include China throughout the 1980s and 1990s, the industrialized nations of the NORTH from early 1900s to the present, and just recently many of the more prominent eastern European, Latin American, and Southeast Asian nations, regions which have exhibited sharp declines in their population growth rates. Although the nation of India would be considered as a stage-two region, it is rapidly moving into stage three. By all indications, the world economy as a whole is currently in the early to middle portion of stage three. It is expected to remain in stage three for roughly 40 to 50 years before proceeding to stage four.

Stage 4: The Low-Stationary Stage

The low-stationary stage is characterized by low crude birth rates and low crude death rates. In this last stage of the demographic transition, crude birth rates continue to decline and approach a stationary crude death rate. As a result, the region's population begins to approach a stable size and economic growth translates into greater economic benefits per capita. Note that only a small proportion of the world's population resides in regions which have entered into this stage. Typically, it is a stage associated with regions of high economic development status—mature and highly integrated economies.

The low-stationary stage is best represented by the ZG population pyramid, should the rate of natural increase fall toward zero (that is, if the average crude birth rate equals the average crude death rate over time). Note, however, that the base of the population pyramid associated with regions in stage four can either be stable or it can shrink through time. Should total fertility rates fall below the replacement level of 2.1 and crude birth rates fall below the crude death rate, the low-stationary stage would be best represented by a negative-growth population pyramid (a pyramid with a shrinking base).

Examples of regions that have entered into the early stage-four status include the high-income, core economies of the United States, Canada, Japan, and Europe. Given the rise of China to economic superpower status, it is not surprising to see that China now stands ready to enter into stage four of its demographic transition.

A CRITIQUE OF DEMOGRAPHIC TRANSITION

The demographic transition model has been criticized on various fronts. While it represents a generalization of the experience of the NORTH when viewed from an historical perspective, many feel that it is not an appropriate model to describe demographic change in regions of the SOUTH. Critics argue that the history of the industrialized NORTH offers a less-than-perfect representation of the complex economic transformations of today's developing regions. They contend that the economic preconditions necessary for modernization and demographic transition are different today than they were long ago. Those preconditions also vary markedly from region to region.

It is important to note that degree of indebtedness of nations in the SOUTH is a condition that did not exist in the core economies during their industrialization and modernization. Borrowing to reduce debt burden or financing projects to correct an ever-enlarging array of development problems continues to undermine the modernization process. Reactive development policies have reduced the likelihood of sustained economic growth and demographic transition. Although development loans are aimed at serving the "common interests" of the NORTH and SOUTH, the diversity of interests in the world economy prevents developing nations from mobilizing sufficient capital and political clout to force economic reforms that would increase their leverage in the world economy and allow them to be managers of their own destinies. To accelerate the pace of development in the SOUTH, and the demographic transition of

its nations, efforts must be redirected toward strategies which promote SOUTH-to-SOUTH linkages in ways that are nonthreatening to the financial security of the NORTH (Loxley, 1986).

The underlying cause-and-effect relationships, which drive demographic change, provide another source of criticism of the demographic transition model. Note that it has been recognized that the cause-and-effect sequences inherent to the European experience are quite different from those associated with the developing SOUTH. Those differences are summarized in Table 7.11.

The demographic transition model suffers from another major shortcoming: The impact of cultural differences are not considered. For instance, current total fertility rates in the SOUTH are much higher today than they were in premodern Europe. This difference may be attributed to a host of factors including social and religious beliefs, marriage customs, traditional attitudes toward marriage, preferences on family size, and resistance to modernization. People may eagerly accept innovations that improve their health and increase their longevity, but they may not give up their traditional viewpoints on family, reproduction, or society. The traditions of having large families is likely to be retained in many peripheral regions.

Note that it is highly probable that crude birth rates will remain high throughout the SOUTH to counter recent increases in mortality rates as the medical communities

| **TABLE 7.11** | **Cause and Effect Sequences of Demographic Transition** |

For Europe (historically) and the NORTH:

1. Socioeconomic changes affected mortality rates up until World War I.

2. Socioeconomic changes influenced crude birth rates.

3. Resulting declines, first in mortality and subsequently in fertility, produced the patterns observed in the classic demographic transitions model, with modern medical science as a factor in only the final stages.

For the SOUTH:

1. Prior to World War II, disease-ridden societies displayed the medieval European balance of high crude birth rates checked by high crude death rates.

2. Medical science has influenced mortality rates since World War II and, as a result, the rate of their decline in crude death rates has been much greater than in the European experience.

3. Socioeconomic change and modernization will determine any further falls in crude death rates.

4. Socioeconomic development will influence crude birth rates and induce their decline in areas with least cultural resistance.

5. The demographic transition will be brought about by the interaction of (2), (3), and (4).

Source: After Simpson, 1994, p. 34, with modifications.

combat a resurgence in highly contagious diseases like tuberculosis, hepatitis, and acquired immune deficiency syndrome (AIDS) and various infections that stem from the human immunodeficiency virus (HIV). These and other diseases are already considered to be in epidemic proportions throughout the SOUTH. In addition, the rise of sexually transmittable diseases (STDs) has been a growing concern for women. STDs have been linked to the recent rise in reproductive tract infections and cervical cancer. Slowing the spread of STDs and HIV may prove difficult despite the wide availability of condoms (a proven and effective control when properly used) given that religious, social, and cultural barriers exist. Efforts to combat these diseases have been hampered by the increasing numbers of people engaging in unsafe sexual activities (Figure 7.15).

Promiscuity of men in the male-dominated societies of sub-Saharan Africa and elsewhere continues to pose a development problem. Although culturally accepted, provided that the men are discrete, promiscuity is generally not viewed as acceptable behavior for women. As a result, males and females suspicious of their partner may be insulted should either suggest the use of a condom, implying that one or the other has been unfaithful to the partnership. By suggesting the use of a condom, a woman is viewed as either admitting to having a disease or worried that her partner may have recently contracted a disease. Consequently, women in male-dominated societies continue to be intimidated and stigmatized by their male partners. They opt to run the risk of pregnancy and disease rather than risk the wrath of their partner by suggesting the use of a condom (Sachs, 1994, p. 18).

Promiscuity and unsafe sex has led to an increase in the incidence of children with AIDS. Approximately 10% of all HIV transmissions are now of the mother-to-child variety—the so-called "pediatric AIDS phenomenon." This percentage is much higher in Africa, as fertility rates remain well above the world average. As more infants become infected with HIV, infant and child mortality rates will continue to rise. In addition, the spread of HIV/AIDS among adults is likely to force many unwanted socioeconomic changes. For example, many children will grow up without parents and parental guidance as adults succumb to the terminal effects of this disease. The disruption in the exchange of knowledge, skills, and customs handed down from generation to generation will cause many to lose a valuable part of their cultural heritage. The cost of caring for the sick will be staggering, not to mention the fact that the loss of men and women in their prime working years will translate into enormous losses of productivity and limit the size of many regional labor forces.

Although the spread of HIV/AIDS is higher in urban communities than in rural areas, its impact has left a devastating impression on the social fabric of so many regions throughout the SOUTH. Average life expectancies are projected to drop by as much as 30 years in nations

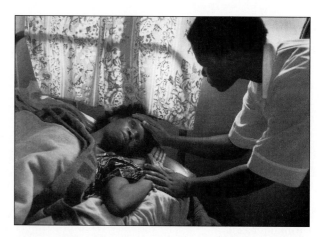

Figure 7.15 The AIDS phenomenon continues to grow to epidemic proportions throughout the SOUTH, leaving a devastating toll on those left behind. Here a nurse tends to an AIDS patient in Masaka, Uganda.
7.15: ©Sean Sprague/Stock Boston

like Zambia, Zimbabwe, Nigeria, and Thailand as AIDS runs its course. Remarkably, China continues to be largely unaffected by the world AIDS crisis. Variability in the demographic, socioeconomic, and cultural characteristics of nations and the impact of AIDS and other diseases does not support the same demographic transition in all regions.

Recent increases in mortality aside, there exists tremendous variability in the nature of demographic transitions for nations in the developed and developing worlds (see Figures 7.16, 7.17, and 7.18). In particular, there exists a large "gap" between crude birth rates and crude death rates for nations in the SOUTH. This gap suggests that many of these regions are still growing exponentially and are in stage two of their demographic transition. With the subtle rise in crude death rates due to the spread of disease and the general decline in crude birth rates, the nations of the SOUTH are expected to move into stage three. The larger the gap, the longer it is likely to take to enter stage three. Nonetheless, the gap between CBR and CDR is likely to close over the next 30 years.

World Health Organization estimates in 1993 placed sub-Saharan Africa as the region hardest hit by HIV/AIDS, with more than 8 million of the 13 million reported cases worldwide. Latin America and Southeast Asia were tied for a distant second place, each with an estimated 1.5 million or more cases (outnumbering the more than 1 million HIV cases estimated in North America in that same year).

Demographers insist that AIDS will take an enormous toll on human life and that its long-term economic impact will be staggering. According to the Center for International Research, U.S. Bureau of the Census, population and economic growth rates will be severely affected in nations such as Central African Republic, Congo, Kenya,

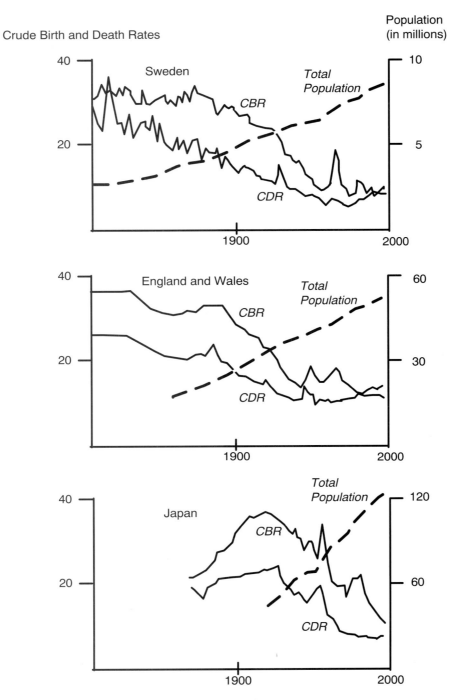

Figure 7.16 Demographic transitions in more-industrialized countries.
Source: Meadows et al., 1992, p. 30.

Rwanda, Uganda, and Zimbabwe. Many sub-Saharan African nations are now expected to exhibit population growth rates between 1% and 1.5% by the year 2010, as compared to once-projected growth rates of 2.5% and higher made prior to the AIDS crisis. The high incidence of AIDS and other communicable diseases has changed the nature of demographic transition in the SOUTH.

As discussed by Teitelbaum (1987), demographic transition theory fails to provide adequate answers to two fundamental questions. First, there is the question of sufficiency. Will the moderate levels of economic growth and development to which the nations of the SOUTH may realistically aspire be "sufficient" to set into motion the forces that will result in a natural fertility decline? Second, there is a question of timing. If declines are observed, will they occur soon enough and/or will the rate be rapid enough to compensate for the high fertility levels of nations in the SOUTH when compared to those of

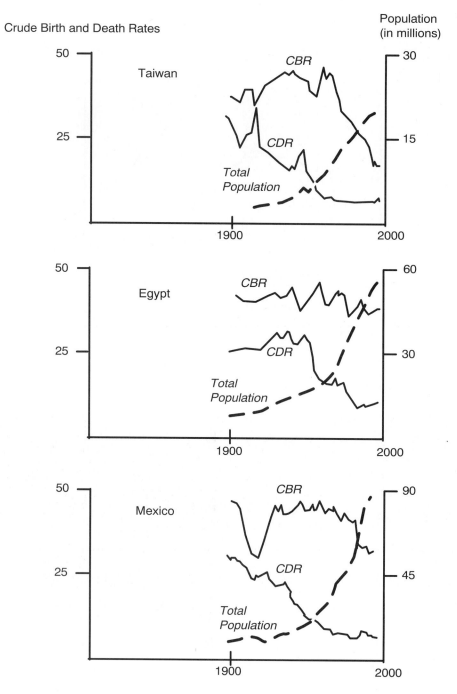

Figure 7.17 Demographic transitions in less-industrialized countries.
Source: Meadows et al., 1992, p. 31.

transitional Europe. In light of the AIDS phenomenon, a third question arises. Will increases in crude death rates be of a magnitude which would cause significant declines in the rates of natural increase as to avert excessive population growth (or overpopulation) prior to nations entering and passing through stage three of their demographic transition? While the long-term impacts of the AIDS epidemic and other growing health risks on world population growth are yet unknown, it is likely that they will introduce a subtle twist to the standard demographic transitions model (as crude death rates rise throughout stage three).

MIGRATION AS A PROCESS

As discernable differences in the economic development status of nations continue, the movement of people in the

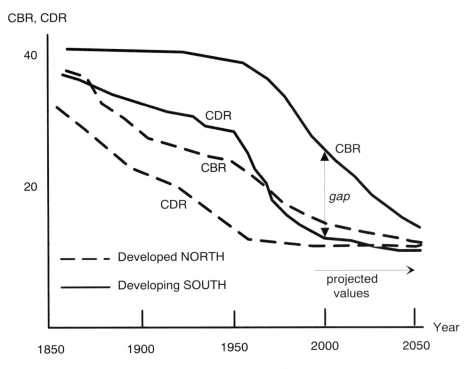

Figure 7.18 Demographic transition: developed NORTH versus underdeveloped SOUTH.

Sources: Based on data collected from W*orld Development Report* (World Bank, 1992); *World Resources 1992-93* (1992); and *World Resources 1994-95* (1994).

world economy (that is, migration) will play an increasingly important role in influencing nations' demographic profiles. In the broadest sense, migration may be defined as the temporary, seasonal, or permanent movement of people to a new location, settlement, region, or nation. For the purposes of this discussion, the term **immigration** refers to *the movement of people between two nations (that is, the movement of people across international borders or boundaries).* Consider the immigration of individuals to the United States from Mexico as an example of a movement across an international boundary. *People engaging in the process of relocating across international borders* are referred to as **immigrants** (Figure 7.19).

This should not be confused with the intraregional movement of people, a process known as "in-migration." Consider, for example, the in-migration of elderly retirees to the state of Florida from Ohio or Pennsylvania. While "in-migrants" relocate from one place or region to another, they remain within the same nation.

Migration may also be thought of as a process by which human beings relocate in response to various "push" and "pull" factors. **Push factors** are related to *forces which increase the likelihood of an outflow of people from a given region,* whereas **pull factors** are related to *forces which increase the likelihood of an inflow of people to a given region.* Push and pull factors affect migration decisions and behavior as human beings respond to the perceived net benefits of moving based upon available information. At the international scale, push and pull

Figure 7.19 Immigration continues to change the face and make-up of nations around the world. Here a group of new immigrants become U.S. citizens.

7.19: ©Michael Newman/Photo Edit

factors are extremely important to the migration process and largely determine whether migration is "forced" or "voluntary." Push factors (at a given location) include adverse or intolerable living conditions arising from a depressed regional economy, civil disorder, war, famine, scarcity of resources, underdevelopment, or a rapidly diminishing quality of life. Pull factors include favorable sociopolitical or economic conditions in relation to the perceived opportunities or stability of an area. In short, push and pull factors may be defined in environmental, political, and/or economic terms.

Note that the likelihood of movement between regions increases as regional economic disparities increase. Migration is, therefore, viewed as a natural reaction to regional economic divergence and uneven development. Migration is the way in which human beings attempt to rationalize location decisions and/or reorganize themselves in space as to maximize their welfare as based upon their perceived net benefits of moving in comparison to staying. The efficiency of migration as a process is constrained by the amount and accuracy of information available on existing conditions and opportunities in other regions and the relative competitiveness of those regions (as destinations) to attract prospective migrants. Migration is a process that is constrained by government and institutional barriers.

Push and pull factors of the political variety are not separate, nor distinguishable in many cases, from economic push and pull factors. Consider the eagerness of many East Germans to flee their once Soviet-dominated state for the safe haven of West Germany prior to the downing of the Berlin Wall and the end of the Cold War. It was the pull of democracy, freedom, and economic opportunity that forced many to be shot, captured, or persecuted while attempting to escape the sterile and unforgiving grip

of command economy. The post Cold War reunification of Germany brought about a strong migration from east to west, a by-product of a pent-up desire to migrate as motivated by both political and economic forces.

THE WORLD REFUGEE PROBLEM

Sometimes people are forced to migrate from a troubled region or nation in response to life-threatening conditions or political upheaval. The victims of this type of forced migration are known as "refugees." Specifically, a **refugee** refers to *any person who has been displaced from their homeland and forced to seek asylum outside his/her country of origin for sociopolitical reasons.* Refugees are generally unable or unwilling to return home in the likely event they would face persecution or death for reasons related to religious or political preference, party or group membership, ethnicity, or nationality. Most refugees are victims of deteriorating political economies, poverty and deprivation, ethnic and religious conflict, and environmental degradation in relation to natural and man-made hazards. These people look to the international community for assistance as they temporarily or permanently relocate to more stable geographic areas (Figure 7.20).

Figure 7.20   Burundian Hutu refugees at a camp near Ngara, Tanzania; part of a growing world refugee problem that has presented an enormous development challenge.

7.20: ©Sean Sprague/Stock Boston

Consider the 30% of Afghanistan's population that was forced to move to Iran and Pakistan, in response to the invasion of the Soviet military in 1979. As sporadic fighting and bloody skirmishes waged over territorial control between defending Afghan rebels and Soviet forces, refugees refused to return home for fear of getting caught up in the crossfire. After the withdrawal of Soviet troops, many were reluctant to return to their homeland for fear of yet another invasion or political persecution.

Large numbers of refugees have fled the many east, central, and southern African nations plagued by violence. Consider the various political factions that attempted to establish dominance and control in war-torn Somalia, forcing refugees to seek asylum in nearby nations. The hoards of Somalian refugees in neighboring Kenya and Sudan are testimony to the legacy of war and fear. A lengthy war in Mozambique, caused roughly a million refugees to migrate to the tiny nation of Malawi in search of safe haven. Deteriorating economic and political conditions throughout the SOUTH are likely to generate unprecedented numbers of refugees in future years.

Intolerable living conditions and a lack of sufficient economic opportunities have led many to seek refuge (legally or illegally) in regions perceived as offering better opportunities. Each year large numbers of immigrants flow out of Mexico, heading north to border regions in the southwestern United States in search of opportunities. While the influx of Mexican workers to the U.S. economy has alleviated unemployment pressures in Mexico and allowed immigrants to support themselves and their families back home, it is not without controversy or cost. A strong public outcry in opposition to illegal immigration continues, as U.S. taxpayers demonstrate their unwillingness to support illegal aliens.

Fleeing the grip of anarchy, the search for freedom and a voice in government have led many to seek stable political climates. The exodus of Haitians and Cubans bound for American shores are prime examples of politically induced migration. Forced migration from a given region in response to deteriorating physical or environment conditions is also not uncommon. Consider the many droughts that have occurred in the Sahel region of Africa. Shortages of water, food, and subsistence opportunities have forced many people from nations like Mali, Ethiopia, Sudan, and Tanzania, to relocate to settlements or refugee camps at great distances from their homeland.

Heightened ethnic tensions in recent years have led to civil war and human rights atrocities in Burma, Burundi, and Rwanda. In the former Yugoslavia, a nation once "united," the rekindling of hatred between Serbs and Croats has created an ethnic division that has escalated into a bloody and ongoing confrontation. Ethnic conflicts such as these may be near impossible to resolve through regional economic development efforts. Intervention from the international community can temporarily bring peace to regions ravaged by civil war, but in many cases outside intervention does little to resolve the conflict. The recent conflict between NATO and Serbian forces in Kosovo has forced hundreds of thousands of ethnic Albanian refugees to flee to surrounding regions to escape Serbian aggression.

The total number of refugees in 1991 was estimated at about 15 million people worldwide. That number continues to rise with the escalation of violence in peripheral regions. In 1995, there were well over 19 million refugees. By the year 2000, the number of refugees is expected to climb toward the 25 million mark.

A regional breakdown of the origins of refugees worldwide is given in Table 7.12. Note that South and West Asia and Africa represent the two largest sources or origins of refugees. Table 7.13 highlights major countries of refuge (destinations), with only those nations harboring more than 90,000 refugees listed. Note that the nation of Iran tops the list, as it has absorbed more than 4 million refugees from the many surrounding war-torn nations.

A listing of the major conflicts from 1990 to 1995 is provided in Table 7.14. These conflicts are responsible for generating the vast majority of the world's refugees and are useful in helping to explain the geographic distribution of refugees as shown in Table 7.13.

The rising flood of refugees worldwide presents a formidable global development problem. According to Newland (1994), the manageability of the worldwide refugee problem depends on four policy elements:

1. an insistence that refugees should not be returned to the countries where they face persecution;
2. the provision of hard, fair, and swift evaluation procedures for those seeking asylum to prevent backlogs and delays that could prove costly to those wishing to leave a troubled region;
3. a strong and unwavering financial commitment from the international community (and core economies) to alleviate the monetary burden of those regions or nations absorbing the greatest number of refugees; and
4. the introduction of proactive policy measures to overt a regional or worldwide crisis situation and

TABLE 7.12	Geographic Origins of Refugees Worldwide (1991)
Region	**Percentage**
Latin America	1.5%
East Asia	3.6%
Middle East	21.9%
Africa	30.1%
South and West Asia	42.9%

Source: The U.S. Committee for Refugees Report (1991).

TABLE 7.13	Major Countries of Refuge (Destinations, as of January 1, 1993)

Country	Number of Refugees
Iran	4,150,700
Pakistan	1,629,200
Malawi	1,058,500
Germany	827,100
Boznia and Herzegovina	810,000
Sudan	725,600
Croatia	648,000
Canada	568,200
Guinea	478,500
United States	473,000
Ethiopia	431,800
Kenya	401,900
Zaire	391,100
Mexico	361,000
Sweden	324,500
Armenia	300,000
Tanzania	292,100
China	288,100
Burundi	271,700
India	258,400
Azerbaijan	246,000
Bangladesh	245,000
Guatemala	222,900
Algeria	219,300
Uganda	196,300
France	182,600
Côte D'Ivoire	174,100
Zambia	142,100
Zimbabwe	137,200
Kuwait	124,900
Costa Rica	114,400
Honduras	100,100
Liberia	100,000
Iraq	95,000

Source: K. Newland, "Refugees: The Rising Flood" *World Watch* 7, (May-June, 1994), pp. 14–15.

TABLE 7.14	Major Conflicts, 1990–1995

Region	Nation	Type	Factors
Central America	Guatemala	I	a
	El Salvador	I	c
	Nicaragua	I	c
South America	Peru	I	c
Africa	Algeria	I	b
	Liberia	I	a
	Sudan	I	b
	Eritrea	I	a
	Ethiopia	I	d
	Somalia	I	a
	Rwanda	I	a
	Burundi	I	a
	Angola	I	c
	Mozambique	I	c
	South Africa	I	a
Europe, Middle East, and Asia	Slovenia	I	a
	Croatia	I	a
	Bosnia, Herzegovina	I	a
	Yugoslavia	I	a
	Moldova	I	a
	Turkey	I	a
	Georgia	I	a
	Azerbaijan	II	a
	Armenia	II	a
	Lebanon	III	d
	Israel	I	b
	Iraq	III	d
	Kuwait	II	d
	Yemen	I	c
	Tajikistan	I	d
	Afghanistan	I	a
	Pakistan	II	c
	India	II	c
	Myanmar	I	a
	Sri Lanka	I	a
	Cambodia	I	a
	Indonesia (East Timor)	I	a

Key:

I—Civil	a—ethnic
II—International	b—religious
III—Both I and II	c—political
	d—multiple/undifferentiated

Source: *Goode's World Atlas* (19th edition), p. 59.

prevent refugee flows from reaching a level that is not manageable.

In addition, a clear-cut statement must be issued by each nation regarding their stance on refugee absorption to identify those places which are willing and able to offer safe asylum to future refugees and those which are "closed" or have limited absorption rates.

The effective implementation of policy to counter the growing refugee problem will require enormous sacrifice, effort, and cooperation on the part of international gov-

ernments and world leaders. It is essential that an international consensus be reached on the policy elements outlined on the previous page, as no one country or region can pursue them successfully acting in isolation of all others.

Certainly, one can argue that such a requirement also applies for other major global development issues that desire our utmost and immediate attention (such as poverty, hunger, and overpopulation). There is a definite need to construct a comprehensive human development platform that targets not only the symptoms of social, economic, political, and environmental turmoil, but one that addresses the very roots of development problems.

While it has been the nations of the SOUTH that have absorbed the greatest number of political and environmental refugees, the so-called "refugee problem" is a global problem.

> The rapid growth in the global refugee problem is symptomatic of the inadequacy of our (development) efforts to date. Refugees are the human fallout of the explosive tensions that have been allowed to accumulate around economic maldistribution, environmental degradation, militarization, and a host of other ills. They are not the only people who have a stake in addressing the causes of their plight. (Newland, 1994, p. 20).

Promoting the existence of safe, stable, and non-threatening political environments, which allow people to exercise various freedoms or participate actively in the political process, is key to the development process. It is the pull of democracy and the benefits of political stability that encourage many to gravitate toward the countries with strong multi-party systems and away from military regimes and/or single-party authoritarian rule. The desire for personal and economic freedom continues to erode the power base of those favoring command economy.

Non-Forced Migration and the Principle of Gravity

Push and pull factors are major determinants of human migration, yet movement in the world economy shows regularities that are easily explained by the principle of "gravity." Traced to the work of Ravenstein (1885), and founded in the principles of Newtonian physics, migration between any two regions is a process that exhibits distinct regularities. The **gravity model** states that *the flow of people between a region i and a region j, I_{ij}, is said to be directly proportional to the size or mass of region i (M_i), directly proportional to the size or mass of region j (M_j), and inversely proportional to the distance between region i and region j (d_{ij}).* Moreover, distance d_{ij} is raised to an exponent ß—a parameter that denotes the "friction of distance" or the difficulty in overcoming physical separation. Note that in the earliest gravity-type applications, it was assumed that ß = 2; that is, movement between an origin i and a destination j was directly proportional to the masses of i and j and inversely proportional to the distance between i and j "squared," as in the case of Newton's famous gravity law. This assumption was later relaxed in models applied to human systems, where flows occur using many different types of transportation technologies and modes (and where ß is time- and mode-specific).

A general gravity formula which describes or predicts migration or the movement of people between any two places or regions (i and j) may be written as follows

$$I_{ij} = k M_i M_j / d_{ij}^{\beta}.$$

Note that k is a system-wide coefficient known as the "constant of proportionality." This constant is used to adjust predicted values to bring them in line with observed flows.

An Application of the Gravity Model

Suppose that 650,500 people move between region i and region j during a given time period. Suppose that the sizes of regions i and j are 68,000,000 and 250,000,000 respectively, and that the distance between the geographic centers of these regions (as perceived by the migrants) is approximately 500 miles. Also, suppose that it has been established (from previous studies) that the constant of proportionality k = .00001, and the friction-of-distance parameter ß is approximately 2. The gravity model would then produce a reasonable estimate of flows, as it would predict that there were 680,000 people moving between region i and j as

$$I_{ij} = (.00001) * 68,000,000 * 250,000,000 / 500^2$$
$$= 680,000.$$

Note that the gravity model has slightly overestimated the flow between region i and region j in this case. System-wide, over- and underestimation will even out such that total outflows from all "generating" regions will equal total inflows from all "receiving" regions, and the sum of squared error—the difference between observed and predicted flows squared for all regional pairings—would be minimized provided reasonable estimates for k and ß were found.

The gravity model can be used to describe the intensity of flows of people or commodities between regions of various sizes (where the propensity of interaction increases with region size) taking into account the "attenuating" effects of distance—how increasing distance or separation lessens the likelihood of interaction. In general, there is a greater likelihood of observing flows between two large regions than between two small regions when those regions are of an equal distance from one another. As there exists a "hierarchy" of regions—regions ranked in terms of their influence or leverage—we would expect that the flow or interaction intensity between any two regions increases as the size or influence of a region increases. In reference to the world economy, flows are most likely to be observed "across" the top of the hierarchy of nations—between core economies. It is also common to observe large flows between regions which are proximate or near to one another rather than between regions separated by a great distance. If flows are observed over great distances, it usually means that the sending and/or receiving regions are large or influential or that the friction of distance is low (that is, there is little resistance to traveling across space). Note also that if relatively small flows or virtually no interaction occurs between two large and proximate regions, then it is likely that the friction of distance is high or there are barriers which hinder or obstruct movement between those regions.

Consider the case of labor migration in the United States as observed during the late 1980s. Figure 7.21 shows the top-twelve destinations of labor migrants leaving the East North Central census region (which includes the states of Wisconsin, Michigan, Illinois, Indiana, and Ohio), and relocating to states outside the region from 1985 to 1988. Figure 7.22 shows the top-twelve destinations of labor leaving the New England census region over the same time period. Note that the top labor migration flows are either to states that are in close proximity to the region of origin or large states at a fairly great distance away. In either case, the gravity model provides a reasonable explanation for the labor migration pattern, which seems to be highly affected by the size of origins and destinations and the distance between them.

A Critique of the Gravity Model

In its attempt to replicate the movement of people, the gravity model suffers from several shortcomings. First, it is a static representation of a dynamic process. It fails to account for the historical aspects of the migration process and the nongeographical factors which influence past and present migration streams. Second, parameter estimates for k and ß can be viewed only as gross approximations that, when viewed in dynamic terms, must lack inherent stability (that is, they are parameters that will change over time once people migrate and reorganize themselves in space, and/or the initial conditions within the system change). Consider the fact that people have become increasingly more mobile over time due to improvements in transport technology. Third, while the gravity model is useful to "describe" flows, its utility as a "predictive" model is questionable given that so many unknowns exist (like the true value of ß or the behavioral and perceptual aspects of migrants). Last, the gravity model fails to capture the essence of Ullman's principles of spatial interaction (as discussed in Chapter 1). In other words, the gravity model does not establish the existence of complementary relations and the transferability of human agents amongst origins and destinations; nor does it account for the effects of a "competitive" spatial structure—how flows are influenced by the relative location of competing destinations and intervening regions. Despite its shortcomings, the gravity model does provide a reasonable explanation for most geographic flow patterns.

The Twelve Generalities of Voluntary Migration

Notwithstanding the obvious movement of people to escape political or economic hardship, a large number of people migrate within nations or between nations for other reasons, including better climate, more amenities, greater employment or economic opportunities, or speculation that a new location may offer a sense of adventure or a chance at making a new start. Despite the unpredictable aspects of the human relocation process, empirical studies on the subject have helped produce twelve generalities of voluntary migration (as discussed by Norton, 1992, p. 117, and others):

1. *The large majority of migrants travel only a short distance when relocating in the world economy.* Due to the "friction of distance" in travel, it is both difficult and costly to move. The tyranny of geographic separation is an impediment to mobility. Together, the friction of distance and tyranny of separation force the large majority

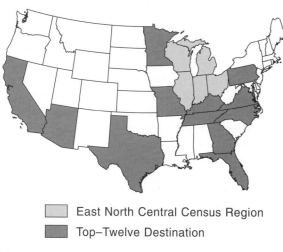

◻ East North Central Census Region

◼ Top–Twelve Destination

Figure 7.21 Top-twelve labor flows from the East North Central census region (1985–1988).

◻ New England Census Region

◼ Top–Twelve Destination

Figure 7.22 Top-twelve labor flows from the New England census region (1985–1988).

of migrants to move only a short distance away from a given place or homeland. While there are many **prospective migrants**—*those that have a desire to relocate,* many prospective migrants never move for one reason or another. The number of prospective migrants is large when compared to the number of actual migrants. Prospective migrants that remain in one place may be classified as "stayers." As a formal definition, **stayers** *are prospective migrants that weigh the net benefits of relocating and opt not to move, or those that are physically constrained to remain in a given place or location as they lack the necessary resources or means to move.*

Advancements in communication and transport technology have lessened the tyranny of geographic separation, allowing a deeper convergence of time and space. We can now move people around the globe in a matter of hours. We can move and diffuse information to the far corners of the earth in the blink of an eye. Increasing "time-space convergence" has undoubtedly lessened the friction of distance, improved information flows, reduced uncertainty, and increased the overall mobility of human agents in modern society. The recent surge of migration worldwide is an outcome of more people (and more prospective migrants) and enhanced mobility. Intensification of migration is an outcome of improved access to modern transport technology and information. The increasing mobility of people in developing regions has been called a "mobility transition," a by-product of development and modernization (Zelinsky, 1971).

2. *Migrants traveling over long distances or internationally tend to favor destinations that are large and typically world centers of commerce or industry.* This preference reflects the fact that more information is generally available about dominant centers in the world economy. At the global level, world cities are magnets for prospective international migrants as they offer the greatest amount of economic and employment opportunities.

3. *Migration is a step-by-step process in which migrants first relocate to large urban areas (or ports of entry) and later to places that are either across or farther down the "urban hierarchy."* For example, a person moving from Australia to the United States may first locate in a large port city such as San Francisco. After becoming familiar with the city and its surrounding region, they may choose to move to a secondary location, perhaps to a nearby city or suburb. A move from San Francisco to Los Angeles, would be considered as a move across the urban hierarchy. A move to San Mateo, California, would be considered as a move down the urban hierarchy. Step-by-step migration is an outcome of a ongoing search process that ends when a person finds a place or location which maximizes his or her perceived and attainable level of welfare.

4. *Each migration stream produces a potentially compensating counter stream of migrants who move in the opposite direction.* In other words, for every migration flow there exists an opposite migration stream (though relatively smaller in magnitude than the primary flow). The compensating counter stream is sometimes referred to as **return migration**—*a movement back to a point of origin due to an unsuccessful relocation or reassessment of the net benefits of the relocation.*

5. *The rural-to-urban migration stream is more commonly observed than the urban-to-rural migration stream.* In other words, we are less likely to see a migration flow from an urban area to rural area than we are to see a migration flow from a rural area to an urban area. Migration is thought to be logical in the sense that larger settlements tend to offer a greater variety of amenities, opportunities, services, and living conditions.

The recent urban-to-suburban migration wave is not thought of as a compensating or counter flow, but simply an indication that many urban areas have grown beyond a size which is considered to be "optimal" from a socioeconomic standpoint. In theory, there exists an **optimal city size**—*a city size that promotes the efficient or equitable distribution of public goods and services, in a setting where the costs of congestion do not exceed the benefits of agglomeration.* While this may differ from place to place depending on the physical layout or structure of an urban landscape and the spatial distribution of urban dwellers, establishments, employment opportunities, etc., the concept of optimal city size helps us to form an understanding of the importance of cities as centers of functional regions. The very reasons why cities emerge on the landscape to begin with—the convenience of agglomeration and the mutualistic nature of production and consumption—are the reasons why rural-to-urban migration flows are more likely than urban-to-rural flows.

The modest increase in urban-to-rural migration, nonetheless, is a logical response to increasing urban population density and congestion. Moreover, many cities have simply grown old. Inner-city decay, poverty, unemployment, crime, and deteriorating infrastructure have combined to reduce the net benefits of agglomeration and overall value of the urban experience. These conditions have enticed the young and affluent (the most mobile members of society) to flee the entrapments of a counterproductive urban landscape.

The outward movement of people from central cities to the urban fringe has brought about the erosion of the inner-city tax base and a worsening of the very conditions that induce out-migration in the first place. The urban poor, however, can seldom afford the luxury of relocating to more desirable surroundings. As such, the "stayers" far outnumber those actually leaving. Nonetheless, large cities continue to be magnets for those coming from rural areas where high-wage employment opportunities are scarce. Some see the money- and technology-driven urban environments as places which offer nothing more than an illusion of a higher quality of life.

6. *Age and gender play prominent roles in the migration process.* Mobility and the propensity to migrate are things which decline with age (as a result of various sociological and economic factors). As families establish roots in a given area and/or as people show loyalty to a given employer or region, they are less likely to move with time. The number and strength of social-economic relationships increase as people age, decreasing the likelihood that they will move due to "inertia."

Differences due to gender have also been observed. For example, females are likely to be part of an intra-regional migration process, whereas males are more likely to venture beyond a regional jurisdiction as part of an inter-regional migration stream. Most commonly, the international migrant is both young and male (and typically of adult status).

7. *Whole families seldom migrate out of the country of their birth (unless forced).* Voluntary international or overseas migration is most commonly a males-only phenomenon, despite the growing number of female migrants in recent years. Difference in mobility in relation to both gender and age generally preclude whole families from voluntarily uprooting themselves and relocating to a new and potentially unfamiliar nation, unless forced to do so (as in the case of refugees). The vast majority of families in the world economy will live and die in the country of their birth.

8. *During the height of industrialization and urbanization, large settlements in most developing regions will tend to grow more by migration than by natural increase for a period of time.* Rapid economic growth and industrialization will result in a rapid increase in the demand for labor. As a result, urban areas undergoing industrialization will tend to absorb labor from the hinterland and surrounding regions. Thus, the accompanying increase in population will be predominantly driven by in-migration. This is true even in mature economies, as many large urban areas continue to display a momentum of growth that is driven by a strong and steady stream of in-migrants.

9. *The volume and intensity of migration increases with the intensity of industrial growth and the expansion of the regional production network.* Intensification of production and rapid urbanization tends to attract migrants from great distances as both real and perceived economic and employment opportunities arise on the landscape. The size of the migration stream is directly proportional to the rate of industrialization and network formation.

10. *Regional migration in developing regions reflects the logical redistribution of labor as economies transform and modernize.* Intensification of agricultural sectors, the production of agricultural surplus, and/or the natural gravitation of people to large urban settlements tends to redistribute labor from areas of low population density (rural agricultural settlements) to areas of high population density (urban settlements). This pattern reflects the impact of economic development and the associated changes in labor force participation by sector as economies transform and diversify.

11. *The greater the flow of people between two complimentary regions, the greater the likelihood of future migration flows between those same regions.* In other words, the greater the **migrant stock**—*the cumulative number of successful migrants from a given region of origin to a specific destination,* the greater the information transfer between those regions and the less uncertainty as to economic conditions and opportunities at that destination. The lesser the uncertainty and the greater the migrant stock, the greater the likelihood of observing successful migration flow in the future between the two regions. Once established, a migration stream will continue until a significant share of migration becomes unsuccessful or as prospective migrants are informed of the reduced benefits of a move due to changing conditions.

Consider the recent decline in U.S. labor flows from the Snowbelt to the Sunbelt. This situation is partly due to slowdowns in the rate of economic and employment growth in the South and the stabilization and reemergence of industry throughout the North. Note also that competing and intervening forces on the economic landscape may combine to reduce the longevity of any given migration stream as decision-making processes of migrants involve the consideration of alternative destinations and the opportunity costs of unsuccessful migration.

12. *Forced migrations are generally sociopolitically or environmentally motivated, whereas the voluntary or non-forced migration is motivated by economics.*

SOME OBSERVATIONS ON THE IMPACT OF IMMIGRATION

Migration continues to play an important role in changing the age, race, and ethnic composition of regional populations. Consider the changing demographic composition of the United States from 1980 and 1990 as shown in Table 7.15. During this period the United States absorbed almost 7 million immigrants, mostly from Asia and Latin America. Note that the largest population increases were associated with people from Asian and Pacific Island

TABLE 7.15	**The Racial Composition of the United States, 1980 and 1990**

| | **Percentage** | | |
Category	**1980**	**1990**	**Change**
White	83.1%	81.0%	−
Black	11.7%	12.1%	+
Hispanic Origin	3.0%	3.9%	+
Asian or Pacific Islander	1.5%	2.9%	++
American Indian, Eskimo, or Aleut	0.8%	0.6%	−

(Percentages do not add to exactly 100% due to errors from rounding.)

Total Population	226,545,805	248,709,873	(+10%)
White	188,371,622	199,686,070	(+6%)
Black	26,495,025	29,986,060	(+13%)
Hispanic Origin	6,758,319	9,804,847	(+45%)
Asian or Pacific Islander	3,500,439	7,273,662	(+108%)
American Indian, Eskimo, or Aleut	1,420,400	1,959,234	(+38%)

(Percentage changes shown in parentheses rounded to nearest percent.)

Source: U.S. Census Bureau, as illustrated in *Business Week,* (July 13, 1992).

nations and people of Hispanic origin. These increases were largely due to the effects of immigration.

Forecasts indicate that by the year 2050, it is likely that the U.S. population will be roughly 50% nonwhite. By that same year, the number of Asian and Pacific Islanders, now the fastest growing segment of the U.S. population, will increase to 12% of the total population. As America's population climbs to more than 380 million by 2025, the number of Americans of Asian and Pacific Island descent will climb to 45 million. Demographic shifts in the U.S. population will bring about many changes in consumer preferences and the marketing of products, as well as the way in which the "average" American is portrayed.

Table 7.16 provides a comparison of the historical patterns of U.S. immigration by region of origin and the two big migration waves observed since the Industrial Revolution. The first major wave lasted from 1901 to 1920 and accounted for well over 14 million legal immigrants entering the United States. The overwhelming majority of those immigrants were of European origins.

The more recent migration wave, from 1971 to 1990, accounted for approximately 10 million legal immigrants. In contrast to the earlier migration wave, the vast majority of immigrants were of Latin American and Asian descent. When measured in absolute terms, these two immigration waves are of comparable size when one considers that there were more than 3 million illegal immigrants entering the United States from 1970 to 1990. See Figure 7.23.

It is interesting to note that the latest immigration wave is easily explained by the gravity principle for

TABLE 7.16	**Two Major Migration Waves**

Time Period	Origin of Immigrants	Percent of Immigrants
Melting pot period: "Give me your tired, your poor, your huddled masses yearning to breathe free . . ."		
1901–1920	N. and W. Europe	41%
	S. and E. Europe	44%
	Canada	6%
	Asia	4%
	Latin America	4%
	Other	1%
Increasing Cultural Diversity Period		
1971–1990	N. and W. Europe	5%
	S. and E. Europe	8%
	Canada	2%
	Asia	37%
	Latin America	45%
	Other	3%

Source: Bergman and McKnight (1993), p. 201.

Period	**Total U.S. Immigration (in millions)**
Legal Immigrants	
(1901–1920)	14.5
(1971–1990)	9.8
Illegal Immigrants in U.S. (1995)	3.5*

Source: From Rubenstein (1992) , p. 104, with additions.

* Estimated.

several reasons. First, the shear size and the rapid population growth in many Asian and Latin American nations has resulted in a greater potential for these areas to send migrants. Second, the glaring economic disparities between core and peripheral economies are largely responsible for creating push and pull factors that make the United States a preferred destination in the world economy. Third, the United States is absorbing people from greater distances due to increases in mobility due to improvements in transoceanic transport technology and commercial air travel. Last, but not least, educational and employment possibilities in the United States and an open-door policy have attracted top international students and labor from the leading nations of the SOUTH. The lure of democracy, capitalism, and an ever-expanding horizon of economic opportunity in the world's leading core economy continues to attract the crème de la crème of the world's labor force.

Several observations can be made regarding the likely impacts of the most recent U.S. migration wave. First and foremost, there has been a noticeable increase in the quality of the U.S. labor force. Consider the case of Asian-Americans, who account for the largest share of legal immigrants recently entering the United States. As a group, Asian-Americans tend to be more educated, more affluent, and more skilled than the average U.S. citizen.

Viewed in this way, immigration can be a powerful domestic economic policy tool and an inexpensive and effective way of improving the quality of a region's labor force in a short period of time. Importing highly skilled workers from abroad allows a region to up-skill its labor force, enhance productivity, and improve its competitive position in the world economy. It is interesting to note that in 1990 the U.S. government redrafted national immigration legislation to give preference to those prospective immigrants that were highly skilled, well educated in math and sciences, and/or affluent. This legislation has made it easier for foreign students to remain in the United States upon completion of their degrees.

Advocates of the open-door policy say that the recent migration wave has greatly enhanced the competitive advantage of the United States as it faces the challenge of remaining competitive throughout the next millennium. The cosmopolitan and highly skilled labor force is likely to keep U.S. industry on the cutting-edge of product innovation as it vies for a share of the world's emerging markets. Critics of liberal migration policies point out that imported labor is likely to steal jobs from native-born Americans and drive down wages in labor markets as the supply of labor (both skilled and unskilled) increases faster than demand.

Liberal immigration policies have, however, helped reduce labor shortages in many regions and sectors in the

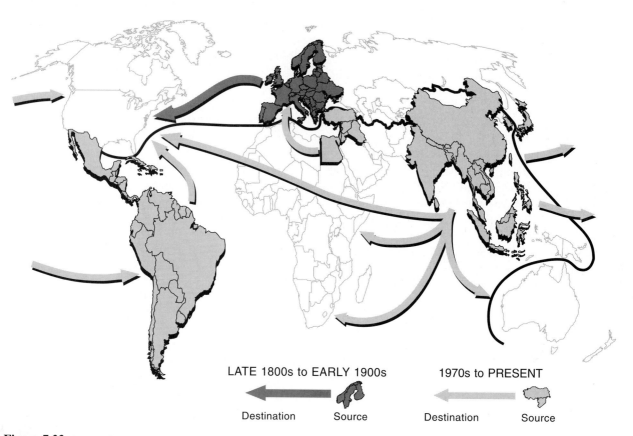

LATE 1800s to EARLY 1900s   1970s to PRESENT

Destination   Source   Destination   Source

Figure 7.23 International migration streams by region.

world economy. There are many benefits for nations willing to absorb "guest workers" from abroad. **Guest workers** may be defined as *foreign labor that migrates to a host nation in response to labor shortages in low-skill, low-status sectors or jobs.* Guest workers in west and central Europe come from a wide variety of places including West Asia (primarily Turkey), the Middle East, North Africa, and other less-affluent eastern European nations. These workers fill voids in local labor markets by working low-wage, service-oriented jobs. Although it is expected that they will stay only a short time as part of a transient labor force, many have gone on to marry, raise children, and become citizens in the countries were they have obtained work. Employment opportunities for guest workers typically include street repair, general maintenance, sanitation and garbage collecting, and positions which support and sustain various transportation systems (including cab and bus drivers and railroad personnel). Guest workers have become a permanent fixture of the social and industrial infrastructure of Europe despite the fact that strict rules have made it difficult for many of them to obtain citizenship (Figure 7.24).

There is a downside to the global reorganization of labor, however. The exodus of skilled and unskilled labor from underdeveloped regions has had a devastating effect on the economies of the SOUTH. As the "brain drain" continues, it reduces the short-term competitiveness and the long-term development potential of affected regions. Liberal immigration policies worldwide have worked against global development efforts. Core economies of the NORTH tend to be net importers of skilled labor and

Figure 7.24 Voluntary migration: A family of immigrants in search of a better way of life.

7.24: ©Owen Franken/Stock Boston

guest workers and non-core economies continue to be net exporters of labor and guest workers in the now global labor market.

Immigration has changed the face of host nations, bringing together people from a wide range of backgrounds and cultures. The exposure to different cultures, ideas, and traditions has yielded immeasurable benefits for society at large. It is an enhanced quality of life that only ethnic, social, and cultural diversity could bring about. Nonetheless, as regions become more culturally diverse they run the risk of encountering increasing conflicts, tension, and divisions between ethnic groups. Consider the impact of Cuban immigration to Florida during the 1970s and 1980s, a factor that was largely responsible for a rapid shift in the composition of South Florida's population.

The Cuban invasion resulted in many uprisings between Cuban immigrants and Florida's black and non-Hispanic populations. Much of the conflict centered on the notion that immigrants placed undue pressures on the regional labor market, under the assumption that only a limited number of employment opportunities existed in the region. Many feared that Cuban immigrants would take away their jobs and force a decline in wages (in retail and service sectors) as immigrants were willing to work for less money. Job competition and job displacement concerns have recently become a hot issue in the southwestern United States, an area greatly affected by both legal and illegal immigration from Mexico.

Despite increasing competition in labor markets, immigration has led to the expansion of urban areas and an expansion of the public sector. This is particularly true in the case of gateway cities and areas which act as ports of entry for legal and illegal immigrants. In general, immigration brings about an increase in the demand for services associated with larger populations. Increasing demand for services and increased retail activity stimulate the economies of host regions, creating job opportunities that would not have occurred in the absence of immigration. Viewed in this way, immigration helps to revitalize an economy.

Yet, immigration in the absence of sufficient employment opportunities places an enormous burden on the local, regional, and national governments, which must support those who fail to find work. Nonetheless, population and economic growth in core economies continues to create a wealth of employment opportunities in low-wage, labor-intensive industries including restaurant and food services, the manufacturing of clothing and shoes, transportation services, and agricultural sectors which require large amounts of hand-processing input (for example, truck farming operations, which emphasize production of fruits and vegetables). Thus, it has not been difficult for legal or illegal immigrants to find work.

In contrast to legal immigrants, illegal immigrants possess few skills and are relatively uneducated or illiterate. Many U.S. taxpayers are calling for strict measures to control the stream of illegal aliens entering the United States from Mexico and elsewhere. Many have demanded that state and federal governments increase efforts to deport illegals and/or deter them from entering the United States by denying them access to publically funded services and programs. The assumed burden of immigration is highly variable over space, however. The largest public outcry has come from regions absorbing the greatest number of illegal immigrants, states such as California, Arizona, Florida, and Texas.

Notwithstanding problems associated with illegal immigration, there have been serious concerns over the rising tide of refugees in core nations. With little chance of successfully integrating into formal economic sectors of the U.S. economy, refugees have become part of the cycle or culture of poverty. Refugees continue to cost host regions big money, as they join the growing sea of welfare recipients in the United States and elsewhere.

The percentage of refugees in America receiving some form of public assistance is substantially larger than the percentage of immigrants or native-born Americans (see Table 7.17). This is due to the fact that refugees are eligible for public assistance benefits that exceed that which are available to the average U.S. citizens. In addition, refugees impose a relatively heavy net cost to host governments given that they tend to earn less income and subsequently pay less taxes. Although active in informal sectors of the economy, many refugees do not gain the necessary skills essential to improve their employability and status. Thus, the majority of refugees are condemned to a life of economic hardship. The refugee dilemma in the United States has also greatly influenced public opinion on the pros and cons of immigration. This dilemma has made it especially hard on prospective immigrants seeking to gain legal entry into the United States. Note that about half of all Americans believe that the majority of incoming foreigners ultimately end up on welfare, despite evidence to the contrary.

Controversies aside, immigration is becoming an increasingly important component of regional and national population change. Although 60% of the population change in the United States is explained by the forces of natural increase (births less deaths), immigration now accounts for 40% of U.S. population increase, up from 11% during the 1960s. Data on the increasing importance of immigration as a component of U.S. population change are presented in Table 7.18. Note that immigration is likely to account for 43% of the U.S. population increase in the not-so-distant future.

Immigration has also provided important benefits to urban economies, despite the strain placed on health and education systems. This is especially true for larger urban areas (the preferred destination for immigrants). In gateway cities such as Miami, New York, San Francisco, Los Angeles, and San Diego, immigrants have begun to create a flurry of economy activity, starting their own businesses

TABLE 7.17	Refugees Receiving Public Assistance in America (1990)

Country of Origin	Percentage of Households on Welfare
Cambodia*	48.8%
Laos*	46.3%
Dominican Republic	27.9%
Vietnam*	25.8%
Former Soviet Union*	16.3%
Cuba*	16.0%
Mexico	11.3%
China	10.4%
Haiti	9.1%
S. Korea	8.1%
Jamaica	7.5%
Colombia	7.5%
El Salvador	7.3%
India	3.4%
Taiwan	3.3%
Legal refugees	16.1%
Legal (non-refugee) immigrants	7.8%
Native-born Americans	7.4%

Source: *Business Week,* "It's Really Two Immigrant Economies," by M.J. Mandel, (June 20, 1994), pp. 74–78.
* Represents a major sources of refugees.

TABLE 7.18	The Importance of Immigration as a Component of U.S. Population Change

Period	Immigrants Share of Total Population Growth
1960–1970	11%
1970–1980	33%
1980–1990	38%
1990–2000	40%*
2000–2010	43%**

Source: U.S. Department of Commerce, Bureau of the Census (1995).
* Estimated.
** Forecast.

and working to improve their local communities. The recent wave of immigrants from Latin America and Asia has provided many American cities with a much-needed boost. Without immigration, the nation's ten largest cities would have shrunk in size by more than 6% during the 1980s, instead of growing by an estimated 5% (Figure 7.25).

According to Briggs and Moore (1994), immigrants have been responsible for starting tens of thousands of new enterprises in urban America. Moreover, there is

Figure 7.25 An immigrant-owned-and-operated establishment in Boston's Chinatown. Immigrants have brought tens of thousands of jobs to America's inner cities and hope for the future.

7.25: ©Michael Dwyer/Stock Boston.

evidence that immigrants create at least as many jobs as they take away, due to the expansion effect they have on the economy (Brown and Shue, 1983). Moreover, immigrants' contributions to the U.S. tax base greatly exceed the associated costs of public services and government assistance provided (Briggs and Moore, 1994). Currently, immigrants pay an estimated $100 billion in taxes on average per year in comparison to the mere $5 billion in welfare benefits they receive. Despite the fact that there is no hard evidence to support the contention that legal immigrants are a drain on America's economic system, there is strong support for immigration policy reform. With growing disparities between core and peripheral economies, immigration will remain a sensitive issue. More and more, immigrants will be viewed as economic assets as nations continue to compete for labor in the global labor market. As a result, governments are likely to impose tougher immigration and nationalization policies to limit the entry of refugees and deter immigrants who are less affluent or less skilled.

In the United States, immigration reform must be concentrated on controlling the flow of illegal immigrants (mostly from Mexico). Currently, there are an estimated 4 million illegal immigrants in the United States. Though hard working and productive, they are generally less skilled and less educated than the average American worker. Each year, an additional 300,000 illegal immigrants cross into the United States from Mexico. As there is a small return flow, the net inflow of illegal aliens to the United States is currently estimated at around 250,000 per year. Since the early 1990s, U.S. taxpayers have forked out an estimated $3 billion a year to provide the necessary education infrastructure to satisfy the education requirements of the children of illegal immigrants. Bil-

lions of dollars per year are being spent for health care to meet the needs of illegals, not to mention the cost of social programs and other services.

Note, however, that illegal immigration has provided a small yet significant stimulus to the U.S. economy. Illegals do pay sales tax on items purchased in the United States and aid in the recirculation of money. Estimated sales tax revenues are in the billions of dollars. This revenue is able to support a small yet important number of public-sector jobs, not to mention the multiplier effects, which are known to ripple through the economy. In light of these facts, illegal immigrants are truly only a minor financial drain on the system. From a legal standpoint, opponents of illegal immigration are quick to point out that their unlawful presence in the United States is hard to justify.

Though illegal immigration has been blamed for increasing job displacement, illegal aliens do fill minor gaps in the labor force, taking on the least desirable and lowest-paying jobs. Yet, the flow of illegal aliens continues to increase the supply of labor in low-wage sectors of the economy. Intensification of competition for a limited number of jobs in areas where the percentage of illegals is high contributes to wage suppression and inequality in the regional economy as a whole.

Many policy analysts seem to think that employers of illegal immigrants should become the target of attack and not the illegal immigrants themselves. They feel that the most effective immigration policies must work to decrease the demand for below-minimum-wage labor. Although reductions in the demand for low-wage labor will lessen the pull economies have on labor, it does little to address the many push factors that exist in the disadvantaged nations. Most likely, the presence of both exploitable labor and exploiters is a condition that arises "simultaneously" (that is, without a time-lag) as both labor and capital have become highly mobile in the world economy. As the debate continues over the problem and its solution, illegal immigrants continue to solidify their ties to host nations. A large percentage of illegals are having children in places outside their homeland. This has further complicated immigration policy, as there now exists this added dimension of finding ways to deal with the transnational aspects of families with multiple citizenships.

The migration process itself shows little potential to become an "equalizer" in the development process. Given the abundant nature of cheap labor in the world economy, it is unlikely that the movement and redistribution of labor are anything more than simple geographic expressions of changing conditions in juxtaposed regional labor markets. The natural relocation of unskilled labor to nearby core economies in North America and Europe only serves to reinforce the dominant position of the core, as labor actively flows toward locations where it can be exploited. Firms and industry continue to retain a distinct advantage having the option to either utilize cheap incoming labor or move to alternative cost-saving regions virtually anywhere on the globe. Unskilled labor, by contrast, will typically move to nearby locations. As such, the elevated mobility of capital is a force that continues to perpetuate regional inequality and uneven development.

CONCLUDING REMARKS

The world's population engine continues to run in overdrive, yet there are indications that relief is forthcoming. Just as the industrial revolution was viewed as a major stimulus to world population growth, many analysts see recent economic slowdowns in Asia and Latin America and the world economy as leading to further declines in regional and global population growth. Demographic trends such as declining total fertility rates are signs that the population engine is beginning to run out of steam. As the world population proceeds through its demographic transition, analysts contend that the world population explosions of stages two and three will take an enormous toll. Excessive regional population growth has already been responsible for irreversible changes to the environment. While most scientists do not believe that the world is in danger of exceeding its carrying capacity, many do question whether 11 billion people by the year 2100 is beyond a "global optimum."

While there is no assurance that zero population growth will ever become a reality, demographic trends are consistent with that outcome. Failure to underscore the urgency of many regional population crises and the need to expedite development are the biggest blunders world leaders can make. As many regional and global development efforts have been thwarted by civil and ethnic conflict around the globe, the need for peace is unmistakable.

> The longer population growth continues, the more committed humanity becomes to a particular set of problems: more rapid depletion of resources, greater pressures on the environment, more dependence on rapid technological development to solve these problems, fewer options, and perhaps continued postponement of the resolution of the other problems including those resulting from past growth. The sooner population growth ceases, the more time humanity has to redress the mistakes of past growth, the more resources it has to decide how it wants to live in the future. These benefits must be balanced against the costs of overriding individual preferences where those preferences do not naturally lead to a cessation of population growth. This balancing can only be achieved in the political arena. (Ridker, 1992, p. 14)

Initiatives to curb population growth and form clearcut policies on the world refugee problem, along with strategies for dealing with illiteracy, hunger, malnutrition, access to safe drinking water, and other important underdevelopment problems were the focus of discussion and debate at the sixth United Nations International Conference on Population and Development held in Cairo, Egypt, in September of 1994. At that conference, 108 delegations, representing a wide range of political, cultural, and regional backgrounds and affiliations came together to iron out a population management agenda. Although proposed solutions and recommendations were multidimensional in scope, offering a wide range of sociocultural, political, and religious viewpoints on the subject of population containment, most participants leaned toward economic development as the universally acceptable remedy for a regional population crisis.

A three-part strategy was proposed to curb population growth and improve "human security." Note that it involves increasing access to survival resources and investment in human capital. Plans called for (1) increasing education and employment opportunities and access to capital, and improving the voice of women in the home and workplace (something which requires educating men about the importance of joint or household decision-making in matters of family size and economics); (2) making family planning available to those who want it but currently do not have access to it for whatever reason; and (3) offering a broad array of health services, including reproductive and medical care for women and infants, treatment for sexually transmitted disease, as well as family-planning counseling for adolescents and adults.

The tremendous success of this approach in places like Bangladesh, Iran, and Indonesia, is a testimony to the type of results that can be achieved when population containment policies become part of a regional development agenda. Note that of the world's poorest nations, Bangladesh is one of the relatively few nations to register sharp declines in its fertility rate, largely attributed to a concerted effort on the part of successive government regimes to concentrate on national family planning, despite opposition from Muslim leaders. Initiatives to control population in the SOUTH (while under religious attack) have not supported coercive sterilization, forced birth control, or gender-selective abortion. These efforts have largely focused on education, dissemination of information, social and economic reform, and the empowerment of women. In many instances, it has become a game of semantics. Use of the phrase "family planning" is sometimes acceptable, even in cultures where there is staunch opposition to "fertility regulation." Anti-abortionists demand that family planning be limited in scope (focusing on proactive rather than reactive measures). In an ironic twist of fate, the Vatican and Islamic fundamentalists have become allies in what they envision as a war against immorality and permissiveness. Despite efforts of social conservatives in core economies, the use of contraceptives has been gaining popularity worldwide. Abortion is now "legal" in 172 countries. Nevertheless, the widespread commitment to population containment through social and economic reform has become the hallmark of the global development agenda, where "economic development" is the best line of attack.

The threat of poverty and hunger remain equally important issues. Geometric-like population growth in the SOUTH is expected to bring about food and grain shortages of immense proportions in the not-so-distant future. Production deficits will be the greatest on the continent of Africa, which will require approximately ten times its current level of imports by the year 2050. Population growth will leave nations such as Ethiopia and Nigeria with unrivaled food production deficits. Projected food deficits in nations like China, India, and Iran are likely to spark intense competition for agricultural surplus from the core. Inevitably, grain prices will move upward. This will leave less-affluent nations unable to afford additional food imports. Many experts see the forthcoming regional food scarcity problems as a world development problem (Brown, 1994). Regional economic development and food security policies cannot, however, be drafted without provisions for population containment and strategies to empower people. While many analysts do not foresee a world moving beyond its human carrying capacity, many do see a world that is divided in terms of its economic and demographic experience. Divisions notwithstanding, the nations of the world must come together to find workable solutions to help accelerate the rate of demographic transition in the SOUTH.

KEY WORDS AND PHRASES

base (of a population pyramid)   150
census   148
cohort   150
cohort-dependency ratio   156
crude birth rate (CBR)   146
crude death rate (CDR)   146
declining-growth (DG) population
    pyramid   151
demographic crisis   145
demographic profile   150
demographic transition   160
early-expanding stage   162
fertility   145
gravity model   171
guest workers   177
high-stationary stage   160
immigrants   167

immigration   167
infant mortality rate (IMR)   149
late-expanding stage   162
life expectancy   149
low-stationary stage   162
maternal mortality rate (MMR)
    149
migrant stock   174
migration   145
mortality   145
natural increase   145
negative-growth (NG) population
    pyramid   154
net migration   145
optimal city size   173
population change equation   145
population pyramids   150

prospective migrant   173
pull factors   167
push factors   167
rapid-growth (RG) population
    pyramid   150
rate of natural increase (RNI)   146
refugee   168
replacement rate (RR)   154
return migration   173
slow-growth (SG) population
    pyramid   152
stayers   173
total fertility rate (TFR)   149
zero-growth (ZG) population
    pyramid   154
zero population growth (ZPG)
    148

REFERENCES

Ashford, L.S. "New Perspectives on Population—Lessons from Cairo." *Population Bulletin* 50 (March 1995), Issue #1.

Banister, J. *China's Changing Population.* Palo Alto, CA: Stanford University Press (1987).

Bergman, E.F., and T.L. McKnight. *Introduction to Geography.* Englewood Cliffs, NJ: Prentice-Hall (1993), Chapter 6, pp. 167–205.

Briggs, V.M., and S. Moore. *Still an Open Door?* Washington, D.C.: American University Press (1994).

Brown, L.R. "Who Will Feed China?" *World Watch* 7, (September-October 1994), pp. 10–22.

Brown, P., and H. Shue. *The Border That Joins.* Totowa, NJ: Rowman and Littlefield (1983).

Business Week. "The Immigrants." (July 13, 1992), pp. 114–121.

Business Week. "A Spicier Stew in the Melting Pot," by B. Bemner and J. Weber, (December 21, 1992), pp. 29–30.

Business Week. "Too Many People," by E. Smith, M. Cohen, and E. Malkin, (August 29, 1994), pp. 64–66.

Cannon, T., and A. Jenkins (eds.). *The Geography of Contemporary China.* London, U.K.: Routledge, Chapman and Hall (1990).

Clark, W.A.V. *Human Migration.* Volume 7. Newbury Park, CA: Sage (1985).

Cross, S., and A. Whiteside (eds.). *Facing Up to Aids: The Socio-Economic Impact in Southern*

Africa. New York: St. Martins' Press (1993).

Dicken, P., and P.E. Lloyd. *Location in Space: Theoretical Perspectives in Economic Geography* (3rd edition). New York: Harper & Row (1990).

Ehrlich, P.R., and A. Ehrlich. *The Population Explosion.* New York: Simon and Schuster (1990).

Fik, T.J., R.G. Amey, and G.F. Mulligan. "Labor Migration amongst Hierarchically Competing and Intervening Origins and Destinations." *Environment and Planning,* A 24, (1992), pp. 1271–1290.

Goode's World Atlas. (19th edition). Chicago: Rand McNally (1995).

Grigg, D. "Ravenstein and the Laws of Migration," *Journal of Historical Geography* 3 (1977), pp. 41–54.

Hardin, G. *Living within Limits: Ecology, Economics, and Population Taboos.* New York: Oxford University Press (1993).

Harries, K.D., and R.E. Norris. *Human Geography: Culture, Interaction, and Economy.* Columbus, OH: Merrill (1986), Chapter 2: World Population, pp. 23–49.

Haynes, K.E., and A. S. Fotheringham. *Gravity and Spatial Interaction Models.* Volume 2. Newbury Park, CA, (1984).

Jackson, R.H., and L.E. Hudman. *World Regional Geography: Issues for Today.* New York: J. Wiley & Sons (1990), Chapter 3, pp. 65–85.

Jacobson, J.L. "The Other Epidemic." *World Watch* 5 (May-June 1992), pp. 10–17.

Jensen, L. *The New Immigration.* New York: Greenwood Press (1989).

Kane, H. "A Deluge of Refugees." *World Watch* 5 (November-December 1992), pp. 32–33.

Lewis, G.J. *Human Migration.* New York: St. Martin's Press (1982).

Lewis, M., G. Kenney, A. Dor, and R. Dighe. *Aids in Developing Countries.* Washington, D.C.: Urban Institute Press (1989).

Linge, G.J.R., and D.K. Forbes (eds.). *China's Spatial Economy.* New York: Oxford University Press (1990).

Loxley, J. *Debt and Disorder: External Financing for Development.* Boulder, CO: Westview Press (1986).

Miller, G.T., Jr. *Environmental Science: Working with the Earth.* Belmont, CA: Wadsworth (1995), pp. 130–162.

Neman, J., and G. Matzke. *Population: Patterns, Dynamics, and Prospects.* Englewood Cliffs, NJ: Prentice Hall (1984).

New York Times International. "Bangladesh, Still Poor, Cuts Birth Rate Sharply," by J.F. Burns (September 13, 1994), p. A5.

Newland, K. "Refugees: The Rising Flood." *World Watch* 7 (May-June 1994), pp. 10–20.

Norton, W. *Human Geography.* Oxford: Oxford University Press (1992).

Ogden, P.E. *Migration and Geographical Change.* Cambridge: Cambridge University Press (1984).

Ravenstein, E.G. "The Laws of Migration." *Journal of the Royal Statistical Society.* 48 (1885), pp. 167–235.

Ridker, R.G. "Population Issues." In *Global Development and the Environment: Perspectives on Sustainability,* J. Darmstadter (ed.). Washington, D.C.: Resources for the Future (1992), pp. 7–14.

Rubenstein, J.M. *An Introduction to Human Geography: The Cultural Landscape* (3rd edition). New York: Macmillan (1992).

Sachs, A. "Men, Sex, and Parenthood in an Overpopulating World." *World Watch* 7 (March-April 1994), pp. 12–19.

Schoub, B.D. *AIDS and HIV in Perspective: A Guide to Understanding the Virus and Its Consequences.* New York: Cambridge University Press (1994).

Shannon, G., G. Pyle, and R.L. Bashshur. *The Geography of Aids: Origins and Course of an Epidemic.* New York: Guilford Press (1991).

Simon, J.L. *The Economic Consequences of Immigration.* Cambridge, MA: Basil Blackwell (1989).

Simpson, E.S., *The Developing World* (2nd edition). London: Longman (1994).

Soldo, B.J., and E.M. Agree. "America's Elderly." *Population Bulletin* 43, no. 3, Washington, D.C.: Population Reference Bureau (September 1988).

Teitelbaum, M.S. "Relevance of Demographic Transition Theory for Developing Countries." In S.W. Menard and E.W. Moen (eds.). *Perspectives on Population.* New York: Oxford University Press (1987).

Thomas, R.W., and R.J. Huggett. *Modeling in Geography: A Mathematical Approach.* Totowa, NJ: Barnes and Noble (1980).

United Nations. *Case Studies in Population Policy: China.* New York: United Nations Publications (1989).

U.S. News and World Report. "Population Wars." (September 12, 1994) pp. 54–56.

U.S. News and World Report. "10 Billion for Dinner, Please," by S. Budiansky, (September 12, 1994), pp. 57–62.

Vernez, G., and D. Ronfeldt. *The Current Situation in Mexican Immigration.* Santa Monica, CA: RAND (1991).

Ward, G.C. "India: Fifty Years of Independence." *National Geographic* (May 1997), Vol. 191 (5), pp. 2–57.

Way, P.O., and K.A. Stanecki. *The Impact of HIV/AIDS on World Population.* Washington, D.C.: U.S. Department of Commerce, Economic Statistics Administration, Bureau of the Census (1994).

Weeks, J.R. *Population: An Introduction to Concepts and Issues* (4th edition). Belmont, CA: Wadsworth (1989).

World Development Report 1990: Poverty. (published for the World Bank) New York: Oxford University Press (1990).

World Development Report 1991: The Challenge of Development. (published for the World Bank) New York: Oxford University Press (1991).

World Development Report 1995: Workers in an Integrating World. (published for the World Bank) New York: Oxford University Press (1995).

World Resources 1994–95: People and the Environment. New York: Oxford University Press (1994).

Yi, Z. *Family Dynamics in China.* Madison, WI: Wisconsin University Press (1991).

Zelinsky, W. "The Hypothesis of the Mobility Transition." *Geographical Review* 61 (1971), pp. 219–249.

8

EXTRACTIVE RESOURCES AND HUMAN-ENVIRONMENT RELATIONS

PART I: AGRICULTURE

Agricultural production systems are integral and indispensable components of regional economies of the SOUTH. These systems are largely oriented toward subsistence activities, where people rely on nature and the land to satisfy basic needs for food, clothing, fuel, and shelter. In light of this reliance, there has been a growing awareness that human beings, as manipulators of the environment, must take special care to manage agricultural resources. Proper management is becoming increasingly difficult in areas that are overworked and degraded.

An **agricultural system** may be defined as *any mutually interacting or interdependent group of agents, factors, and/or institutions involved in the planting and tending of crops (cultivation) and/or the caring for and raising of livestock (animal husbandry) for the purposes of producing food and fiber.* We may distinguish between **commercial agriculture**—agricultural systems which produce exportable "surplus," where a large portion of output is sold to consumers outside the production region—and **subsistence agriculture**—agricultural systems in which output is predominantly consumed within the production region.

Agricultural products from the SOUTH represent an important export base to many underdeveloped regions. As increasing pressures are placed on agricultural systems to meet local and regional demand, export possibilities are declining over time. Though most agricultural production in the SOUTH is subsistence oriented, there is

a significant amount of commercial production (that is, "cash crops" for export). Given the ongoing population explosion in the SOUTH, the future of cash crops and **non-food farming**—*the production of agricultural commodities for uses other than food (for example, cotton, sisal, hemp, flax)*—remains questionable.

Agricultural production patterns are greatly influenced by economic, sociocultural, and environmental considerations (see Table 8.1). The choice of what to produce and the success of production are highly influenced by the physical setting, circumstance, and the dynamics of local and regional economies (Tarrant, 1991). While it takes more than just land to ensure success in agriculture, the success of the world's agricultural systems depends largely on the human agents which oversee those systems. Given recent concerns about environmental degradation, resource mismanagement, land scarcity, and declining physiological densities, many questions arise as to the "sustainability" of food production systems in the SOUTH.

The term **sustainability** was originally introduced as a region-based concept. It was used *in reference to any production regime or reproducible resource that could be managed and sustained over an indefinite period of time.* This definition was later expanded by ecologists to include concerns over the environment and the interrelationships between living organisms (with considerations for nonliving components as well) in ways that would preserve the quality, status, and function of a regional system. Economic geographers have discussed sustainability in terms of the economic factors which sustain growth and development and promote improvements in the human condition or quality of life. Human geographers, sociologists, and anthropologists have incorporated considerations for human-environment relations, as well as the socioeconomic, physical, and cultural aspects of production systems in a working definition of sustainability.

While the use of the term sustainability varies widely among members of the scientific community, there is a commonality. Scientists insist that studies in sustainability take an interdisciplinary approach when attempting to understand evolution and change in regional production systems. This includes examination of the links between economic activities, environmental conditions, and human behavior and development (Brown et al., 1991; Toman, 1992) and the constraints imposed by biophysical limitations, resource availability, natural hazards, and the dynamics of human-environment relations (Jones, 1991).

According to Brundtland (1987), "sustainable development" refers to action and activities which satisfy the needs of human economic agents in the present without compromising the abilities of future generations to satisfy their own needs. For our purposes, we will define **sustainable development** *as the responsible management of production systems and/or extractive resources in ways that would allow production and consumption to be carried out indefinitely without debilitating the environment and without imposing impediments to progress in the human condition both now and in the future.*

Notwithstanding debate over the usage of the word sustainability and the lack of a universal definition, let us define **sustainable agriculture**—*as an agricultural system in which a sufficient level of per capita output or a given level of exports or surplus can be sustained or increased over an indefinite period of time through proper management of local and regional production factors and with little or no consequence to the environment.* Before questioning the sustainability of the major agricultural systems of the SOUTH, let us review some important terminology.

Note that we may distinguish between "intensive" and "extensive" agricultural activities. Table 8.2 highlights the attributes of intensive and extensive production systems, distinctions that will aid in the classification of agricultural regions. Based on the terms intensive, extensive, commercial, and subsistence, we may describe four types of agricultural systems. These systems are differentiated by yield, production intensity, behavior, end-use of products, and geographic attributes (see Table 8.3).

In virtually every underdeveloped country, the vast majority of the economically active population, in some way, shape, or form are involved in subsistence activities. The overwhelming majority of people in underdeveloped regions of the SOUTH are associated with one of the following production systems: extensive subsistence (ES), extensive commercial (EC), and intensive subsistence (IS) agriculture. Although some specialty and intensive commercial (IC) agriculture can be found in scattered pockets throughout the SOUTH, the total output from specialty systems are small in comparison to the overall

TABLE 8.1	Factors Which Affect Agricultural Outputs	
Economic Factors	**Sociocultural Factors**	**Environmental Factors**
Location of market(s)	Personal or group preferences	Relief
Availability of labor	Demographic change	Climate
Access to capital	Land tenure	Soils and geology
Availability of technology	Traditions and values	Susceptibility to hazards
Government policies	Attitudes toward risk	Degradation
Subsidies and regulation	Local knowledge of farmers	Amount of arable land
Financial resources	Attitudes toward technology	Transportation infrastructure
Efficiency of production system	Organizational aspects of agricultural system	
	Modes of transport	

Source: Tarrant (1991), p. 20, with additions.

TABLE 8.2	Intensive and Extensive Production Systems: A Comparison and Generalization
Production Systems	

Intensive	**Extensive**
High labor input per land unit	Low labor input per land unit
Relatively small land holdings	Large land holdings or coverage of vast areas
High levels of capital input per land unit	Low levels of capital input per land unit
High yields per land unit	Low yields per land unit
Commercial and/or subsistence oriented depending upon location	Commercial and/or subsistence oriented depending upon location
Specialized production zones (as determined by climate and location)	

Note: "Capital inputs" refer to any tools, implements, devices, structures, chemicals, fertilizer, machinery, or animals used in the production process.

TABLE 8.3	Four General Types of Agricultural Systems

Extensive Subsistence (ES) Agriculture

—a production system that is entirely subsistence oriented

—low overall production volume on a per land unit basis

—products extracted directly from nature

—output is limited by external and natural conditions

—agricultural products harvested for "use value" only

—variations in output due to changing ecology and physical setting

—overall output limited by regional carrying capacity

—humans exhibit little or no control over the land or the natural environment

—movement of people is common in this system to minimize the impact on the environment or as a reaction to changing environmental conditions

—most common to low-density regions and extreme environments (arid lands, mountainous terrains, etc.) of Africa, the Middle East, and North and Central Asia, and remote tropical areas of South America and Africa

Extensive Commercial (EC) Agriculture

—a commercial system characterized by large-scale, production-for-profit operations

—high overall regional production volume, but low output per land unit

—production costs are kept to a minimum due to size of operations (scale economies)

—manipulation of land and environment is necessary and commonplace

—emphasis is on surplus production and the growing of "cash crops" for export

—most common to the less remote tropical areas of Latin America, Africa, and Southeast Asia, and covering large land areas in Europe, North America, and Oceania

Intensive Subsistence (IS) Agriculture

—a production system for food and fiber that is subsistence oriented

—extremely high volume of output per land unit

—maximization of output regardless of cost

—high amount of labor input

—little concern over profit or exchange value of products produced

—less "economically efficient" than commercial systems

—a system that is highly susceptible to environmental degradation and natural hazards

—largely located in highly populated areas in rural Asia, along major river systems and their floodplains

Intensive Commercial (IC) Agriculture

—a production system for food and fiber that promotes maximization of output in relation to known or projected cost constraints

—a commercial system that is highly sensitive to regional and global market forces

—end-products produced for their exchange value

—a system which promotes economic efficiency in production and the maximization of profit

—most commonly associated with the premier agricultural lands of North America, as well as west and central Europe

Note: As a subgroup of IC agriculture, **"Specialty Agriculture"** (SA) is defined as *a production system that is highly specialized in terms of output and largely associated with unique climatic conditions or physical settings that favor a specific type of crop or activity.* Specialty agricultural zones are found relatively close to market (or in areas that have access to markets via a transportation system).

level of output generated in intensive subsistence and extensive commercial systems.

Let us proceed to discuss each system in turn.

EXTENSIVE SUBSISTENCE AGRICULTURE

Five hundred years ago, roughly two-thirds of humanity was engaged in subsistence activities in one form or another. Today the world is still oriented toward subsistence, yet the patterns of production are more regionally distinct. Extensive subsistence (ES) agriculture is a system that accounts for less than 5% of the world's population (Table 8.4), despite the fact that ES activities cover vast portions of the earth's land surface. Note that ES activities can only support a limited population base. Hence, these activities are usually found in areas of extremely low population density in the less-industrialized regions of the SOUTH.

Extensive subsistence agriculture is the predominant agricultural system of the world's major population voids, where population density is generally less than three people per square mile. ES agriculture is most generally associated with regions that are either extreme cold and dry (with low temperatures and low precipitation) or tropical (with high temperatures and high precipitation). ES agriculture is associated with low levels of output per worker and little or no surplus production beyond the subsistence group or region. In addition, ES agriculture is characterized by the use of simple farming tools, cultures with very little land ownership, and settlements that are largely "nonsedentary" (that is, where movement from place to place is commonly observed).

Extensive subsistence activities, although declining in terms of their overall economic significance, will most likely continue for two major reasons: (a) they have been spared from the intrusive forces of industrialization and modernization because of their isolation or remoteness; and (b) the extreme climatic conditions preclude competition from other agricultural or nonagricultural systems.

TABLE 8.4	Estimated Number of People Engaged in Extensive Subsistence Activities Worldwide

Activity	Percent of World Population	
	1990	2000*
Hunting and gathering societies	.04%	.02%
Nomadic pastoralism	.30%	.25%
Shifting cultivation	5.0%	4.6%

Source: U.N., Food and Agricultural Organization (1990).

* Forecast based on population growth trends.

We may identify three distinct types of extensive subsistence activities: hunting and gathering, nomadic pastoralism, and shifting cultivation.

Hunting and Gathering

In **hunting and gathering societies,** there is little manipulation of the environment beyond setting traps or nets to capture animals or the use of simple tools for hunting and harvesting food directly from nature. These societies subsist on animal meat and animal by-products, fish, various insects, fruits, berries, honey, roots and root crops, leaves, and food and fiber collected from trees, shrubs, and small plants. There is virtually no seed planting or cultivation of the soil, and no attempt to improve habitat or production conditions. It is a system in which nature provides everything that is needed to subsist. Hunting and gathering societies can be found over many diverse areas of the earth's less inhabitable land surface (Figure 8.1).

Some hunting and gathering societies in equatorial regions supplement food production by growing limited amounts of staple crops such as corn, beans, or cassava—a starchy tuberous root crop, high in carbohydrates, that is ground into flour and roasted, having a consistency of corn meal. Supplemental production is necessary to ensure that adequate supplies of food are on hand when harvesting from nature becomes less bountiful (say, during the dry season in tropical zones). In addition, a limited amount of animal rearing may occur throughout hunting and gathering regions.

Consider the peasant farmers living along the Amazon River Basin in central South America. During the wet season (roughly December through May), the region may receive up to 100" of rainfall, leaving much of the land and floodplain covered by water. Inhabitants of this region are exposed to a rich and varied harvest from nature with over 2000 species of fish and an abundance of diverse plant life. During the dry season (June through November), the waters of the floodplain recede and consumption patterns change to fruits, berries, wild strains and varieties of crops planted in upland and lowland fields. In addition, many farmers keep a limited number of animals on hand (for example, swine) to be consumed when fish stocks are low. With proper management of resources and the environment, subsistence practices in many of the more remote tropical areas could easily be classified as sustainable (Figure 8.2).

Some hunting and gathering societies, however, live dangerously close to the edge of starvation. Basic survival skills in many traditional hunting and gathering societies have been eroding through time, as these cultures have been transformed through contact with the modern outside world. This has created a situation in which many people have become dependent on outside regions for food and other necessities. Many once traditional hunting and gathering societies must now rely on various commercial activities to sustain themselves.

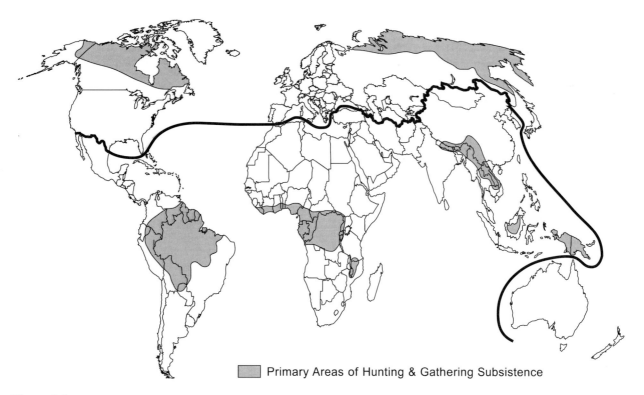

Primary Areas of Hunting & Gathering Subsistence

Figure 8.1 Subsistence agriculture: hunting and gathering.

Figure 8.2   Floodplain housing in Brazil's Amazon River Basin.
8.2: ©Anna E. Zuckerman/Photo Edit

Figure 8.3   The existence of modern technology in the Arctic circle region. Here a man takes his family for a ride on an ATV in Toksook Bay, Alaska.
8.3: ©Paul Souders/Gamma Liaison

Consider the case of the Alaskan Eskimos and other indigenous groups of the Arctic Circle region. Contact with the "white-man," mostly during the post-World War II whale-trade days (when the market for whale oil and other by-products were lucrative), has had tremendous repercussions on their society. For example, the introduction of the rifle as a means by which to hunt and exploit nature's abundant resources led to the deterioration of traditional hunting skills. It also reduced the likelihood that traditional skills would be handed down from generation to generation. The use of the rifle has entirely redefined the art of hunting in many cultures, creating further dependence on the modern world for ammunition, parts, and maintenance of this hunting tool.

The erosion of valuable hunting and survival skills and the replacement of old methods and implements (sustainable in their own right) with more modern approaches and technologies (poorly applied in many cases), have not only altered a way of life, but have threatened the longevity of this system. A once harmonious relationship between humans and nature has been replaced by a society that is highly dependent upon outside regions. A less-than-lucrative fur trade and a highly uncertain tourist and recreational industry have left many to question the sustainability of settlements in the region. The very future of the Eskimo remains in doubt as a result of modernization. Although it is likely that most of the larger settlements throughout the Arctic Circle will remain physically intact, many will shrink in number as people, primarily the young, out-migrate in search of a better way of life (Figure 8.3).

Nomadic Pastoralism

Nomadism refers to *a wandering lifestyle with little or no reliance upon sedentary cultivation or a given location*. It is based upon the purposeful movement of people in their attempt to survive and live off of the land. Migration is motivated by a constant search for water and sustenance in harsh and seemingly unforgiving environments.

Pastoralism *is the practice of breeding and rearing various herbivorous animals*. Hence, **nomadic pastoralism** may be defined as *the migratory practice of animal caretaking (that is, the tending of domesticated animals) in response to the availability of water and forage and changing climatic, environmental, or ecological conditions*.

Note that pastoralism is not unique to remote or peripheral regions of the world. In fact, pastoralist activities can be found in virtually every major agricultural system on the earth's land surface. Note that several types of pastoralist activities have been identified, including **commercial pastoralism**—*the herding of cattle and sheep conducted on the large grasslands of the world* (including segments of New Zealand, the rich Argentinean Pampas, and large tracts of land in the western United States and central Europe). As a commercial activity, these systems are primarily associated with surplus animal production for meat, hide, and fiber markets.

Subsistence herding or nomadic pastoralism, by contrast, takes place in fairly remote areas of the world; those with less-favorable production environments (Figure 8.4). Nomadic pastoralists do not produce in response to market signals, they simply produce enough to subsist. Though subsistence herding is viewed as a primitive exercise by Western standards, it can be argued that nomadic pastoralists possess a very high degree of sophistication given their ability to sustain themselves under adverse conditions.

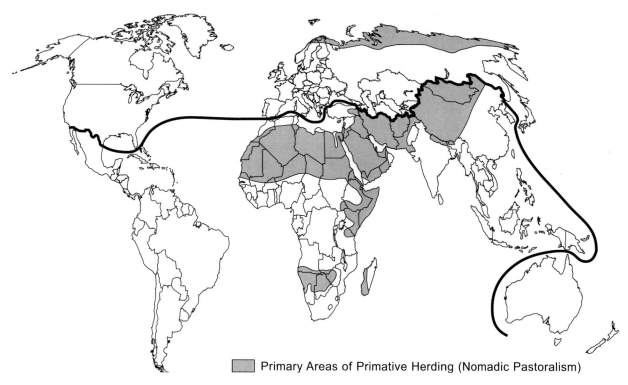

□ Primary Areas of Primative Herding (Nomadic Pastoralism)

Figure 8.4 Subsistence agriculture: primitive herding (nomadic pastoralism).

The pastoral practice of **transhumance**—*the moving of animals seasonally or periodically between two different climatic regimes or elevations*—can be found in and along many highland regions of the world. Animals are herded to the highlands or mountainous areas during the summer and transferred down to lowland areas or valley pastures during the winter. The practice is readily observed in the commercial grazing regions of the Alps (south central Europe, forming the arc from southern France to Albania), the Pyrenees (running between France and Spain, from the Bay of Biscay to the Mediterranean Sea), and the Rocky Mountains (extending from northern Mexico to Alaska, along the Continental Divide), as well as the subsistence regions of the Andes (west coastal areas of South America), the Himalayas (in south central Asia), and mountainous regions of East Africa.

Two of the more important groups of nomadic pastoralists in terms of numbers are: (1) the Bedouin tribes of the great Sahara, and Arabian, Syrian, and Northern African deserts; and (2) the seminomadic peoples of east central and south central Africa (Kenya, Tanzania, Somalia, Ethiopia), known as the Masai, who tend to engage in migratory activities and sedentary farming practices.

Bedouin tribes cluster in clans—a small group of people, typically an extended family. Clans migrate from place to place according to their animals' need for water and forage. Hence, all migratory patterns are purposeful and productive to the clan. Animals consist primarily of

goats, sheep, and camels, and occasionally some horses. The products in this system include animal milk and by-products such as cheese and meat. Clothing and shelter (tents) are made from animal hair, wool, and skins. Tools, toys, jewelry, and trinkets are commonly constructed from animal bones. Dried animal refuse (dung) is used for cooking fuel (Figure 8.5).

Despite the primitive appearance of nomadic pastoralists, it is not uncommon to see them taking advantage of modern technology. As with other subsistence groups, contact with the outside world and the subsequent adoption of modern ideas and innovations have occurred. Contact with contrasting societies has reshaped Bedouin culture in ways that seem odd to those adhering to more traditional practices and values. For example, it is not uncommon for the more affluent members of a clan to use trucks to carry water to herds when necessary. Many herders can now be seen with modern gadgets such as radios and Sony Walkmans. Wealth and status in this system, however, are judged by the number of animals a person owns; but this too is changing as the Bedouins have become exposed to Western ideology and technology.

Bedouin migration behavior is logically in tune with the forces of nature, and highly influenced by the availability of water and forage. The migratory practices common to this system are problematic for several reasons, however. Bedouins are independent people who tend to ignore political boundaries during their migratory cycles.

Figure 8.5 Nomadic pastoralists moving across the arid lands of the Middle East, where migration patterns are influenced by the availability of water and forage.

8.5: ©A. Ramey/Photo Edit

They may cross over several jurisdictions or political boundaries in their search for water and forage. Governments are not particularly fond of having people wandering back and forth across national borders, and, as a result, have made various attempts to force nomads to settle in one place.

Many efforts to "settle" nomadic pastoralists have met with stiff resistance, as sedentary lifestyles are not compatible with this system. Government-sponsored projects to create permanent settlements for nomadic pastoralists have had only limited success.

Perhaps the biggest obstacle to initiatives aimed at settling nomadic groups is the fact that loyalty of nomadic pastoralists is not so much to a particular country, but to their clans or tribes and sometimes to a particular region which has served them well. Transition from a nomadic culture to a sedentary lifestyle, however, is nearly complete in many parts of the world (including the former Soviet Union, Mongolia, China, and portions of the Middle East).

Though nomadic pastoralism represents a symbiotic relationship with nature, and therefore is a system that is sustainable, it is a traditional way of life that is continuously being threatened by economic and political forces. In general, nomadic pastoralism continues to decline in its importance for four major reasons. First, there has been enormous pressure from governments to create permanent settlements for nomadic people. Second, nomadic pastoralism is becoming increasingly incompatible with other sedentary land-use activities. Migration patterns have recently begun to intrude on irrigated agriculture, commercial grazing lands, and areas involved in mining operations or petroleum exploration. Third, the diffusion of modern culture and the exchange of ideas and technology have placed the system at risk. The system faces impending labor shortages as many of the younger members

of nomadic society opt to abandon a traditional lifestyle in favor of exploring alternative livelihoods and locations. Last, the natural forces of desertification and environmental degradation have dramatically reduced the regional carrying capacities of many arid lands.

Shifting Cultivation and Alternatives

Shifting cultivation is a practice that defines an agricultural production system that accounts for more people than any other extensive subsistence activity. According to recent estimates by the Food and Agricultural Organization (FAO), more than 270 million people are supported by this activity in areas that cover almost 20% of the earth's land surface. Geographically, we can identify three broad (equatorial bound) regions where shifting cultivation takes place: (1) Central Africa, covering large portions of Kenya, Zaire, Congo, Cameroon, and other tropical land areas; (2) a large area that stretches the vast tropical lands of Central America, across the Amazon Basin—from the Andes mountain chain to the Atlantic Ocean; and (3) a zone of dense tropical forestland that runs through Southeast Asia, with extensions to adjacent offshore islands (Figure 8.6).

Specifically, **shifting cultivation** *is a practice characterized by a predictable and destructive production cycle in which land is first cleared, crops are planted and harvested seasonally, and the land is abandoned once parcels cease to be productive in terms of yield.* Declining productivity is followed by the movement of the production system to another location where the cycle begins again. All in all, shifting cultivation plays a major role in feeding a fairly large number of people in the tropical regions of Africa and Asia.

Subsistence farmers today live much the same as their ancestors did thousands of years ago. Food preferences have remained fairly stable over time, as most farmers cultivate plants that were domesticated from wild plants native to the region. Crops commonly found in this system include: maize (the leading grain in the Western Hemisphere, a crop first domesticated in Mexico); rice (staple crop of Southeast Asia, originally from Thailand); millet (a small-seeded cereal grain and forage grass known for its abundant foliage and fibrous root system; its origin is traced to Uganda); and a variety of crops such as peanuts, beans, bananas, and tomatoes (a crop that originated in the highland of Peru). In addition, sugar cane (originally from the islands of Southeast Asia) and starchy "root crops" such as yams and cassava are favorites among subsistence farmers. Local varieties of yams were domesticated separately in Asia, Africa, and the Americas; cassava, on the other hand, is native to Columbia.

Cassava, a preferred crop in many subsistence regions, grows as a shrub with large tuberous roots. It produces a much higher yield of carbohydrates and calories

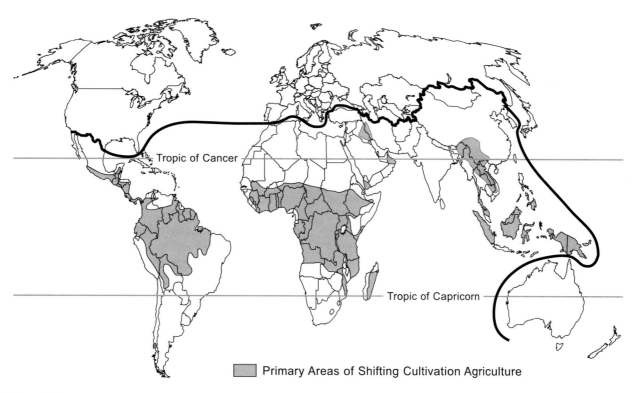

Primary Areas of Shifting Cultivation Agriculture

Figure 8.6 Subsistence Agriculture: Areas of Shifting Cultivation

than either maize or rice, and as a result is second only to yams as a food source per unit of land. Unlike yams, cassava also carries the distinction of being a "cash crop" as it is used as a thickening agent to make starches and tapioca.

Cassava is also more tolerant to drought than yams, and is highly resistant to insects and disease. These qualities led to the rapid diffusion of cassava throughout the tropics, making it a preferred staple crop amongst the world's peasant farmers. The world's largest cassava producers are shown in Table 8.5. Note that the top-six producers of cassava accounted for two-thirds of world output in the early 1990s.

Subsistence farmers engaged in shifting cultivation live in close accord with their environments. The following characteristics apply: (a) they exercise little control over their environments and are largely "adaptive" agents; (b) there is little or no use of draft animals or mechanical power in these systems (primitive tools such as digging sticks and hoes are used for planting and cultivating) and only limited use of other capital inputs such as fertilizers and pesticides; (c) they are people who live hand to mouth; (d) there is little concern for soil maintenance in many areas (as farmers simply change fields when yields become low); (e) there is no genetic engineering of plant stocks or use of hybrids (in some cases this has led to natural inbreeding of crop strains and the phenomenon of mutated crops); and (f) farmers produce relatively low levels of output per unit of land in comparison to other agricultural systems around the world.

TABLE 8.5	World Leaders in Cassava Production (1990–1992)

	Percentage of World Production*
Brazil	16%
Thailand	14%
Nigeria	13%
Zaire	12%
Indonesia	11%
Tanzania	5%

Source: *Goode's World Atlas* (1995), p. 39.

* Out of a total (average) production volume of over 150 million metric tons of cassava worldwide.

Note: A metric ton is roughly 2200 lbs.

In the equatorial regions of the Caribbean, Central America, South America, Africa, and Southeast Asia, the climate is warm year-round, and there is abundant rainfall, conditions that seem ideal for large-scale commercial agriculture at first glance. Yet, they are far from ideal conditions. In tropical rainforest areas, trees manage to flourish despite the poor soils and heavy rain. Their extensive root systems and dense canopy allow the tree to protect the relatively thin soils from the torrential force of the rain. When the natural vegetation is cut down, burned, and removed, the soils are directly exposed to the elements. During the rainy season, the nutrient-rich top-soil is washed away, leaving an exposed and unproductive subsoil.

Shifting cultivation or "slash and burn" agriculture, as it is sometimes called, is the biggest contributor to the loss of topsoil in the SOUTH. As peasant farmers slash and burn the natural vegetation cover in order to expose their fields to sunlight, and destroy the tropical rainforest in the process, they promote **leaching**—*nutrient loss and erosion*—of topsoil. The leaching process causes soils to lose their natural fertility after only a few growing seasons once the natural vegetation cover is removed. The duration of the shifting cultivation cycle is normally less than five years, as farmers are forced to abandon a working parcel due to declining productivity. Over time, the tropical rainforest becomes patterned with small clearings in various stages of recovery from the slash and burn practice. Continuous streams of smoke are commonly observed rising up to the air in virtually all remote tropical regions in close proximity to human settlements. The imprint of devastation on the economic landscape is unmistakable. The average recovery time of burnout land parcels is no less than 30 years. Some parcels never fully return to their original form. At various stages of recovery, the land and its existing vegetation are often set on fire to replace soil nutrients. The ash, however, offers only a temporary solution to the infertility problem. In many parts of the newly homesteaded Brazilian Amazon Basin, farms that were thought of as instruments of development and symbols of food security are now viewed as a testimony of environmental mismanagement.

Note that soils in tropical areas are extremely susceptible to leaching for several reasons. Tropical soils lack significant amounts of organic matter to begin with. For plants to assimilate soil nutrients, those nutrients must be "soluble"; that is, they must be easily dissolved in water. Heavy precipitation causes soluble nutrients to be washed away. The remaining materials, mostly nonsoluble iron and aluminum oxides, result in an infertile reddish colored soil, a soil that maintains a nice appearance, but is virtually worthless in terms of its productivity.

The harvesting of tropical hardwoods for industrial uses and slash and burn agriculture have been identified as the leading contributors to rainforest destruction and associated losses in biodiversity. From 1987 to 1989, it was estimated that 80,000 square miles of rainforest in the Amazon Basin were destroyed by shifting cultivation and hardwood extraction. In addition, over 50,000 square miles of rainforest were destroyed by fire in that same period, as slash and burn agriculturalists migrated to nearby areas in search of higher yields. Recent estimates place losses in excess of 10,000 square miles per year during the 1990s, despite the best efforts of government and industry to control the problem. Slash and burn is, once again, on the rise as local and region population pressures force agriculturalists to produce more food to meet a growing demand. Certainly, this system cannot be classified as something which is sustainable (Figure 8.7).

There are, nonetheless, sustainable alternatives to slash and burn agriculture, methods which are now successfully being used throughout the tropics. One such method is "rotational cultivation." In this system, cultivated fields are rotated (planted or left fallow) from growing season to growing season, as opposed to being used until yields decline. Abandoned fields are allowed to regenerate nutrients naturally. The harsh fertility declines associated with slash and burn are virtually eliminated as the natural vegetation recovers quickly once a parcel is idled.

After harvest, fallow land is left to regenerate itself for a period of anywhere from 7 to 15 years, ranging on the number of fields in the system and the size of the population to be supported. Idle fields are later cleared, fired, and cropped again when their turn comes due in the rotation cycle.

The practice of rotational cultivation results in a more stable human-environment relationship, allowing agriculturalists to support higher population densities with less need for physical relocation of families or a settlement. Obviously, sedentary subsistence activities such as this represent a more progressive agricultural system. When properly managed, these systems may be sustained indefinitely and even produce a modest amount of surplus, which can be used for exchange or export.

Sophisticated soil erosion controls have also been put into place. These controls include, although are not limited to the following: (a) the use of water drainage ditches to control run-off, especially on slopey lands where soils are susceptible to leaching during the rainy

Figure 8.7 Destruction of the tropical rainforest brought about by "slash and burn" agriculture.

8.7: ©Randall Heyman/Stock Boston

season; (b) the use of hedges, wind breaks, and residual vegetation cover to reduce wind erosion and lessen potential of soil leaching; (c) **agroforestry**—*systems which integrate natural vegetation and forest cover within a food production system* (typically the use of trees intermixed with shrubs and crops in working parcel) to help soils retain moisture and nutrients by establishing cover, protection, shade, and water-retentive root structures; and (d) the practice of **interculture**—*the planting and growing of multiple crops in the same field or parcel* (as opposed to monoculture).

Erosion controls and diversification of outputs in subsistence agricultural systems have made them less susceptible to failure and degradation. Science, innovation, and good old-fashioned common sense have combined to produce workable solutions to the problem. The interplanting of crops, such as oranges and banana with manioc/cassava, black pepper, and mango (with suitable erosion controls), in agroforestry settings is now a common practice in places like the Brazilian Amazon.

The switch from nonsustainable practices, like slash and burn, to sustainable alternatives has been significant over the past decade or so. Yet much more needs to be done in the way of modifying the traditions and behaviors of peasant farmers. With increasing market activity and exchange of agricultural products in the SOUTH comes an increasing interaction between those who have been successful in adopting sustainable practices and those who rely on outmoded and environmentally insensitive practices. The geographic diffusion of new production methods and ideas has brought about a newfound optimism for those trying to bring an end to the needless destruction of the rainforest.

In subsistence regions around the world, agricultural goods are often exchanged in what are called **periodic markets**—*an interlocking network of village markets that are active centers of trade and interaction, where people congregate in regular cycles (for example, every third or fourth day) to exchange goods and socialize.* These centers have been instrumental in the diffusion of ideas in subsistence areas. Visits to the periodic markets allow farmers and consumers exposure to local and regional trends in production, the latest farming methods, and changing crop alternatives and preferences. Each market in the network gets their turn to attract people from distances both near and far (allowing people in the largest region to be in the vicinity of at least one market). Goods of various kinds are brought to market for sale or barter by those travelling on foot, on the backs of animals, or by truck (Figure 8.8).

While shifting cultivation in the Amazon Basin area may be labeled as an extensive agricultural activity, there is tremendous variability in the "intensity" of production. Less remote production areas—those in close proximity to markets or large settlements—are greatly affected by market forces. While the emphasis is mostly on subsis-

tence crops, there may be a secondary emphasis on cash crops to sell or exchange at a nearby market. Hence, even remote subsistence regions can take on some of the characteristics of a commercial system (in a spatially limited sense). Though only a modest percentage of what is produced by local farmers in subsistence areas is transported to market, the system itself is still considered as a subsistence system given that little or no surplus will actually leave the larger geographic region.

Thus, within subsistence agricultural regions, we find variations in both **agricultural land use intensity**—*the degree to which available arable land is cropped in a given region (that is, the percentage of arable land in production)*—and **agricultural intensification**—*the degree to which labor and capital inputs are utilized to produce agricultural output on the average working parcel of land and/or the extent to which crop output is diversified in response to any number of economic, sociocultural, and environmental factors which influence local production.* Agricultural intensification is directly related to the size and growth rate of a region's population and the degree to which a region has accessibility to outside market(s). High levels of intensification have been linked to land degradation in areas engaged in nonsustainable agricultural practices.

It is widely accepted that agricultural land use patterns and agricultural intensification are greatly affected by land and labor availability, physical land characteristics, climate, crop preferences, perishability and transportability considerations, prevailing mode(s) of transport, the overall efficiency of the transportation system, agricultural price levels or the exchange value of crops, and the nature of local and regional demand.

Of all the factors which affect production, it is distance to market that plays the leading role in influencing agricultural intensification. The closer farmers are to market(s), the greater the influence of the market(s) and the more likely farmers are to respond to market signals.

Figure 8.8   A typical "periodic market" in Africa where goods, food, and conversation are traded.

8.8: ©Anna E. Zuckerman/Photo Edit

Hence, distance to market is a factor which will greatly affect farmers' decisions to grow specific types of crops. In general, it can be stated that agricultural intensification and crop diversity increase with decreasing distance or increasing accessibility to market. Hence, less-intensive agriculture and specialization is generally observed at greater distances to market. In addition, it can be noted that the size of the average working parcel of land is smaller the closer to market. Typically, small farming operations tend to be more intense in terms of agricultural practices.

Predictability in cropping patterns and agricultural land use patterns are attributed to the work of the Prussian land economist Johann Heinrich von Thünen and his book entitled *Der isolierte Staat*, published in 1826. **von Thünenian analysis** *focuses on the value of arable land (in a production activity) in relation to agricultural product prices, demand, yield, transport costs, and relative location.* By observing the logic behind agricultural land use patterns in southern Germany, von Thünen posited that the location of market(s) played a dominant role in guiding production decisions. Moreover, von Thünen theorized that (over time) agricultural land would generally gravitate toward its **highest and best use**—*that activity in which it is most profitable, most competitive, or best suited for given localized conditions and constraints.* The principle of highest and best use can be applied to virtually any market-oriented agricultural or nonagricultural land use system.

Perishability and transportability are important considerations when examining the production potential of a given crop in a given area. It is not uncommon to see a wide variety of diverse and highly perishable crops (for example, vegetables and fruits) being grown at a short distance to market, whereas more transportable varieties of crops and those less perishable or bulkier (for example, maize, wheat, or potatoes) being grown at greater distances from market. Note also that the degree to which farmers have access to one or more markets is dependent upon the nature of the transportation system and the available modes of transport. The road network and road quality play important parts in influencing production patterns, farming behavior, and the movement of agricultural products. In addition, the physical environment can have a profound impact on production patterns.

Consider the observable differences in the crop selection of farmers in lowland versus upland areas of the Amazon. In particular, several observation can be made for farmers with access to the Monte Alegre market. First, there is a greater tendency toward crop specialization on the floodplain, whereas upland areas tend to be more crop diverse. Second, farmers on the floodplain tend to lean more toward marketable products such as short-season and highly perishable vegetables, which can be quickly and easily transported by boat. As upland transport (typically by truck) must cover rough and rugged terrain to reach market, the production of fragile crops such as melons are more apt to be produced in the floodplain. Upland crop production is dominated by more transportable crops such as manioc, oranges, rice, beans, and bananas. Third, intensification and diversification of production are not only affected by access to market, but are also influenced by constraints placed upon the system by the biophysical environment. Residents of the floodplain must face flooding, excessive rains, and pest management problems, while farmers in upland areas are more concerned with occurrences of drought (Swales, 1998).

INTENSIVE SUBSISTENCE AGRICULTURE

Sustainable agriculture can take on many forms. The success of any agricultural system is not just a question of whether a system is highly productive but whether that system has the ability to consistently and continually provide adequate amounts of food and fiber to meet an increasing demand. Of all the agricultural systems in the world today, perhaps the greatest burden has been placed on a system which supports the most populated and fastest growing regions of the world. In South and East Asia, "intensive subsistence" (IS) agriculture is, and will remain, the principal food production system. No other system in the world produces more on a per land unit basis than IS agriculture. It is a system in which farmers' main objective is to maximize output regardless of cost.

Leading nations in this system include China, India, Thailand, Bangladesh, Vietnam, Indonesia, Philippines, Laos, Cambodia, Myanmar, and Malaysia, spanning a geographic area that contains the two largest population concentrations on the planet (see Figure 8.9). Approximately two-thirds of the world's farmers are involved in IS agriculture. It is a system that helps feed roughly 60% of the world's population. Intensive subsistence agriculture is characterized by little or no surplus production; that is, virtually all that is produced is consumed within the IS production region. If agricultural exports are observed, they are usually earmarked for deficit-producing subsistence regions in close proximity to the source (typically other highly populated and nearby areas). IS agriculturalists are savvy and sophisticated producers. They exert an extraordinary amount of influence and mastery over the physical environment. Although not economically efficient, this system is responsible for producing the highest yields of any farming system in the world.

In general, intensive subsistence agriculture is a system typified by agents with small land holdings, extremely high amounts of labor input per unit land, and relatively high amounts of capital inputs (in the form of simple farming tools and draft animals for plowing).

The dominant crop in this system is rice, a crop that was traditionally produced in monocultural settings.

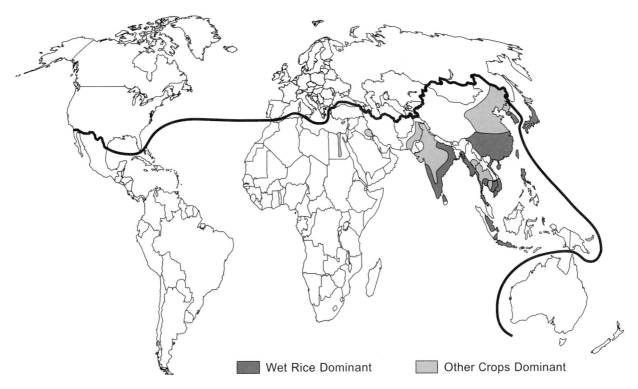

Wet Rice Dominant   Other Crops Dominant

Figure 8.9 Subsistence agriculture: intensive subsistence.

Typically, "paddy" or wet rice is grown in small plots (typically one acre or less) on submerged parcels of land along the major floodplains, deltas, and valleys of the major river systems found throughout South and East Asia. These fertile areas provide a constant supply of nutrient-rich "alluvial" soils, from deposits of sediments that are continuously transported down the rivers and tributaries during the rainy season. Topsoil depth can reach as high as three feet. The nutrient-rich topsoil rests on an underlying clay subsoil that prevents water seepage. Hence, water levels remain high throughout the growing season (Figure 8.10).

Premier rice growing regions are found along river systems in eastern India and Bangladesh (the Ganges and the Brahmaputra rivers) and areas associated with the major floodplains of China's three longest rivers—the Yangtze, Huang He, and Xi Jiang. In the rural agricultural districts associated with these river systems, population densities often exceed 2500 people per square mile.

Consider that only 10% of China's total land area can be used for agricultural purposes. Nearly all this land is currently under cultivation. China is the undisputed world leader in irrigated agriculture. Roughly 50% of China's arable land is irrigated. The burden placed on China's agricultural system is enormous when one considers the fact that China has less total arable land area than the United States (96 million hectares as compared to 189 million hectares) and almost five times the U.S. popula-

Figure 8.10 A peasant farmer working in a rice paddy in China on the floodplain of a major river system.
8.10: ©Paul Conklin/Photo Edit

tion. It is no surprise that virtually all the food produced in China is consumed within its borders.

Note that approximately 80% of China's cropland (which includes irrigated farmland) is devoted to food crops. Rice cultivation accounts for just over 30% of the total cultivated area (and the greatest amount of output in terms of tonnage and value). China's second-most important food crop, wheat, occupies large tracts of land along the North China Plain and the fertile valleys of the Wei and Fen rivers. Other crops in China's intensive subsistence farming regions include sorghum and millet (two important food and feed crops), corn, oats, sweet potatoes, and a wide variety of fruits and vegetables. China is also a leading producer of soybeans, peanuts, and sugar.

With the intensified use of fertilizer in IS systems and the emphasis on output maximization, it is not uncommon to see several crops being planted and harvested along the floodplains. The growing season for wet rice is approximately four to six months in length. This allows for **multiple cropping**—*multiple sequences of planting and harvesting of a single crop in a given year on the same parcel*. Hence, two or three rice crops may be harvested in a given year as the replanting of rice occurs directly after harvest. **Intercropping**—*the simultaneous cultivation of two or more crops on the same parcel* is also observed in some regions. For instance, aquaculture or fish farming in rice paddies is not an uncommon practice. Intercropping also involves the cultivation of other crops on the rice paddies' edge.

The overall production output of IS systems is staggering. Note that rice yields are typically over 3500 lbs per acre. Differences in agricultural output are largely a function of variations in fertilizer use. The leading rice-producing nations of the world are listed in Table 8.6. These regions are also leading producers of fertilizer.

Note that rice production is highly correlated with population size or demand. The eight leading rice-producing nations account for approximately 85% of world rice production. Note that two of the top-four rice exporters in the world are not top producers (see Table 8.7). With the exception of Thailand, the highly populated nations of South and Southeast Asia export very little rice. In fact, total rice exports represent only a small fraction of the rice that is produced within this system. Worldwide, rice exports are less than 3% of total production. Together, Thailand and the United States accounted

for about half of the world's exportable rice surplus in the early 1990s.

It is interesting to note that distinct locational patterns emerge in the Asian IS system (see Figure 8.9). Wet rice production is largely restricted to areas along major floodplains and deltas. Just outside the floodplain, in higher elevations, "upland" or dry rice is grown. Upland rice production has left a distinct mark on the landscape as it is generally grown on terraces cut along the edge of hillsides and slopey valley regions. Upland rice effectively competes with crops such as maize, wheat, and corn in these areas. In geographic areas located outside major river systems (for example, India's Northwest and Central Punjab Region and Northeastern China) we find a wide range of crops including wheat, cane sugar, sesame and sunflower seeds, coconuts, millet, sorghum, and tea.

All in all, intensive subsistence agriculture produces very little exportable surplus. Growth of food output in many of these regions may not be able to keep pace with the rapidly increasing demand for food. With virtually all arable land currently in production in these systems, expansion must come from further intensification. Proper management of land-based resources in the SOUTH requires that these systems do not attempt to exceed intensification thresholds as they attempt to meet the growing demand for food. Should forthcoming production deficits be satisfied by outside suppliers, the prospects for sustainability of this system are excellent.

SPECIALTY AGRICULTURE

There are several significant pockets of "specialty agriculture" in the SOUTH. Specialty agriculture is a system which covers limited geographic areas. A large proportion of the crops produced in these specialty production zones are earmarked as exports to markets in the NORTH. Of the specialty agricultural systems that do exist, two stand out in terms of their overall importance: (1) Mediterranean agriculture; and (2) natural fibers production.

Mediterranean agriculture is a system that is largely defined by climatic conditions. This activity is located in

TABLE 8.6	World's Leading Rice Producers (1990–1992)
Nation	**Percentage of World Production***
China	36%
India	21%
Indonesia	9%
Bangladesh	5%
Vietnam	4%
Thailand	4%
Malaysia	3%
Japan	3%

Source: *Goode's World Atlas* (1995), p. 38.

* Note that average annual world production of rice (1990–1992) was over 500 million metric tons.

TABLE 8.7	World Leaders in Rice Exports (1990–1992)
Exporting Nation	**Percentage of World Exports***
Thailand	33%
United States	17%
Vietnam	11%
Pakistan	8%

Source: *Goode's World Atlas* (1995), p. 38.

* World exports of rice average just under 14 million metric tons per year (1990–1992).

zones that are primarily suited for grape and citrus production. Geographic locations include the Central Valley of California, central Chile, the Mediterranean Basin region of southern Europe, Northern Africa and the Middle East, and the coastal regions of South Africa (see Figure 8.11). Main attributes of this system include nearness to a large body of water, areas that receive maximum precipitation during the winter months (with 10"– 40" of rainfall annually), regions with fairly mild temperatures year-round and a continuous growing season. In addition, Mediterranean agriculture is typified by dry summer seasons with plenty of sunshine prior to harvest (Figure 8.11).

Production includes grapes (as part of a growing viticulture industry) for wines, table use, and raisins. Grapes do well in Mediterranean climates for several reasons. The deep-rooted structure of grape vines provide access to groundwater during the dry season. Aboveground, the vines also provide shade and keep soil cool, minimizing moisture loss during the hot summer months. Grapes also have the ability to maintain a favorable sugar-to-acid balance, even in droughty environments. Note that the quality of harvest is extremely sensitive to microclimatic variations, not to mention soil type, slope and sun-angle

orientation of land parcels, and growing methods. Citrus crops are also found in this system (including oranges, grapefruit, lemons, and limes). Other big crops or activities in the Mediterranean agriculture include the production of olive oil and jarred olives.

Although Mediterranean agriculture is extremely important to the regions contained within this climatic regime (particularly for the limited number of areas in the SOUTH which fall into this system), it represents a very small amount of the total agricultural base of the SOUTH. Of much greater importance to the SOUTH are crops which are grown for fiber. Natural fiber production refers to crops grown for textile markets. These crops are used to make clothes, ropes, sacks, and industrial filaments and materials. Of all natural fibers produced in the world today, cotton is the most significant in terms of output and geographic coverage (see Figure 8.11). World production of cotton stands at over 20 million metric tons per year, with production highly correlated with population and demand. The United States is the world's second largest cotton producing nation, accounting for roughly one-fifth of world output. The nations of China, India, Pakistan, and Uzbekistan account for well over 50% of the world cotton production.

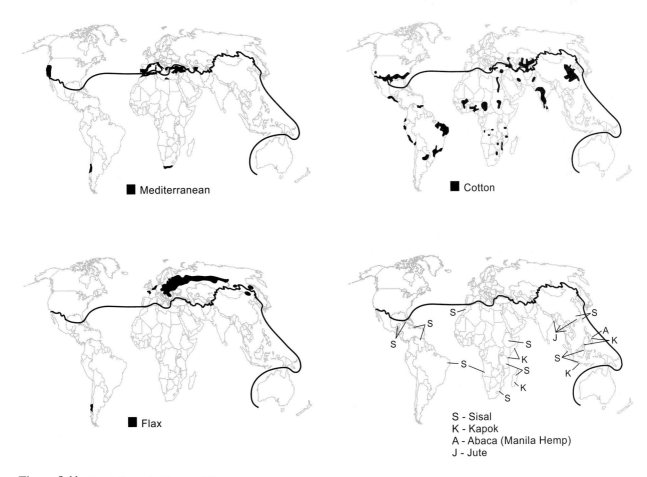

Figure 8.11 Specialty agriculture and fibers.

Another important crop grown for fiber is jute. It is the second-largest fiber crop next to cotton in the world today. Jute is used to make products such as ropes, sackcloth, and carpet backing. The top producers of jute include India, Bangladesh, China, and Thailand. These four nations account for well over 90% of the world's jute production, which now totals well over 3 million metric tons per year.

Other natural fibers produced in the SOUTH include sisal, a durable white fiber that is used to make cords and binder twine. Sisal is grown throughout southern and eastern Africa, Central America, Brazil, China, and Southeast Asia. Production of kapok, from the silky fibers that dress the ceiba tree, is a crop used as fillers for mattresses, cushions, sleeping bags, and life preservers. Kapok is primarily grown in West Africa and Southeast Asia. Another important fiber crop is flax. Flax is a long, slender plant with silky fibers that is used to make linen. Its seeds are also used for oil and cattle feed. Mostly grown in eastern Europe and the former Soviet Union, there are significant producing regions in Chile, Argentina, and China. The production of abaca (or manila hemp) is also important to this system. Abaca is a fiber obtained from the leafstalk of a banana plant native to the Philippines, now mostly grown in Southeast Asia and Central America. It is used to make cloth, durable cords, mats, and rope. Last, but not least, is silk, a lustrous and tough elastic fiber spun from the protective cocoon (larvae) of silkworms and used for thread, yarn, and fabric. The production of silk is a highly specialized activity that continues to thrive in India (Figure 8.12).

Given population pressures in the SOUTH and the proliferation of "synthetic fibers," which are mostly produced in the NORTH, food crops have become increasingly competitive in fiber production systems of the SOUTH. Despite increased competition and a growing world demand for fibers and textiles, prices in natural fiber markets have remained relatively low. This is due to the emergence of product substitutes—cellulosic fibers like rayon and acetate, and noncellulosic fibers like nylon and polyester—and bountiful annual harvests of natural fibers throughout the developing world. Nonetheless, the fiber industries of the SOUTH constitute a sustainable export activity that will continue to support industrial growth and modernization.

The increasing demand and production of natural and man-made fibers and the growth of a globalizing textile industry have become increasingly important to economies like India, China, Pakistan, and Bangladesh. Governments have encouraged and supported the expansion of textiles at three levels:

1. "cottage industries"—home-spun or village-based operations involved in the local production of yarn and handloom cloth;
2. "labor-intensive" garment production—large-scale operations that take advantage of economies of scale and low-wage labor to produce a myriad of hand-made cloth and clothing; and
3. "fully mechanized" spinning and weaving operations—ultra-large-scale facilities that produce in high volume for both regional and overseas markets.

Despite the need to intensify food production efforts in many regions of the SOUTH, fiber production has become an essential ingredient to the agricultural-led industrialization and development of many peripheral and semiperipheral economies.

EXTENSIVE COMMERCIAL AGRICULTURE IN THE SOUTH

In contrast to subsistence activities, commercial agriculture in the tropics represents an important export activity. Crop output in this system comes from two main sources: (1) from small independent peasant farming operations; or (2) larger-scale tropical plantations. While the former system is responsible for producing a diverse mix of crop outputs for both local/regional consumption and export, where crop variety is highly dependent upon such factors as farmers' preferences, tradition, price levels, and the physical setting, the latter system is almost exclusively oriented toward production for export to major world markets (mainly in the NORTH). As many small peasant farming operations engage in both subsistence and commercial production depending upon their relative location to market(s), it is hard to distinguish many extensive subsistence areas from extensive commercial areas. Nonetheless, if a large share of the end product from farming operations in the SOUTH reaches consumers in the NORTH, those operations are considered as part of the SOUTH's extensive commercial agricultural system.

Figure 8.12   Worker spinning fabric in a textile mill in Manali, India.

8.12: ©Jane Tyska/Stock Boston

TROPICAL AGRICULTURE

As a geographically diffuse, diverse, and extensive system, **"tropical plantation agriculture"** has emerged as the premier commercial agricultural activity of the SOUTH. Its roots can be traced to 19th century colonial rule and the establishment of a global market for agricultural outputs. It is a low-latitude activity, generally associated with the world's equatorial regions, that is highly specialized in terms of production orientation and output. The large majority of tropical agriculture is geared toward the production of one or several cash crops destined to be sold in markets located at great distances from the production region (Figure 8.13).

Tropical plantation agriculture is a highly labor-intensive activity, requiring large amounts of hand labor for planting, crop maintenance, pruning, and harvesting.

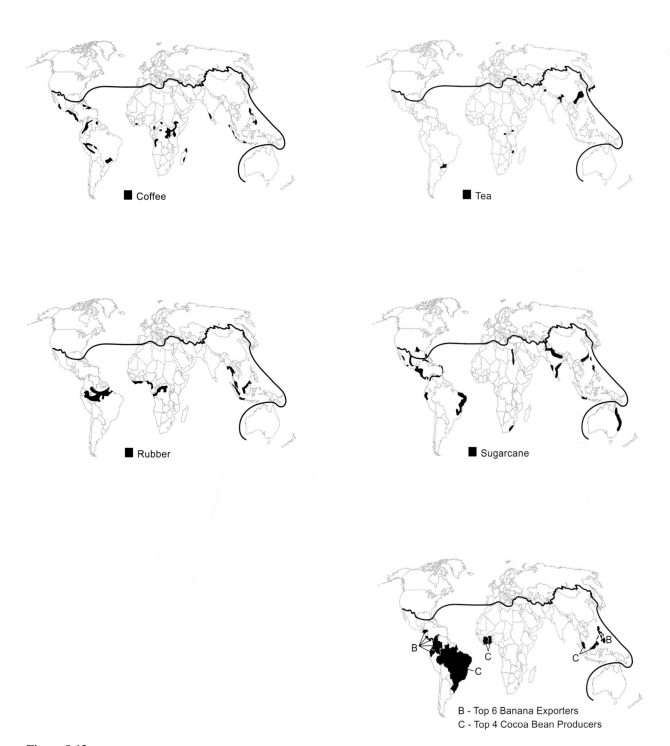

Figure 8.13   Tropical plantation agriculture: major cash crops.

This system is reliant upon an imported and seasonal labor force that is low-skilled and poorly paid. Hence, it is a system where exploitation of labor is commonplace. In addition, it is a spatial extensive system, usually involving large-scale operations (or plantations) concentrated along coastal areas or along major transportation corridors (near railways and navigable riverways).

It is interesting to note that tropical plantation agriculture generally involves production of crops that have been transplanted from another region of the world. Major products from this system include banana, cacao, coffee, copra, palm oil, (natural) rubber, sugar from sugarcane, tea, as well as various spices, nuts, and fruits (Figure 8.14). Top producers of some of the more popular commodities from this system are highlighted in Table 8.8.

Tropical plantation agriculture, the largest cash-crop system in the SOUTH, has a future that is somewhat unpredictable. Much of the uncertainty stems from the fact that (a) economic assessments on the viability of this system must be made on a crop-specific basis given the dynamic nature of market forces which influence the price, profitability, and production of various tropical plantation crops; (b) prices of commodities associated with this system have plummeted over that past decade (a by-product of increasing supply and increased competition among production regions); (c) mismanagement of genetic resources has threatened the long-term viability of several major plantation crops (including coffee); and (d) synthetic and natural substitutes for many tropical plantation crops have been successfully developed and marketed throughout the world.

Several of the leading cash crops in the tropics have direct product substitutes that are now competitively produced by core economies. For example, sugar from sugar beets is widely produced throughout Europe and the United States, not to mention the fact that product substitutes (like synthetic sweeteners) are now competitive in the sugar market. Big sugar beet producers include Germany, France, Ukraine, the United States, and Italy (which account for half of world sugar beet production). Despite the emerging sugar beet industry, sugar from sugarcane still represents an important export crop for nations like Cuba and Brazil.

In many of the tropical regions, production is biased toward a single tropical crop. For example, consider the many tropical plantations in Central and South America where banana is the preferred crop and export. As such, the leading banana-producing nations have earned the reputation of being part of the so-called "Banana Republics." Specialization in a relatively low-value export, however, has meant low overall export revenue. The

Figure 8.14 Peasant farmer harvesting sugarcane in Cuba.

8.14: ©Paul Conklin/Photo Edit

TABLE 8.8	**Top Producers of Selected Tropical Plantation Crops (based on average annual production from 1990–1992)**

	Top-four Producing/Exporting Nations by Rank (% of world production shown in parentheses)			
Crop	**1**	**2**	**3**	**4**
Banana exporters	Ecuador (24.3)	Costa Rica (15.5)	Colombia (13.4)	Honduras (7.6)
Cacao	Côte d'Ivoire (31.6)	Brazil (14.1)	Ghana (11.2)	Malaysia (9.6)
Coffee	Brazil (23.7)	Colombia (15.8)	Indonesia (6.9)	Mexico (5.4)
Copra	Philippines (40.0)	Indonesia (26.7)	India (8.8)	Mexico (3.7)
Palm oil	Malaysia (51.5)	Indonesia (22.8)	Nigeria (7.1)	Thailand (2.0)
Rubber (natural)	Indonesia (25.1)	Malaysia (24.5)	Thailand (23.8)	India (6.4)
Sugar from cane	India (17.5)	Brazil (12.2)	Cuba (10.2)	China (7.8)
Tea	India (28.4)	China (22.4)	Sri Lanka (8.8)	Indonesia (6.3)

Source: *Goode's World Atlas* (1995), pp. 36–39, p. 41, and p. 45.

Caribbean Basin has a similar problem, as the dominant plantation crop in the islands is cane sugar, another low-value commodity.

World price levels for sugar have remained low due to the proliferation of the sugar beet industry and a reduced demand for imported sugar in core economies as a result of government subsidies, quotas, and price supports to protect domestic sugar producers from (low-cost) outside competition (as in the case of the United States and Europe). Note, however, that the artificially high price of sugar in the U.S. market and the increasing demand for sugar led to the development of sweeteners, both natural and artificial. In 1987, "high-fructose corn sweeteners" became the sweetener used in most processed products, capturing more than 50% of the U.S. market. The demand for sugar from sugarcane has also declined due to the growing use of artificial sweeteners such as saccharin and NutraSweet. This is especially true now that there is more awareness about one's diet intake of "empty calories." Artificial sweeteners combined with fat substitutes make it possible to produce fat-free, sugar-free, low-calorie ice cream that is hard to distinguish from the real thing. The success of product substitutes has dramatically reduced the power of sellers in the marketplace, a situation that can have grave consequences on economies which are heavily reliant on crops like sugar as an export (such as nations in the Caribbean). Given recent international agreement to reduce trade barriers and eliminate subsidies on agricultural products, it would appear that there might be a growing market again for imported sugar in the United States and Europe. However, with the increasing use of artificial sweeteners and corn syrup, there will be less and less demand for sugar from cane. Note also that most tropical environments are not particularly suitable for growing corn, so these regions cannot compete in the corn syrup market. Even as the markets of core economies continue to open up, the world price of sugar is not likely to rise above its current level. If prices of sugar remain stable, then there is no significant economic benefit for sugarcane-producing regions to expand sugar production and exports.

As tropical plantation exports go, banana offers little as a money-making commodity given its relatively low value per unit weight. Native to southern Asia, and later transplanted to the Western hemisphere in the early 16th century, banana serves as both a cash crop and subsistence food crop in many tropical areas, particularly in Latin America. Production of banana is a highly labor-intensive activity in terms of harvesting and planting, as each banana tree can yield only one crop in its lifetime. Top banana producers and exporters include Ecuador, Costa Rica, Columbia, and Honduras. Although banana prices are likely to remain low for some time to come, the future export growth potential of this crop is bright. Banana is a favorite in modern Western society and no product substitutes exist.

Note that the development of synthetic substitutes has, however, slowed the growth of demand for natural rubber. Rubber from the rubber tree (native to the Amazon Basin, and smuggled out to Southeast Asia where it now predominates) is a tropical plantation crop that seems to being doing well despite the expansion of the synthetic rubber market. Roughly 80% of the world's natural rubber now comes from just four nations: Malaysia, Indonesia, Thailand, and India. Minor producers of natural rubber include the Philippines and Sri Lanka.

The natural rubber industry has shown modest yet significant growth. The increasing demand for latex and the enlarging market for radial tires in Asia and elsewhere (in response to the growing demand for automobiles and trucks) is likely to keep this industry afloat for some time to come. A "boom," however, is unlikely for several reasons. First, plantation output cannot keep pace with world demand. Second, synthetic substitutes have been developed and are widely utilized in regions with the largest markets for rubber and rubber products. Big synthetic rubber producers (in descending order of importance) include the former Soviet Union, the United States, Japan, Germany, and France. Together this group accounts for more than 70% of the world's synthetic rubber production, which now stands at over 10 million metric tons annually. Synthetic rubber now accounts for about two-thirds of all rubber produced worldwide. The flood of synthetic rubber in the world market has kept price levels of natural rubber low, minimizing the export-related benefits of rubber as a cash crop (Figure 8.15).

Coffee is a relatively strong cash crop despite its low value per unit weight. Native to the Ethiopian highlands, coffee has emerged as the quintessential cash crop of the

Figure 8.15 Farmer cutting tree to extract natural rubber on a tropical plantation in Sumatra, Indonesia.

8.15: ©A. Ramey/Stock Boston

tropical plantation system. Coffee production is concentrated in Central and South America, Africa, and Southeast Asia, with the majority of output flowing to the United States, Germany, France, Japan, and Italy. These top-five coffee importers are the final destinations of two-thirds of world's coffee (Figure 8.16).

Despite a seemingly strong world coffee trade, numerous questions have surfaced regarding coffee's future potential as a revenue generator as coffee prices have remained low in the world market. Low coffee prices are likely to continue as there is intense competition between existing producers around the globe. Also, the existence of product substitutes such as tea and synthetic caffeine continues to diminish the overall importance of this key agricultural export of the SOUTH. Nonetheless, the coffee habits of Americans and Europeans have remained strong. The greatest potential lies in the production and exportation of specialty coffees and regional blends for the growing legion of discriminating coffee drinkers in core and semiperipheral economies.

Tea, a crop native to the subtropics of Asia, is now grown in many areas of the SOUTH. It is a crop that re-

Figure 8.16 Workers harvesting coffee beans on a coffee plantation in Chamba Valley near Lusaka, Zambia.

8.16: ©Wolfgang Koetner/Gamma Liaison

quires intensive pruning and significant amounts of hand labor. As the quality of tea is highly dependent upon soil and local conditions, tea production has been favored in only a limited number of geographic locations. Two general types of tea are regularly consumed: processed "black tea" (where tea leaves are picked, withered, fermented, and fired) and "green tea" (where withered tea leaves are steamed to stop fermentation, a transformation process that produces a smooth and mild-tasting tea that is extremely popular in Asia).

All in all, the production and export future of tea is good. Though demand for tea in the NORTH is highly dependent upon coffee prices, as coffee is a direct substitute, the tea industry will remain competitive for a long time to come. Tea is a favorite commodity of consumers in both the NORTH and the SOUTH. It is a particularly strong commodity in Asia. The largest importers of tea include the United Kingdom, Russia, Pakistan, the United States, Egypt, Iran, and Japan.

Cacao is a tropical plantation crop that is in no real danger of losing ground to product substitutes. Cacao, from the dried and partially fermented seeds of fleshy yellow pods of the cacao tree, is a highland tree crop of South American origin. It is now widely produced in West Africa and used for a variety of products including chocolate and cocoa butter. The future export potential of this crop is bright as chocolate is also a favorite commodity of consumers around the world.

Tropical plantation agriculture is also noted for the production of oils for cooking and other purposes. Copra, from the white inner-lining of the coconut, is a South Seas native. It is largely used to make vegetable oil. Leaders in copra production include the Philippines and Indonesia, nations which account for two-thirds of world production.

Palm oil, from the fleshy fruit of the African oil palm tree, is another crop that is likely to do well. It is used as a cooking oil and is a major ingredient in margarine, as well as soaps, candles, and various lubricants. Palm oil production is primarily a Southeast Asian activity. This region accounts for more than three-quarters of the world output. Worldwide, copra and palm oil have promising and competitive futures despite recent trends in the West to switch to healthy alternatives to using vegetable oil in cooking.

All in all, tropical plantation crops offer only a limited export potential for the economies in which these crops are now grown. Although tropical plantation agriculture is a system that can be viewed as sustainable in its own right, as it has flourished for hundreds of years and continues to grow, there is evidence to support the contention that its relevance as an economic activity is waning for several reasons. First, revenues generated by the sale of tropical plantation commodities continue to become a smaller and smaller percentage of the world's total revenue from all production and exchange activities. Second, and despite positive growth in demand and production for tropical products, revenues associated with the sale of these products have declined. This is due to the fact that price levels have dropped faster than production levels have increased during recent times (see data in Tables 8.9 and 8.10).

TABLE 8.9	World Production Trends of Selected Crops		
	Average Production in Metric Tons (×1000)		**Percent Increase in Production**
Crop	**1980–1982**	**1990–1992**	
Key Tropical Plantation Crops			
Cane sugar	70,732	74,245	4.9
Rubber (natural)	3826	5117	33.7
Banana exports	6971	10,255	47.1
Copra	4823	4880	1.1
Palm oil	5605	12,042	114.8
Cacao	1616	2409	49.0
Coffee	5255	6042	14.9
Coffee exports	3841	4936	28.5
Tea	1885	2539	34.6
Other Crops of Interest			
Beet sugar	35,375	38,856	9.8
Rubber (synthetic)	7964	9878	24.0
Cotton	14,647	19,182	30.9

Source: *Goode's World Atlas* (17th edition), 1986 and *Goode's World Atlas* (19th edition), 1995.

TABLE 8.10	World Prices for Selected Commodities			
		Average World Price per Unit in 1990 U.S. $		**Percent Decrease in Price**
Commodity	**Unit**	**1980–1982**	**1990–1992**	
Key Tropical Plantation Crops				
Cane sugar	kg	0.55	0.22	−60.0*
Rubber (natural)	kg	1.80	0.99	−45.0*
Banana	mt	536.0	510.9	−4.7
Copra	mt	531.6	289.4	−45.5*
Palm oil	mt	742.1	330.3	−55.4
Cacao	kg	2.98	1.16	−61.0*
Coffee	kg	4.37	1.70	−61.1*
Tea	kg	2.87	1.90	−33.8
Cotton	kg	2.54	1.55	−38.9*

Source: *World Resources 1994–95* (1994), p. 262.

Note: kg = kilograms; mt = metric tons.

* Those commodities where the absolute value of the percentage decrease in price exceeds the percentage increase in production (as shown in Table 8.9).

Price data from World Bank, compiled from major international marketplaces for standard grades of each commodity.

For instance, while the production of natural rubber (over the ten year period examined) increased by almost 34%, the (per unit) market price for natural rubber has declined by 45%. Ironically, increased production over the last ten or so years has worked against producers and regions which export tropical plantation commodities. Price levels for these commodities have continued on a downward spiral as supplies increase and as the impact of substitutes are fully realized.

It is doubtful whether the cash crop systems of the tropical regions of SOUTH will ever generate sufficient growth in revenue to support more intensified economic development efforts, or to further a revolution in agricultural-led industrialization. Nonetheless, agricultural commodities of the tropics are extremely important generators of foreign currency for many regions of the SOUTH.

Perhaps the greatest threat to the tropical plantation system comes from within the system itself. Many researchers lament the loss of biodiversity of plant species from the rapid destruction of tropical forests. Conservation and preservation efforts to sustain plant resources have been slow in coming to many regions and in some cases nonexistent.

As a result, much of the rich gene pools of traditional and wild varieties of tropical plants have been lost forever. This loss has reduced the possibility of genetically strengthening the more commonly produced and limited varieties now in use—strains that have become increasingly vulnerable to disease and insects. Resource development efforts in the tropics have, therefore, refocused attention on the goal of preserving the quality of gene pools for traditional crop varieties and wild plant populations. Endeavors to minimize the loss of biodiversity (including preservation efforts and gene banks) will allow a now troubled system to avert widespread destruction of crops which are vital to the economic well-being of the SOUTH (Smith, 1987; Smith et al., 1992).

The loss of tropical rainforest areas has not only been linked to nonsustainable agricultural practices, but to industrial logging activity—the clear-cutting of tropical forests by lumbering operations for products to meet a growing world demand. By some calculations, roughly 40% of the world's tropical rainforests have been lost since 1945. At current consumption levels, it is possible that there will be virtually no virgin tropical rainforest left in the world in 60 years. Use of the forests primarily for fuelwood has reduced forest cover in many parts of the world. Consider that Ethiopia's forest cover declined to just 4% of its total land area in 1992, down from 40% in 1940. It is feared that bad management practices will be spread by the seemingly insatiable demand for tropical hardwoods.

Deforestation—the destruction of forest cover and structure—is a world resource problem. Deforestation has resulted in habitat disruption, species endangerment and extinction, a loss of a forest's bioveristy, and changes in climate and hydrological systems (Figure 8.17).

Slightly more than half of the world's wood is produced to meet the demand for fuelwood and charcoal, a trend that is highly population-driven. A little over one-quarter of wood production is for sawlogs and wood veneer (for building materials and furniture), and just under one-fifth is used for pulpwood, paper, and other industrial uses. Table 8.11 gives a breakdown of world leaders in cut roundwood production.

Figure 8.17 Hardwood logging in the tropics has contributed to the demise of the rainforest and threatens the loss of biodiversity and natural habitat.

8.17: ©Kevin Horan/Stock Boston

TABLE 8.11	World's Leading Roundwood Producers	
Nation/Region	Percentage of World Output	Predominant Forest Regions*
United States	14.7	a,b,c
Former USSR	10.8	a,b,c
China	8.1	b,c
India	7.9	d
Brazil	7.6	d
Canada	5.2	a,c
Indonesia	4.8	d
Nigeria	3.1	d

Source: *Goode's World Atlas* (19th edition), 1995, p. 44.

* Forest regions:

a = conifer (softwoods)

b = temperate (hardwoods)

c = mixed (softwoods and hardwoods)

d = tropical hardwoods

Note: World output of cut roundwood = 3,462,350,000 metric tons.

While **sustained yield** practices are in place in most developed regions—*where the rate of cut does not exceed the rate of growth*, many underdeveloped regions continue to utilize forest resources at rates which are far beyond the regeneration rates of the affected forests. This is particularly true in tropical hardwood areas where hardwood forest growth is slow in comparison to "coniferous" and "mixed forest" systems.

Logging activities have been responsible for the rapid depletion of tropical hardwoods in India, Brazil, Indonesia, and Nigeria at a rate that would hardly allow this activity to be classified as sustainable. Much of the primary forests in western Africa have already been removed. With the increasing demand for tropical hardwoods in core economies, it is likely that the remaining and abundant tropical hardwoods of many central and southern Africa nations will be the next to be targeted for extraction.

Large hardwood reserves still exist, however, in Zaire, Cameroon, the Congo, and the Central African Republic. The continued removal of hardwood from forest systems in Africa, Central America, and Asia, as a way in which to secure export revenue, is happening at a rate which far exceeds the rate of regeneration and growth. By promoting irresponsible hardwood production, underdeveloped nations are not only squandering a valuable natural resource base, but they are accelerating the loss of valuable plant and animal species as they destroy the ecosystems in which these species thrive. As scientists are just beginning to understand the depth and potential of forest resources, the loss of biodiversity poses an unnecessary risk to the world economy and humanity.

Tropical forests continue to yield new discoveries and mysteries. Many chemicals and drugs obtained from plants in remote tropical areas cannot yet be created, duplicated, or manufactured in laboratory settings. Science has yet to find uses for many of these chemicals, many of which may hold the key to curing diseases that continue to plague humankind. It is easy to argue that the short-term and geographically limited export benefits of non-sustainable forest-based industries are simply not worth the cost of eroding tropical resources (with potentials still unknown). Government, industry, and the scientific community must continue to work together to champion sustainability of the tropics and its diverse resource base. The continued destruction of the world's forest systems has many yet unknown environmental, economic, and social consequences.

AGRICULTURAL GEOGRAPHY: A RESEARCH AGENDA

The field of agricultural geography has gained considerable attention over the past few years. Leaders in the field have opted to take a more holistic approach to examining issues related to sustainability of production systems. The components of agricultural systems and the changes in those systems have been influenced by many factors (physical, economic, social, cultural, political, and spatial/relational). With the recent intensification of extensive systems comes the need to redefine the nature of sustainability in systems which emphasize both subsistence and commercial activities.

Up to now, the study of agricultural systems has focused on 15 major areas of interest as discussed by Grigg (1995):

1. *The biology of agriculture.* There has been a resurgence in the interest over the biological characteristics of plants and animals. It is now well recognized that biology plays an important role in determining the functional efficiency and adaptability of species in various environments. This focus has enabled scientists to elaborate on the complex interrelationships and interdependencies between the flora and fauna of the physical world and regional food chains.

2. *The geography of climate.* The variability of climate over the earth's surface and its effect on agricultural systems is an important area of research. It has been established that varying climatic conditions are a major determinant to variability in crop distribution and agricultural productivity.

3. *Soil characteristics and land management.* Increasing attention has been given to studies of soil types and characteristics. Research has centered on how soils are affected by climate, the geology and geomorphology of regions, natural vegetation cover and parent materials, etc. Some studies have focused on local farmers' knowledge of soil conditions and how they affect their behavior.

4. *The importance of morphology and the biophysical environment.* The physical features of the earth's surface greatly affect agricultural potential and opportunities. From a suitability standpoint, slope and altitude are two of the most important factors which affect crop production. There is a growing interest in researching the distinctions between farming operations in upland versus lowland areas and how the choice of crops and agricultural practices are affected by the constraints imposed by morphology and the biophysical environment.

5. *Variations in demand.* Local and regional variations in demand for agricultural products are intrinsically tied to the distribution of income and crop preferences. Contextual and cultural features of the human economic landscape are important to the analysis of agricultural systems.

6. *The degree to which markets affect behavior and production.* Market forces are known to affect the behavior of farmers in commercial systems. It is recognized that markets influence the intensity and geography of production. Recently, attention has turned toward investigating the degree to which local, regional, and global markets affect farmers and production in subsistence regions.

7. *Economic development and agricultural change.* Economic development and modernization are known to bring about structural changes in agricultural systems. As such, there is a growing interest in how regional economic transformation affects the agricultural intensification, production, and productivity.

8. *The impact of government in agricultural economy.* It is well known that governments play an important role in economic matters. Recent attention has focused on issues of agricultural policy, protectionism, and land reform.

9. *Locations of markets and the movement of agricultural products.* There is an interest in examining the degree to which declining transportation costs have influenced agricultural production patterns and the movement of agricultural commodities. Advancements in food transport technology, processing, storage, and distribution have lessened the impact of locational separation.

10. *The effects of urban sprawl on rural land values.* The ongoing expansion of urban areas continues to result in a loss of valuable rural farmland. Researchers have begun to examine the implications of land use changes and the degree to which the encroaching urban fringe has affected rural land values and regional production.

11. *Agricultural intensification.* Many studies now focus on the relationship between population density and intensification of agricultural production.

12. *Farm size, farm structure, and tenure.* The size of farming operations, their physical layout, and the nature of land ownership continue to have significant impacts on agricultural productivity and profitability.

13. *The diffusion of agricultural technology.* The adoption of new farming methods is related to the speed at which information and innovations diffuse through an agricultural system. The rate of adoption is known to differ dramatically across cultures and regions, a factor that greatly influences production patterns and productivity. Diffusion of a given production technology is affected by affordability, willingness to adopt (that is, resistance to change), and attitudes toward risk.

14. *The effects of culture and tradition.* Culture, history, and tradition are key to explaining geographic

variations in crop production and the preferences of farmers and consumers toward specific types of crops.

15. *Achieving balance in agro-environment relations.* The sustainability of an agricultural system is highly dependent upon its ability to be nonthreatening to the land or the environment. Research continues to examine the delicate balance between agricultural intensification and environmental degradation in various farming systems.

From the above summary, it is apparent that a diverse set of factors are known to affect agricultural production and sustainability. It is essential, therefore, to entertain a multidimensional and multidisciplinary approach when studying agricultural systems and regional production patterns.

COMMERCIAL AGRICULTURE SYSTEMS: AN OVERVIEW

Commercial agriculture involves the production of agricultural commodities produced exclusively for their "exchange value." Hence, production patterns in commercial systems are greatly influenced by market forces and the geography of demand. While commercial agriculture can be found throughout the world, it is a system that is best associated with the surplus producing regions of the NORTH, although minor exceptions do exist. It is a system best associated with the United States, Canada, nations of central and southern Europe, coastal regions of eastern and southern Australia, New Zealand, and the larger semiperipheral nations of the SOUTH including Brazil, Argentina, and South Africa. In general, commercial agriculture is an activity that is generally associated with fertile farmland in close proximity to large settlements or major markets (Figure 8.18).

Commercial agricultural production intensifies with decreasing distance to market. Production of highly perishable commodities or production activities which require large amounts of labor input per unit land area are generally located at a short distance to market or along transportation corridors that provide easy access to one or more markets. Intensive commercial agriculture includes dairy farming, mixed farming, horticulture and greenhouse farming, and **truck farming**—*production of specialty products, fruits, and vegetables, that are transported to market by truck*) (Figure 8.19).

Extensive commercial activities are those that require more land and less labor input per unit of land (for example, commercial grain production, livestock grazing, ranching activities, and tree farming for pulp and paper production). These activities occur farther from the market, covering extensive geographic areas with relatively low population densities.

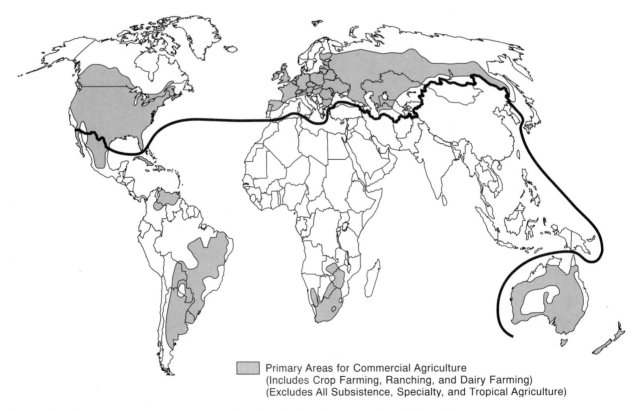

Primary Areas for Commercial Agriculture
(Includes Crop Farming, Ranching, and Dairy Farming)
(Excludes All Subsistence, Specialty, and Tropical Agriculture)

Figure 8.18 Commercial agriculture: production of surplus (basic commodities and livestock).

Overall patterns of production in commercial systems are guided by market forces, the physical setting, productivity and yield, transportability and perishability considerations, and the competitiveness of agricultural activities as defined in terms of their profitability in relation to market conditions, preferences, and expectations. The principles of von Thünen apply well here, as commercial agricultural land is generally allocated toward its "highest and best" use. Agricultural land use patterns are a manifestation of the complex interrelationships between urban and rural land markets and local, regional, and global agricultural markets.

Commercial agriculture is increasingly becoming an **agribusiness** venture—*a system in which agricultural operations are becoming increasingly integrated and market-driven, involving firms which engage in farm production and farm management, as well as food process-ing, storage, distribution, and agricultural research*. It is a system where virtually every aspect of production is controlled from seedling to the marketplace. Agribusiness is now a global activity. Companies like the Archer Daniels Midland corporation have assumed leadership roles in agribusiness, with a growing emphasis on improving food production and quality through science and research.

The Vulnerability of Commercial Agriculture

Commercial food production systems are highly susceptible to unpredictable changes in regional and global markets for agricultural commodities. Yet, expanding commercial food production has meant that market prices of agricultural commodities have remained relatively low over time. As a result, so has farm income. Low prices have been particularly troublesome for U.S. farmers. Although no one factor is to blame for farm problems in the United States, many farmers have been forced out of the industry as a result of low market prices and less-than-adequate income. The demise of the U.S. "family farm" has been hastened by an increased dependence on expensive production inputs (such as fertilizer), increased mechanization and farm debt, chronic excess capacity (the effects of increasing surplus and supply), diseconomies of small farming operations, declining land values, and misguided government policies.

Many small family farming operations continue to be helpless in an industry where agricultural prices, inventories, and commodity exchange are becoming increasingly controlled by large grain companies and agricultural commodities brokers. Increased competition and low prices have allowed only the larger and more efficient producers to survive the ongoing shake-out (Figure 8.20).

Figure 8.19   The New England produce center: Trucks getting ready to haul fresh fruits and vegetables to markets in the Northeast and elsewhere.

8.19: ©Frank Siteman/Stock Boston

Figure 8.20   A large and efficient corporate farming operation in Wisconsin, a contrast to the small and less-competitive family farm.

8.20a: ©William Johnson/Stock Boston; 8.20b: ©David Ulmer/Stock Boston

The recent market trends have led to fewer and larger farms in the United States and elsewhere. Many of the small and less economically efficient family farming units have now either exited the industry or have been bought out by those able to expand and achieve economies of scale in production. The lost agricultural output due to foreclosures of family farms has been more than made up by the larger and "corporate-style" farms, which now predominate the agricultural landscape. The loss of farm income, however, has taken a toll on many farming communities. The demise of the family farm has been swift indeed. Consider that in 1920, there were over 7 million farms in the United States. At that time, the average farm size was approximately 150 acres. Farm income supported a rural population of well over 30 million people (roughly 30% of the U.S. population) just prior to World War II. By the late 1980s, the U.S. farm population declined dramatically, accounting for less than 5 million people (approximately 2% of the U.S. population). By the early 1990s, the number of farms in the United States dropped to 2.1 million with an average size of roughly 450 acres. By the year 2010, it is expected that the number of farms will drop under the 1.8 million mark. The average size of farms at that time is expected to exceed 500 acres.

With proper care and maintenance of the land-based resources and increasing production efficiency, there is little doubt that the commercial agricultural systems of the world are sustainable. As food and agricultural commodities markets expand, the world's most efficient producers will always find ways to generate revenue and turn a profit. Nonetheless, even the most efficient producers will find themselves at risk due to the forces of nature. As crop yields, grain inventories, and price levels can be severely affected by unforgiving weather patterns, floods, droughts, insect infestations, outbreaks of disease, etc., farmers in commercial systems continue to run the risk of losing it all.

World Leaders in Agricultural Production

World prices of agricultural commodities are generally influenced by output levels in the top-producing regions. The world's leading producers and exporters of selected agricultural products are listed in Table 8.12. Note that the majority of commercial agricultural production and exports are largely dominated by just a handful of nations; particularly, the United States, Australia, France, Russia, Brazil, Argentina, and New Zealand. While China remains as a world leader in the production and exportation of several agricultural commodities including grain, swine, sheep, and cattle production, the large majority of China's agricultural outputs tends to be consumed within its borders or within the larger Asian subsistence region.

Fertilizer Production and Use

Not unlike intensive subsistence agriculture, commercial agriculture is highly dependent upon fertilizer as a production input. Farmers in subsistence regions are primarily concerned with using fertilizers to maximize output (with a low sensitivity to cost), whereas the use of fertilizer in commercial systems allows farmers to sustain high yields (on a per land unit basis) in order that they may generate sufficient revenue and profits to remain in business. While organic fertilizers (for example, manure) are important sources of basic plant nutrients (nitrogen, phosphorus, and potassium), inorganic fertilizers have become increasingly popular in regions where access to organic fertilizer is limited due to production or cost constraints.

The leading producers of phosphate rock, the major component in the production of phosphorous-based chemical fertilizers, are the United States, China, Morocco, and Russia (in descending order of importance). Together, these four nations account for more than 75% of world phosphate production. Note that phosphates are an important export base for Morocco, a nation that contains roughly three-quarters of the world's known phosphate rock. The world's leading synthetic nitrogen producers are China, the United States, Russia, and India. Note that these nations are also world leaders in overall agricultural production and the world's largest users of fertilizers. In general, inorganic fertilizer production is positively correlated with the size of the agricultural production system (or demand).

PART II: NATURAL RESOURCES, ENERGY, AND MINERALS

A **resource** may be defined as *anything that is useful or valuable to human beings*. A **natural resource** is *a product from nature that is of value to humans*. Natural resources exist independent of human activity. In general, natural resources may be classified as one of two varieties:

1. **nonrenewable resources**—*resources which are limited in quantity and exhaustible (that is, resources that cannot be replaced by natural processes in a reasonably short period of time)*; or
2. **renewable resources**—*resources which are naturally produced or replenished at a rate that is at least equal to the rate of depletion.*

Note that, by definition, renewable resources can last indefinitely, whereas nonrenewable resources have a limited life-span. Note also that many "potentially" renewable resources may actually have limited life-spans if improperly managed or when the rate of depletion exceeds the rate of replacement. For example, fishing and forestry resources (although commonly referred to as renewable) could conceivably be exhausted should those

TABLE 8.12	Top-Four World Producers and Exports of Selected Agricultural Commodities (1990–1992)

Commodity	Top-Four Producing/Exporting Nations by Rank (% of world production/exports shown in parentheses)				Output* (×1000)
	1	2	3	4	
Wheat	China (17.3)	U.S. (11.5)	India (9.4)	Russia (7.9)	567,555
Wheat exports	U.S. (29.1)	Canada (20.4)	France (16.3)	Australia (9.9)	106,049
Corn	U.S. (42.2)	China (19.5)	Brazil (5.1)	Mexico (2.9)	498,857
Corn exports	U.S. (66.6)	China (10.2)	France (9.0)	Argentina (6.2)	70,084
Potatoes	Russia (13.0)	China (12.2)	Poland (11.2)	U.S. (7.0)	264,658
Soybeans	U.S. (51.1)	Brazil (16.6)	Argentina (10.3)	China (9.4)	108,291
Citrus	Brazil (25.6)	U.S. (13.5)	China (8.2)	Spain (6.2)	77,291
Cattle	India (15.0)	Brazil (11.7)	U.S. (7.7)	China (6.2)	1,284,672
Beef exports	Australia (15.4)	Germany (14.6)	France (9.1)	U.S. (8.1)	4809
Swine	China (43.0)	U.S. (6.4)	Russia (4.4)	Brazil (3.8)	861,267
Sheep	Australia (13.8)	China (9.7)	New Zealand (4.8)	Russia (4.7)	1,164,309
Wool exports	New Zealand (37.9)	Australia (26.1)	U.K. (6.6)	Argentina (4.2)	391,000

Source: *Goode's World Atlas* (19th edition), pp. 36–37, pp. 39–41, and pp. 43–44.

* In metric tons of grain, or number of head.

resources be used or consumed faster than they are replenished.

In terms of resources use and management, the concept of "resource scarcity" can take on a variety of connotations. In actuality, we may distinguish between four different types of scarcity: (1) physical scarcity; (2) geopolitical scarcity; (3) economic scarcity; and (4) environmental scarcity (Rees, 1991, p. 6).

Physical scarcity *is a condition brought about by the exhaustion of nonrenewable resources, production shortfalls due to technological or physical constraints, or the depletion of renewable resources at rates that are nonsustainable.*

Geopolitical scarcity *involves the use of resources or commodity exports as political weapons (embargoes imposed on a nation) or when the location of low-cost resource production shifts to regions that are considered to be hostile.*

Economic scarcity *arises when adequate supply is not available to meet demand at existing market prices. This is a condition which may arise when (a) the demand at current price levels exceeds the ability of suppliers to*

meet that demand (resulting in regional shortages); (b) the needs of a region exceed the ability of that region to pay for those resources; (c) a region is outbid by another region in its attempt to acquire essential resources; and/or (d) specific renewable resources are exhausted due to market failure.

Environmental scarcity *refers to sustainability problems which arise from the disruption of biogeographical cycles and natural processes (usually from human intervention), pollution loads which exceed normal absorptive capacity or tolerance, and the loss of biodiversity, habitat, landscapes, and production capacity associated with environmental change (with consequences that are not always fully understood).*

Note that all four types of scarcity apply to energy resources and energy use.

ENERGY RESOURCES

Energy may be simply defined as the ability to do work or produce. In accordance with the laws of thermodynamics, two conditions apply to **energy use.** First, the "conservation of energy" ensures that energy cannot be created nor destroyed. In other words, energy can be only transformed from one state to another. Second, the impact of "entropy" ensures that energy use is an inefficient process. As the tendency of any system is to move toward increasing disorder (or maximum entropy), the transformation of energy to a less concentrated form will always result in an energy loss. Consider the dissipation of heat in the burning of fuel to power the modern-day combustion engine.

From an economic standpoint, it is desirable to encourage the **efficient use** of energy resources; that is, *uses that offer the greatest economic gain (or smallest loss) per energy unit expended.* This is no easy task when attempting to satisfy the energy needs of the world economy. Nonetheless, it is of the utmost importance that wasteful use of energy be minimized in work and production activities. Subsequently, there has been a renewed interest in **energy conservation**—*the prudent use and management of energy resources for the purpose of maximizing economic and social benefits.*

Note that regional conservation efforts, energy production and consumption, energy cost, and efficiency in use are all things which are highly variable over the economic landscape. For example, while the nations of the SOUTH currently account for roughly 75% of the world's population, they consume less than 30% of the world's energy. Note, however, that energy consumption in the SOUTH is growing at a rate that is approximately seven times faster than in the NORTH, an outcome which is largely driven by rapid economic and population growth.

Reliance on Fossil Fuels

The current energy needs of the world economy are mostly satisfied by the use of **nonrenewables**—*energy resources that are finite in quantity and exhaustible.* At present, roughly 90% of the world's commercial energy comes from nonrenewable sources (see Table 8.13). More specifically, 90% of the world's energy comes from the burning of **fossil fuels**—*materials or deposits in the earth's crust (made up of decomposed plant and animal matter) that, as a result of exposure to heat and pressure over millions of years, have been transformed into various hydrocarbons that can be burned to generate heat.* The three major fossil fuels are coal, crude oil, and natural gas. These fuels are the preferred energy sources of the industrialized and industrializing worlds.

Note that the use of nonrenewable energy is not so much correlated with population as it is with wealth and economic development status. Consider the data on energy consumption per capita as shown in Table 8.14. Note that the per capita energy consumption in the United States is roughly 25 times that of India. Note that in the early 1990s, the United States, Canada, and Japan (na-

TABLE 8.13	A Breakdown of World Commercial Energy Production

Fuel	Percentage of World Commercial Energy Production
Fossil Fuel Production	
Crude Petroleum (from gas and liquids)	39.7
Natural Gas	22.7
Solid Hydrocarbons	27.9
Primary Electricity Production	
Nuclear Power	6.9
Hydro (water power)	2.8

Source: *World Resources 1994–95* (1994), p. 166.
Note: Solid hydrocarbons include bituminous coal, lignite, peat, and deposits in shale.

TABLE 8.14	Regional Differences in Energy Use

Nation	Barrels of Oil Equivalent per Capita (1988)
United States	19.8
W. Germany	8.6
India	0.8

Source: The World Watch Institute (1993).

Commercial Energy Consumption for Selected Geographic Regions 11.037 Billion Metric Tons (coal equivalent) in 1991		
Region/Nation	Percentage of Total World Consumption	Per Capita Consumption*
North America	27.7%	(a)
Europe	24.9%	(b)
Russia	10.6%	(a)
China	8.5%	(c)
Japan	5.3%	(a)
India	2.5%	(d)
Asia (excluding those nations listed above)	11.4%	(d)
South America	2.8%	(c)
Africa	2.2%	(d)

Source: *Goode's World Atlas* (19th edition), p. 51.
* Key: Predominant trend in energy consumption in coal equivalents, kilograms per capita:
(a) 4500–15,000
(b) 1500–4449
(c) 500–1499
(d) less than 500

tions which comprise only 7% of the world's population) consumed approximately one-third of the world's energy. While the developed nations of the NORTH account for roughly 20% of the world's population, they account for the lion's share of total energy consumption. While the industrialized nations of the NORTH are big energy consumers, they have become increasingly aware of the importance of energy conservation. Industrializing nations in the SOUTH, however, tend to be more inefficient users of energy.

Energy sources that fall into the "renewable" category are sources that are produced naturally. These resources, for all intents and purposes, cannot be depleted as they are constantly replenished by natural processes (usually as fast as they are being consumed). Renewable sources, such as wind power, biomass, tidal power, geothermal, hydro/water power and solar power, account for around 10% of the world's current energy needs. As nonrenewable sources become less available or are exhausted, regional production deficits will be made up by renewables and other energy alternatives.

The extent of the life of any energy resource can be measured by the "proven reserves" that are on hand. **Proven reserves** refer to *the amount of identified and usable energy resources that are available (profitably) at current market prices and under prevailing extraction technologies.* Proven reserves are commonly expressed in units that show how much of a resource is left (usually expressed in years) based on the current rates of consumption. At any point in time, the amount of proven reserves depends upon five factors:

1. technological know-how and extraction skills;
2. consumer demand (levels and growth potential);
3. production and processing costs;
4. market price; and
5. the availability and price of substitutes

(Rees, 1990, p. 21).

Recent data on proven reserves of fossil fuels are shown in Table 8.15. According to these estimates, oil is the least abundant of the fossil fuels, whereas coal is the most abundant. Note that the largest share of coal, oil, and natural gas reserves are held by just a handful of countries. The most significant reserves are associated with the members of the **Organization of Petroleum Exporting Countries (OPEC)**—*an oil cartel that actively sets production limits on crude oil.* Members of OPEC decide how much crude oil will be pumped out of the ground over a period of time. By controlling the supply of oil, OPEC can greatly influence the price of crude oil in the world market where demand has remained strong and ever increasing.

Proven reserves suggest that the world will be almost exclusively dependent upon OPEC for crude oil in less than twenty years (under the current rate of consumption and provided that no other significant oil reserves will be

| TABLE 8.15 | Proven Commercial Energy Reserves (1990) |

Region	Coal	Oil	Natural Gas
World	209	45	52
Developing Countries [SOUTH]	163	68	96
Oil-Importing Developing Countries	152	19	58
OPEC*	1511	91	116
North America	242	10	11
Industrialized Nations [NORTH]	239	12	37
OECD	231	10	14

Proven Fossil Fuel Reserves (at current rates of consumption) (shown in years)

Source: *World Resources 1994–95* (1994), p. 167.
* OPEC members: Algeria, Gabon, Indonesia, Iran, Iraq, Kuwait, Libya, Nigeria, Qatar, Saudi Arabia, United Arab Emirates (U.A.E.), and Venezuela.

found elsewhere). Surprisingly, this impending oil scarcity problem has not led to dramatic increases in the market price of oil, nor has it led to a curtailing of demand for crude. In fact, as proven reserves continue to be updated over time (and increase for any number of reasons), and as leading members of OPEC display fairly liberal production policies, virtually flooding the world with oil, crude oil continues to be a real bargain in the world's energy marketplace. In fact, the price of crude oil has remained relatively flat over the last several decades. As OPEC production continues to satisfy the growing demands of the world's leading oil-consuming nations (the United States, Japan, Germany, France, Italy, etc.), there has been a diminished sense of urgency to adopt strict oil conservation policies. This is unlikely to change in the near future given that the market price of oil does not reflect its true value in relation to its economic, physical, environmental, and geopolitical scarcity (Figure 8.21).

Despite the lack of concern over scarcity, the fact remains that oil is a nonrenewable resource. The relatively low price of oil does not mean that oil production is not profitable. Recent prices of a barrel of crude oil in the $16 to $26 range, are well above its production cost in most regions (see Table 8.16). Production cost differences largely reflect the extraction difficulty, quality of crude oil, and the size and accessibility of proven reserves.

Data on oil reserves suggest that the world is expected to run out of crude in about 45 years. Yet, it is difficult to imagine this coming to pass as it is likely that proven reserves are expected to rise as (a) the market price of oil rises; (b) new oil deposits are found; (c) new extraction technologies emerge; and (d) world petroleum demand increases over time. Hence, it is very unlikely that the world will run out of crude oil by the year 2045. It is very likely, however, that crude oil deposits in some regions will be almost completely exhausted over the next several decades.

There is wide speculation that the oil production leaders in OPEC will continue to pump oil out of the ground (seemingly as if there is no tomorrow) to intentionally keep the price of oil low. Many experts agree that the price of crude oil is not at all reflective of its scarcity, nor is it in line with increasing world demand. As the price of oil is key in setting the price of energy in general, cheap oil translates into less incentives for industry and government to allocate research and development expenditures for further oil exploration or for alternative energy sources and energy-saving technology. In effect, cheap oil ensures that nations with significant oil reserves (many times their only natural resource) will have a strong and steady market for oil for some time to come.

Figure 8.21 A typical oil-producing field in North Africa or the Middle East, where oil is plentiful and a chief export.

8.21: ©Jim Holland/Stock Boston

In the 1945–73 period of low and declining energy prices the geography of energy production gradually became dominated by a rapidly increasing flow of oil from the Middle East and a small number of countries elsewhere in the world. . . . The prolific oil resources with extremely low production costs of this small group of countries undermined the economic production of most other sources of energy in most of the rest of the world . . . as capital could not be attracted to enterprises unable to compete with low-cost imported oil. (Odell, 1989, p. 89)

Much to the dismay of OPEC's renegade members (for example, Iraq and Iran) or those with the smallest known reserves, powerful oil-producing nations such as Saudi Arabia and Kuwait (those with the largest oil reserves) have continued to increase oil production to meet the growing demand for oil in both the NORTH and the newly industrializing nations of the SOUTH. Note that the cheap oil allows many nations to make a substantial profit when prices are above production costs. Note also that nations with large reserves generally tend to have lower per barrel production costs, and thus, retain the highest profits. Depressed oil prices will inevitably translate into less overall revenue for nations with smaller reserves (for example, Libya and Nigeria). These nations currently have reserves that are likely to run out long before the price of oil rebounds to a level that more accurately reflects its physical scarcity.

The majority of crude oil consumption is transportation related, generally consumed in the form of gasoline, diesel, and aviation fuel. Cheap and abundant oil has kept the price of gasoline reasonably low in high-consuming regions like North America. Low prices at the pump have allowed consumers to enjoy the luxury of increased purchasing power at a time when the growth in real wages has been sluggish. Moreover, cheap gasoline and diesel fuel has rekindled America's love affair with private transportation, a factor that continues to contribute to urban sprawl and energy waste. Low prices have also thwarted attempts by policy makers to forge a clear-cut

TABLE 8.16	Oil Production Costs for Selected Nations and Regions			
	Less than $2/bbl	**Between $2 and $5/bbl**	**Between $5 and $10/bbl**	**More than $10/bbl**
OPEC	Saudi Arabia	Libya	Nigeria	Gabon
	Kuwait	Venezuela	Indonesia	Algeria
	Iraq	Abu Dhabi		Qatar
	Iran	(U.A.E.)		
Non-OPEC		Mexico	Russia	Egypt
		Malaysia	Northern	U.S.
		Oman	Alaska	North Sea Region

Source: International Energy Agency (1995).

Note: bbl = barrel.

national energy agenda and the underlying need to support renewable energy and alternatives. The switch to renewable energy sources has been slow in coming. This has made the U.S. economy and consumers increasingly vulnerable to forthcoming increases in oil prices or disruptions in supply (as experienced during the 1973 Arab oil embargo, where the price of gasoline skyrocketed and America resorted to gasoline rationing). As national energy policies become increasingly important to matters of national development, many developed nations face a rude awakening should members of OPEC decide to flex their muscle and restrict supply. The recent Gulf War provided us with a reminder of just how important oil is to the United States and the rest of the industrialized world.

Of all the fossil fuels, natural gas is the fastest growing in terms of use. Current trends suggest that natural gas consumption will sustain an average annual growth rate of 2.5% from 1992 to the year 2010. Coal production runs a close second with an annual average growth rate of 2.3% over the same time period. Although natural gas is a clean-burning, low-polluting, and efficient form of energy when compared to oil or coal, it does not have a true "global" market. Unlike crude oil which can be easily shipped by oceanic super-tankers to ports around the world, natural gas is difficult and costly to transport. To make it cost competitive, large volumes of natural gas must be transported in liquid form. This, however, makes it extremely dangerous to handle, transport, and store (especially in and around large urban areas) given its explosive characteristics.

Due to transport limitations, much of the world's natural gas reserves were squandered. Note that, even today, it is still a fairly common practice to "flare" (burn away) residual gas during the extraction and processing of crude oil. The practice of "flaring" has become less visible as the price of energy rises and as transport technologies and infrastructural development (that is, pipelines) allow natural gas to be competitively sold in regional markets that are distant from the source.

Coal is the most abundant of all the fossil fuels with an estimated lifespan of well over 200 years at current rates of consumption. Given its abundance, many industry analysts believe that we are about to embark on a new coal age (once crude oil reserves are depleted). Recent improvements in coal transport technology have kept coal cost competitive. Innovations such as coal "slurry" (and the slurry pipeline)—crushed and concentrated coal that is mixed with water and transported by pipeline—and the "unit train"—custom-designed, high-capacity railroad cars devoted exclusively to the movement of coal (that shuffle between mines and regional distribution or consumption points) have been instrumental in reaching consumers at great distances from coal extraction points. Nonetheless, the extraction and burning of coal is not without its problems. Coal is dangerous to mine, detri-

mental to the local environment, and a high-polluting energy source.

Coal is typically mined in one of three ways:

1. "strip-mining"—the removal of overburden (soil and rock covering) to expose a coal seam near the surface;
2. "auguring"—the use of a drill to bore into a coal seam, allowing it to be removed in the process; or
3. underground shaft mining—the blasting, drilling, and excavating of underground tunnels through which the coal is brought to the surface.

Of the three extractions methods, underground shaft mining is the most dangerous as there is always the possibility of fire and explosion when miners encounter methane gas pockets, which usually occur in close proximity to coal deposits. In addition, there is a high incidence of "black lung" disease among coal miners given their constant exposure to coal dust.

Although coal is an abundant, fairly accessible, and inexpensive energy source, coal mining activities cause several environmental problems including "subsidence" of the land (above or in the vicinity of a mine) and the permanent scaring of the economic landscape. Once coal deposits are extracted, the land is left unproductive and aesthetically unappealing. Restoring the land to its original state is difficult and costly, and most times impossible.

The burning of coal is also problematic as it releases annoying residue and harmful chemical agents and impurities into the atmosphere. Burning coal is known to release sulfur, carbon dioxide, and nitrogen oxide—gases that have been identified as contributors to the recent "global warming" trend. The use of modern "scrubber" technologies has reduced the amount of potentially harmful coal-burning emissions discharged into the atmosphere from industrial smokestacks. The use of this technology, however, is costly to industries that must rely on coal for power or heat.

Coal-fired steel-making facilities and utility companies throughout the developed world, for example, must now comply with tough environmental guidelines regarding sulfur and nitrogen emissions. Regulatory agencies continue to monitor industrial emissions and ensure that emissions and pollutants discharged are kept to a safe level. Carbon-emissions taxes on high-polluting industry have been discussed as a way in which to reduce air pollution and recover the cost to the environment.

Nuclear Energy

Nuclear energy rounds out the league of nonrenewable energy sources. The production of nuclear power relies upon uranium and "nuclear fission" (the splitting of atoms into lighter elements), a process that initiates a violent chain reaction which allows vast amounts of energy to be released. The intense heat from this reaction permits water to be turned into steam, which is used to drive turbines that ultimately generate electrical power.

Plagued by a series of mishaps such as the incidents at Three Mile Island (Pennsylvania) in 1979 and Chernobyl (in the former Soviet Union) in 1986, there has been much concern over the safety of nuclear power. In particular, industry analysts have grown weary over the threat posed by many of the old-style nuclear power plants still in operation today. The nuclear industry, however, is rapidly rebuilding its reputation as a promoter of cheap and safe energy. Modern nuclear reactors have virtually eliminated the possibility of accidents and core meltdowns. The safety records of the modern reactors are once again beginning to restore confidence in nuclear power as an innocuous and reliable alternative energy source.

Currently, there are well over 100 active nuclear power plants operating in the United States. These plants generate about 20% of America's total electricity. Nations like France and Japan continue to show a greater commitment to nuclear energy, a source that now accounts for more than 70% of their electricity needs. The global expansion of the nuclear industry is likely to continue as nuclear power is a clean energy source in comparison to coal or oil. Worldwide, nuclear power now accounts for roughly 20% of all electricity production. Despite the strengths of this clean and efficient energy source, there are two major shortcomings. First, long-term storage of radioactive waste continues to be a problem. Most communities simply do not want a nuclear waste disposal facility to be opened in their backyard, leaving existing waste facilities to be overburdened and shortening their lifespans. Second, nuclear power (at this stage of the technology game) is classified as a nonrenewable energy source, with a finite amount of uranium reserves on the planet. At the current rate of nuclear power production and use, uranium reserves are expected to last roughly 60 to 70 years.

As with most nonrenewable forms of energy, market price increases and technological innovations are likely to lead to future increases in the amount of proven reserves. Combined with conservation efforts, the lifespan of many of the nonrenewables will increase even further in the years to come. Unfortunately, future increase in the cost of energy presents a formidable obstacle to economic growth and industrialization in the SOUTH. Many underdeveloped nations will not be able to afford enough energy to meet their economic development and industrial needs. Even the use of renewables will require a large commitment on the part of the SOUTH to invest in appropriate technologies to utilize the now costly energy alternatives.

The leading producers and proven-reserve holders of the nonrenewable energy sources discussed previously are listed in Table 8.17. Note that the proven reserve data is based upon current prices, extraction technologies, and present rates of consumption. It is interesting to note that well over 40% of the world's remaining oil reserves are held by just three members of OPEC (Saudi Arabia, Iraq, and Kuwait). As the most influential members of OPEC, these nations are likely to have a strangle hold on the world's oil market by the year 2015. Note also that of the major industrialized nations of the world, the United States, Russia, China, and Australia have significant reserves in one or more of the four major nonrenewable energy sources. Outside of OPEC's Middle East contingent, and large semiperipheral nations such as China, India, and Mexico, very few nations in the SOUTH possess significant proven reserves of fossil fuels.

While the developed economies of the NORTH are the world's leading consumers of nonrenewable energy, the greatest growth in nonrenewable energy consumption is actually occurring among the nations of the industrializing SOUTH. The data in Table 8.18 illustrate the recent increase in fossil fuel consumption, a trend that is easily explained by increasing population, industrialization of the SOUTH, increasing trade and commodity flows, increasing demand for electricity, more vehicles, and a greater reliance on private transportation. Fossil fuel per capita is declining in the NORTH as a result of increasing efficiency in use and conservation. Note, however, that there has been a dramatic increase in per capita consumption of fossil fuels in the rapidly populating regions of the SOUTH. In fact, the SOUTH has more than doubled its per capita fossil fuel consumption in the last twenty or so years.

Industry analysts emphasize the importance of promoting energy conservation in the high-growth industrialized corridors of the SOUTH so that the lifespans of nonrenewables will not be prematurely shortened. Not only does the SOUTH face a situation of inadequate supplies of energy to meet the needs of industry, but it continues to deal with widespread inefficiency, use of high-polluting sources of energy (coal and wood), nonsustainable usage rates, and government policies which ultimately subsidize energy consumption. Government involvement for the sake of providing affordable energy, increases the likelihood of waste and mismanagement of energy resources and discourages conservation. A shortage of investment capital in the SOUTH is the single greatest impediment to expanding the much-needed energy infrastructure that will be required to enhance efficiency, lessen the impact on the environment, and allow regions to meet a burgeoning demand (*World Resources 1994–95*, pp. 171–172).

The search for energy alternatives and the push toward energy conservation must become a local, regional, national, and global development priority. Governments in the SOUTH must make a concerted effort to (a) raise energy prices (across the board); (b) limit the use of highly polluting forms of energy by imposing strict and enforceable environmental regulations on industry and

TABLE 8.17	Production and Proven Reserves of Nonrenewable Mineral Fuels*

Leading Nations
(% of world production/reserves shown in parentheses)

Source	1	2	3	4
Coal Production	China (23.8)	U.S. (20.0)	Germany (8.2)	Russia (7.7)
Coal Reserves	U.S. (23.1)	Russia (16.2)	China (11.0)	Australia (8.8)
Petroleum Production	Russia (14.6)	Saudi Arabia (12.9)	U.S (12.3)	Iran (5.5)
Petroleum Reserves	Saudi Arabia (23.8)	Russia (14.3)	Iraq (9.1)	Kuwait (8.7)
Natural Gas Production	Russia (30.9)	U.S. (24.2)	Canada (5.7)	Netherlands (4.0)
Natural Gas Reserves	Russia (36.1)	Iran (13.7)	U.A.E. (4.2)	Qatar (4.1)
Uranium Production	Canada (20.2)	Russia (8.7)	Kazakhstan (8.4)	Uzbekistan (8.1)
Uranium Reserves	Australia (20.3)	former USSR (18.5)	S. Africa (12.4)	U.S. (10.8)

Source: *Goode's World Atlas* (1995), pp. 52–53.

* Based on averages for the period 1990–1992.

TABLE 8.18	Fossil Fuel Consumption and Worldwide Growth of Selected Activities

Fossil Fuel Consumption (in gigajoules per capita)

Region	1966–1970	1976–1980	1986–1990
Industrialized NORTH	142.53	169.52	160.06
Developing SOUTH	8.26	12.91	17.28

Worldwide Consumption/Capacity per Year

Source	1970	1990
Oil	17 billion bbls	24 billion bbls
Coal	2.3 billion tons	5.2 billion tons
Natural Gas	31 trillion cubic ft	70 trillion cubic ft
Electric Generating Capacity	1.1 billion kilowatts	2.6 billion kilowatts
Registered Automobiles	250 million	560 million

Kilometers Driven (estimated)

Passenger Cars	2584 billion	4489 billion
Trucks	666 billion	1536 billion

Source: *World Resources 1994–95* (1994), p. 8; and Meadows et al. (1992), p. 7.

consumers; (c) establish tough efficiency standards for appliances, vehicles, manufacturing equipment, and industry; (d) institute carbon-emissions taxes on industries and consumers using high-polluting fossil fuels; (e) create a system of tax exemptions, incentives, and rebates to offset the increased burden of purchasing costly, yet energy-efficient alternatives; (f) educate people and industry on the benefits of energy conservation and encourage behavior that is conservation oriented; and (g) aggressively promote the development and use of renewable energy to help underdeveloped regions attain a greater degree of self-reliance in energy production.

Long-term sustainable development in all countries requires a gradual move toward renewable sources of energy such as wind, solar, hydro, and biomass, that are more equitably distributed and less environmentally destructive than current fossil fuel sources.

Developing countries (other than oil-producing nations of the Middle East) possess less than 20 percent of the world's crude oil reserves, and only about half of all developing countries have known recoverable reserves. Nonetheless, with the notable exceptions of China and India, which have extensive coal reserves, most developing nations rely on oil, and often siphon off a high percentage of foreign exchange earnings to finance its purchase.

Aggressive development of renewable energy sources offers developing nations the prospect of increasing energy self-reliance, both nationally and locally, and reaping the attendant economic and security benefits. (*World Resources 1994–95*, p. 174)

Given the highly concentrated nature of the world's oil supply, the consumption habits of developed and developing economies, and the finite nature of fossil fuels and uranium, there has been an increasing interest in renewable energy. Let's briefly examine some of the leading renewable energy sources and discuss the implications of their use and their future potential.

Renewable Energy

Solar energy has become one of the more popular supplemental energy sources in the world. It involves *harnessing the power of the sun (solar radiation) to generate heat and produce electricity* and the use of solar-powered, energy-saving innovations. Although commercial solar energy applications are relatively new in the world's energy marketplace, solar energy is truly the definitive source of all the earth's energy.

Improvements in passive-solar home-heating systems and solar-conscious architecture offer new possibilities for conserving energy and stretching an energy budget. The design and use of energy-efficient windows has allowed housing units to utilize the winter sun to warm the interior of a dwelling, and to keep the interior cool during hotter periods by minimizing the effects of the harsh summer sun. Active collection of solar energy involves the use of solar panels composed of "photovoltaic cells"—thin silicon wafers that directly convert the sun's rays to electricity.

Unfortunately, active solar technology is currently just outside the reach of the average consumer, as solar panels and long-term storage devices for active solar applications are currently not cost competitive. The price of solar technology, however, is expected to decline as applications become more commonplace. Note that photovoltaic cells are now widely used in water heaters and pumps, electric lights, calculators, remote telecommunications equipment, and a wide variety of consumer electronic devices. The gain in popularity of solar technology has been somewhat of a quiet yet profound revolution. Solar-powered vehicles are currently being developed and tested, although commercial availability is some time away. Fierce competition exists between solar-powered transporters and other more promising new-age transport technologies, including the electric car and vehicles powered by natural gas.

Even for consumers in up-scale markets, the high cost of solar technology for the home is not justified by the savings. Solar energy is likely to become more cost competitive, however, as the price of oil increases. Despite its current cost disadvantages, scientists continue to work feverishly on solar-powered applications in an industry that is gaining strength in the energy markets of the developed and developing worlds (Figure 8.22).

During the past two decades, the cost of photovoltaic power has fallen from $30 a kilowatt-hour to just

Figure 8.22 Solar panel application in industry. A growing trend toward alternative energy sources.
8.22: ©Tony Freeman/Photo Edit

30 cents. This is still four to six times the cost of power generation from fossil fuels, so further reductions are needed for solar power to be competitive with grid electricity. New applications will spur further cost reductions, which is likely to lead to widespread use of solar cells. As they become more compact and versatile, photovoltaic panels could be used as roofing material on individual homes, bringing about the ultimate decentralization of power generation. Around the same time, perhaps a decade from now, large solar plants could begin to appear in the world's deserts—providing centralized power in the same way as do today's coal and nuclear plants. (Flavin and Lenssen, 1991, p. 12)

Centralized electricity is also available from "solar-thermal" technologies—from systems which utilize large mirrored troughs to reflect the sun's rays onto a liquid-filled tube. This allows the water in the tube to be "super-heated" to produce steam, which drives a series of turbines to generate electric power. These systems can produce electricity at about 8 cents per kilowatt-hour, a cost that is almost competitive with electric power generation from fossil fuels.

Hydroelectric power—*electric power generation from flowing water*—is still the preferred source of electricity in the world today when viewed from an economic and environmental perspective. Hydroelectric power involves the use of natural waterfalls or man-made dams (reservoirs of water that can be later released), where water flow along a vertical drop may be used to spin turbines and generate electricity. Note that hydroelectric power currently satisfies about 10% of U.S. electricity needs (although this is less than 5% of the total U.S. energy budget). In comparison to coal-fired electric power generation, hydroelectric power is one of the cleanest renewable energy forms. Despite the fact that it is one of the less environmentally harmful energy sources, and the fact that many improvements have been made in high-voltage electric power transmission (which now allow consumption regions to be far removed from production sites), there is very little room for expansion of this industry in most regions. Most of the premier hydroelectric sites in the NORTH have already been put into use. Nonetheless, smaller micro-hydro installations (using small-scale dams) will become more economically viable with higher energy prices in the not-so-distant future. Areas of the world that are poor in coal and oil reserves, such as South America, Africa, and Southeast Asia, are likely to turn to hydroelectric power. Currently, these regions have utilized as little as 20% of their hydroelectric power capacities (Figure 8.23).

Though hydroelectric power has a well-earned reputation as a clean energy source, it is not without its problems. For instance, the construction of dams is known to have adverse effects on the environment, altering the natural balance of river systems and their floodplains, not to mention the impact on aquatic life downstream. The installation of hydroelectric power plants is also very costly, usually requiring large government subsidies to fund dam construction projects. The cost-prohibitive aspects of these projects have made hydroelectric power less of an option in many peripheral economies.

Geothermal energy—*the extraction of energy using the earth's natural internal heat, generated from molten rocks well below the surface*—is a highly localized and limited renewable energy source. While the use of hot springs, geysers, hot dry rocks, or geopressurized thermal resources do offer some potential for electric power generation (in areas where subsurface temperatures are high or where deep pockets of water lie trapped below the surface), they are not likely to replace fossil fuels as a primary energy source given that geothermal potential is highly variable from one location to another. The use of geothermal energy in areas residing on geothermal "hotspots" offers much in the way of a supplemental energy source, although these hotspots are sporadically distributed over the earth's surface.

Note that applications of geothermal technology have already proven successful in the generation of elec-

Figure 8.23 Hydroelectric power generation, a dominant source of electricity.

8.23: ©David Barber/Photo Edit

tricity in places like El Salvador, Nicaragua, the United States (mostly in the West), Iceland, Italy, China, and the Philippines. Great potential for the expansion of this renewable energy source exists across the entire island nation of Japan. Great geothermal potential also exists in east-central Africa and portions of Central America. Note that even geothermal energy, a seemingly innocuous source of power, has an environmental drawback. Rising water and steam tend to bring underground sulfur to the surface, limiting the use of this technology in ecologically sensitive areas (Flavin and Lenssen, 1991).

Wind power—*harnessing the power of moving air to create electric power*—is yet another promising energy alternative. Using the force of the wind to spin turbines (located atop giant windmill-like structures) to generate electricity has gained general acceptance as a supplemental energy source. The success of wind power has been overshadowed by its one major drawback. Applications of wind power are geographically limited as they require a steady wind flow to make them commercially viable. Large-scale electric-power generation from "wind farms" — large tracts of land devoted exclusively to the generation of electric power from wind turbines, such as the famous installation at Altamont Pass (California), could be used in both on- and off-shore areas where wind speed is relatively high and uniform (Figure 8.24).

Wind power is now becoming increasingly viable as a supplemental source, as modern turbines can now produce electricity for as little as 7 cents per kilowatt-hour (in 1995 dollars). Technological advancement in this area and increased demand for electricity are likely to cut the costs of electric power generation in half by the turn of the century. Much potential exists for expanding this industry throughout the Great Plains region of the United States, across North Africa and "the Horn" of East Africa, and in off-shore platforms in the windswept North and Baltic Seas.

Figure 8.24 Spinning turbines producing electric power at the famous wind farm at Altamont Pass, California.
8.24: ©Peter Menzel/Stock Boston.

Biomass—*solid waste and combustible organic material*—continues to gain popularity as an important supplemental energy source. Biomass includes energy or fuels derived from the harvesting of woodcrops, forestry and agricultural waste, landfill gas, refuse and municipal waste. Biomass use is a way to capture and utilize the solar energy stored in wood, vegetation, and other products. Conversion of biomass to liquid fuels like "ethanol," derived from corn and sugarcane, shows much promise as a transportation fuel supplement. Almost 10% of the gasoline sold in the United States contains "gasohol"—a fuel mixture that contains ethanol.

Although biomass may not be the answer to all of the world's future energy needs, as it is largely defined in terms of local and regional applicability, it will undoubtedly prove to be an important supplemental energy source as the cost of waste materials declines. Biomass, however, does carry some environmental risks. It is known to be a relatively high-polluting energy source in terms of pollutants released upon incineration.

Up to now, renewable energy sources (such as those discussed previously) have played only a minor role in the world's energy markets. Yet, it is only a matter of time before they are taken more seriously. Perhaps the biggest challenge is finding innovative ways to utilize renewables in the rapidly industrializing and populating regions of the SOUTH.

> With the exception of hydroelectric power, renewable energy technologies have had a hard time penetrating the energy sectors of developing nations. There are several reasons, including lack of commitment on the part of energy planners, technical failures attributed to poor capacity for local maintenance, and the cost of many new technologies, especially relative to the price of oil. (*World Resources 1994–95*, p. 174)

Renewables have been unable to compete in energy markets which fail to account for "externalities"—external costs or spillover effects that are not figured into the market price of energy. This is a situation that is unlikely to change unless the economic and environmental damages caused by high-polluting, nonrenewable energy sources are accounted for.

Note that technological advancements in renewable energy, funded largely by the private sector, have made the pursuit of environmentally sensitive energy options a reality. Technology has allowed costs to come down to levels that are almost competitive with nonrenewables. A cost comparison of the leading renewable energy sources is shown in Table 8.19.

Economics will continue to play a prominent role in determining the expansion capability of many renewables. As the price of oil will most assuredly rise over the next twenty or so years, renewables are likely to become more cost competitive in many regional energy markets. Note that the amount of energy saved by "energy conservation" was the number one source of new energy during the 1980s and 1990s. As consumers change their attitudes and behaviors, and continue to use less energy and/or make strides toward promoting more efficient use of energy resources, conservation will play a significant role in redefining the longevity of nonrenewable energy sources.

Before renewable energy systems are more fully utilized, it is likely that the world will enter into another Coal Age, much as it did at the onset of the Industrial Revolution. It is bothersome, however, to think that the most-abundant coal reserves are of the highest-polluting forms. In particular, there is an abundance of the softer, low-carbon grades of coal, namely, bituminous, subbituminous, and lignite. Low-grade coals are notorious for their high sulfur content when burned.

Note that several important alternative energy prospects remain. Research in the nuclear industry continues to explore the possibility of nuclear "fusion"—the union of atomic nuclei that results in an enormous release of energy. Should scientists find a way to fuse particles at normal (or near normal) temperatures, fusion would be a potential limitless source of energy. Many researchers are excited about the possibilities of harnessing and exploiting ocean currents and "tidal wave" power. Tidal power could be an effective supplemental energy source for markets located along coastal areas, where population densities and the demand for electricity are high. Synthetic fuel or "synfuel"—*fuel from coal gasification and liquification*—is likely to play an increasingly important role as crude oil reserves become scarce. Oil production from hydrocarbon deposits in shale rock (that is, "shale oil") and tar deposits in sand (that is, "tar sand") is likely to increase over time. With abundant coal reserves and substantial hydrocarbon deposits throughout the United States and Canada, coal-based gasoline with be competitive will gasoline from crude as the price of energy rises. The discovery of new petroleum reserves from continued exploration of off-shore locations offers yet further po-

TABLE 8.19	Cost of Renewable Electricity (1980–2030)			
	Cost per Kilowatt-Hour in Cents			
Technology	**1980**	**1988**	**2000**	**2030**
Wind Power	32	8	5	3
Geothermal	4	4	4	3
Photovoltaic Cells	339	30	10	4
Solar Thermal Applications (minimum cost)	24	8	6	5
Biomass	5	5	–	–

Source: Flavin and Lenssen (1991), p. 12.

tential to extend proven crude oil reserves. In short, extraction and processing of hydrocarbons will ensure that fuel will be available for at least another century.

Competition among alternative fuels is likely to heat up as the commercial viability of vehicles that run on hydrogen gas gain serious consideration in high-energy-consuming regions in the developed world. Although compressed hydrogen gas is a dangerous substance, the combustion of hydrogen and oxygen emits an environmentally safe by-product—water. For hydrogen-gas-fueled vehicles to gain serious acceptance, new engine materials and designs must be employed and energy costs must rise considerably—but it could be only a matter of time.

> Eventually, hydrogen fuel could be even more prevalent than oil is today. The gas could become cost-competitive as a transportation fuel within the next two decades. Solar and wind-derived electricity at 5 cents per kilowatt-hour—achievable by the late 1990s—could produce hydrogen that could sell at the pump for about the equivalent of a $3 gallon of gasoline. While that is more than Americans are now charged to fuel their cars, it is less than the price most Europeans pay. (Flavin and Lenssen, 1991, p. 14)

While production breakthroughs are still some time away in alternative energy industries, many renewable energies do show great promise. Price competitiveness, in many instances, is the only thing holding back commercial expansion of several renewable (and sustainable) energy sources. In addition, there is a good deal of optimism over the ability of science and technology to redefine the world's extractive energy potential. New and exciting energy sources and energy-saving devices will not only improve energy conservation efforts, but they will allow us to meet the immediate and long-term energy needs of economies in the NORTH and the industrializing SOUTH. It is especially critical that developing regions adopt "energy-smart" industrialization and modernization

policies—policies which seek to improve living standards, while minimizing degradation of the environment and reducing long-term dependence on nonrenewable energy sources.

Regional variations in the use and scarcity of nonrenewable energy resources notwithstanding, there is a misconception that the world may run out of energy in the near future. Although the world is heavily reliant upon nonrenewable forms of energy at present, the energy well is far from dry when one considers the numerous energy alternatives that do exist. The bottom line, however, is that energy will cost substantially more in the future than it does now.

Minerals

As discussed earlier, many nations of the SOUTH rely on just a few major exports as a means to secure foreign exchange to internally fund development. Hence, the exportation of one or several key minerals is something that is commonly observed. As with most extractive resources, mineral exports are unreliable revenue generators for several reasons. First, low and sometimes declining price levels have reduced export revenue potential. Note that most minerals are readily available in regional and global markets. Second, restrictions in supply by major producing regions do not necessarily help to raise the market price of minerals. This is due to the practice of **stockpiling**—*keeping inventories of precious commodities for later use.* Stockpiles of key minerals are held by core and semiperipheral nations in amounts that are above and beyond their present needs. Stockpiling protects these nations against possible shortages which may arise from an embargo or war.

Stockpiling of oil and **strategic minerals**—*substances that are critical for the defense of a nation or its key industries, yet only available from external suppliers*—is an inventory control used to ensure that adequate supplies will be on hand should the usual supply lines be cut off. In particular, it is common to observe the stockpiling of "ferro-alloys" like cobalt, manganese, chromite, etc.—metals used in the production of steel and military armaments. If production of a strategic mineral or metal is voluntarily restricted and market price of that commodity begins to rise, stockpiled minerals and metals are dumped on the world market, forcing prices to spiral downward (to preexisting levels or lower). Nations which engage in stockpiling activities, and not regions which supply the commodities, ultimately control the market price.

The leading producers of selected minerals are shown in Table 8.20. Notice that much of the production/extraction of these commodities takes place in the SOUTH. Refinement, use, and the majority of value added, however, occurs in the NORTH. As with other primary sector commodities, market prices of minerals have

TABLE 8.20	World Leaders in Mineral Extraction and Processing (1990–1992)

Leading Producers
(% of world production/reserves shown in parentheses)

Commodity	1	2	3	4
Copper (mined)	Chile (19.3)	U.S. (18.1)	Canada (8.6)	Zambia (4.7)
Copper (refined)	U.S. (19.0)	Chile (11.3)	Japan (10.0)	Germany (5.0)
Bauxite and Alumina	Australia (37.9)	Guinea (13.7)	Jamaica (10.5)	Brazil (9.6)
Aluminum from alumina	U.S. (21.0)	Russia (14.8)	Canada (9.2)	Australia (6.3)
Tin	China (21.2)	Brazil (16.3)	Indonesia (14.1)	Malaysia (10.5)
Iron-Ore	Brazil (19.1)	Australia (14.0)	China (10.8)	Russia (10.0)
Ferro-Alloys:				
Chromite	S. Africa (35.1)	Kazakhstan (29.9)	India (9.9)	Turkey (6.8)
Cobalt	Zaire (40.9)	Zambia (24.8)	Canada (7.8)	Russia (7.4)
Manganese	Ukraine (24.3)	S. Africa (19.1)	Gabon (11.3)	Brazil (10.1)
Tungsten	China (58.4)	Russia (15.6)	Austria (3.4)	Portugal (3.2)
Vanadium	S. Africa (49.7)	Russia (25.6)	China (14.3)	U.S. (7.7)
Crude Steel Production	Japan (14.3)	EU top 4* (14.0)	U.S. (11.4)	Russia (10.4)

Source: *Goode's World Atlas* (19th edition), p. 46 and pp. 48–49.
* Germany, Italy, France, and the United Kingdom.

TABLE 8.21	World Prices of Select Commodities (1980 and 1992)

Per Unit Price in 1990 (U.S. $)

Commodity	Unit	1980	1992
Copper	mt	3031.80	2140.30
Tin	kg	22.84	5.63
Iron-Ore	mt fe	39.03	29.65
Bauxite	mt	57.25	30.02
Aluminum	mt	2023.10	1176.90
Nickel	mt	9057.50	6569.00
Manganese Ore	10 kg	2.18	2.06*
Phosphate Rock	mt	64.90	39.20

Source: *World Resources 1994–95* (1994), p. 262.
* Price in 1985.
Note: kg = kilograms; mt = metric tons; fe = iron.
Price data from World Bank compiled from major international marketplaces for standard grades of each commodity.

higher prices and more cost-effective extraction methods, and further geological exploration should greatly extend mineral reserves. Commodity substitutes—the use of synthetic materials, plastics, or other more abundant materials—could add years, and potentially decades, to the life of most mineral reserves. Recycling, recovery, and reuse of minerals will play an increasingly important role in extending the life of reserves as these activities become more commercially viable. As with most extractive resources, sustainability of mineral industries has become a matter of economics (Cutter et al., 1991).

CONCLUDING REMARKS

The economics of extractive resources has become a global issue. While the extractive resource potential of regions varies greatly over the earth's surface, it is the guiding and powerful force of the global market that continues to influence production, consumption, and exchange of primary sector commodities. Nations with favorable resource endowments, however, have not exhibited a clear advantage over nations with little extractive resource potential. Many nations have been able to overcome resource deficiencies by relying on their wealth, economic clout, and/or their production and trade advantages in secondary and tertiary sectors. In the next chapter, we will focus on the importance of economic clout as we examine the geography of manufacturing and investment.

shown a tendency to decline in value, despite the increase in demand and use (see price data for selected commodities in Table 8.21).

Note that economically recoverable reserves (that is, proven reserves) of most minerals suggest that there is enough on hand to meet world demand for several decades. Should demand for these commodities rise with world population, reserves may be diminished somewhat faster than previously thought. Nonetheless, changing economic and technological conditions, particularly

REFERENCES

Adams, W.M. "Sustainable Development?" In *Geographies of Global Change: Remapping the World in the Late Twentieth Century*, (R.J. Johnston, P.J. Taylor, and M.J. Watts (eds.). Oxford, U.K.: Blackwell (1995), pp. 354–374.

Brown, L.R., C. Flavin, and S. Postel (eds.). *Saving the Planet: How to Shape an Environmentally Sustainable Global Economy.* New York: W.W. Norton (1991).

Brundtland, H. *Our Common Future.* Oxford: Oxford University Press, published for the World Commission on Environment and Development (1987).

Cutter, S.L., H.L. Renwick, and W.H. Renwick. *Exploitation, Conservation, Preservation: A Geographical Perspective on Natural Resource Use* (2nd edition). New York: J. Wiley & Sons (1991).

Darmstadter, J. (ed.). *Global Development and the Environment: Perspectives on Sustainability.* Washington, D.C: Resource for the Future (1992).

de Blij, H.J., and P.O. Muller. *Geography: Realms, Regions, and Concepts* (7th edition). New York: J. Wiley & Sons (1994), pp. 439–440.

de Souza, A.R., and F.P. Stutz. *The World Economy: Resources, Location, Trade, and Development.* New York: MacMillan (1994).

Flavin, C. "Facing Up to the Risks of Climate Change." In *State of the World 1996*, L.R. Brown (ed.). New York: W.W. Norton (1999), pp. 21–39.

Flavin, C., and N. Lenssen. "Here Comes the Sun." *World Watch* 4 (September–October, 1991), pp. 10–18.

Goode's World Atlas (17th edition), E.B. Espenshade, Jr. (ed.). Chicago: Rand McNally (1986).

Goode's World Atlas (19th edition), E.B. Espenshade, Jr., and J.C. Hudson (eds.). Chicago: Rand McNally (1995).

Grigg, D. *An Introduction to Agricultural Geography* (2nd

edition). New York: Routledge (1995).

Hamilton, I. (ed.). *Resources and Industry.* New York: Oxford University Press (1992).

Hartshorn, T.A., and J.W. Alexander. *Economic Geography* (3rd edition). Englewood Cliffs, NJ: Prentice Hall (1988), Chapters 3, 4, and 8, pp. 27–47; pp. 102–114.

International Energy Agency. *Oil, Gas, and Coal Supply Outlook.* Paris, France: OECD (1995).

Jackson, R.H., and L.E. Hudman. *World Regional Geography: Issues for Today.* New York: J. Wiley & Sons (1990), Chapter 4, pp. 91–100.

Jones, D.K.C. "Environmental Hazards." In *Global Change and Challenges: Geography for the 1990s*, R. Bennett, and R. Estall (eds.). London: Routledge (1991), pp. 27–56.

Knutson, R.D., J.B. Penn, and W.T Boehm. *Agricultural and Food Policy* (2nd edition). Englewood Cliffs, NJ: Prentice Hall (1990).

Long, R.E. (ed.). *Energy and Conservation*. New York: H.W. Wilson (1989).

Marshall, B. (ed.). *The Real World*. Boston, MA: Houghton Mifflin (1991).

Meadows, D.H., D.L. Meadows, and J. Randers. *Beyond the Limits: Confronting Global Collapse, Envisioning a Sustainable Future*. Post Mills, VT: Chelsea Green (1992).

Merrick, D., and R. Marshall (eds.). *Energy: Present and Future Options*. New York: J. Wiley & Sons (1981).

Miller, G.T., Jr. *Environmental Science: Working with the Earth* (5th edition). Belmont, CA: Wadsworth Publishing Co. (1995).

Odell, P.R. "Draining the World of Energy." In *A World in Crisis?* R.J. Johnston, and P.J. Taylor (eds.). Oxford, U.K.: Basil Blackwell (1989), pp. 79–100.

Rees, J. "Resources and the Environment." In *Global Change and Challenges: Geography for the 1990s*. R. Bennett, and R. Estall (eds.). London: Routledge (1991), pp. 5–26.

Rees, J. *Natural Resources: Allocation, Economics, and Policy* (2nd edition). New York: Routledge (1990).

Rosenberg, N.J. "Greenhouse Warming: Causes, Effects, and Control." *Renewable Resources Journal* 6 (Autumn 1988), pp. 4–8.

Seitz, J.L. *Global Issues: An Introduction*. Oxford: Blackwell (1995).

Smith, N.J.H. "Genebanks: A Global Payoff." *The Professional Geographer* 39 (February 1987), pp. 1–8.

Smith, N.J.H., E.A.S. Serro, P.T. Alvim, and I.C. Falesi. *Amazonia: Resiliency and Dynamism of the Land and Its People*. New York, NY: United Nation University Press (1995).

Smith, N.J.H., J.T. Williams, D.L. Plucknett, and J.P. Talbot. *Tropical Forests and Their Crops*. Ithaca, NY: Cornell University Press (1992).

Swales, S.E. "Dynamics of Land Use and Agricultural Practices on the Uplands and Adjacent Floodplain in the Lower Amazon." Ph.D. dissertation. University of Florida, Department of Geography, Gainesville, (October 1998).

Tarrant, J. (ed.). *Farming and Food*. New York: Oxford University Press (1991).

Toke, D. *Green Energy*. London, U.K.: The Marlin Press (1990).

Toman, M.A. "The Difficulty in Defining Sustainability." In *Global Development and the Environment: Perspectives on Sustainability*. Darmstadter, J. (ed.). Washington, D.C: Resource for the Future (1992), pp. 15–23.

Union of Concerned Scientists. *Cool Energy: The Renewable Solution to Global Warming*. Cambridge, MA: UCS Publications (1990).

U.S. Congress, Office of Technology Assessment. *Renewing Our Energy Future*. Washington, D.C.: U.S. Government Printing Office (1995).

von Thünen, J.H. *Isolated State: An English Edition of* Der isolierte Staat (orginally 1826). Translated by C.M. Wartenbberg (P.G. Hall, editor). New York: Pergamon Press (1966).

World Resources 1994–95: People and the Environment, published for the World Resource Institute in collaboration with the United Nations Environment and Development Programs. New York: Oxford University Press (1994).

MANUFACTURING, INVESTMENT, AND TRADE: PATTERNS AND TRENDS

E conomic Clout

Like most human economic activities, manufacturing is unevenly distributed over the earth's inhabitable land surface, yet concentrated in some areas. The presence or absence of manufacturing activities tells us many things about the economic development status and leverage of a region in the world economy. Manufacturing is a major contributor to a region's economic growth and development. Regions with large manufacturing capacities hold the highest development status and, therefore, possess tremendous economic clout.

Manufacturing regions have also emerged as international centers of finance in terms of generating and absorbing foreign direct investment (FDI). Though manufacturing infrastructure occupies only a small portion of the world's economic landscape, in terms of land use, the relevance of this activity to regional economies cannot be underestimated (Wheeler et al., 1998). At the global level, it is manufacturing that is the driving and dominant force behind international trade and commerce (Dicken, 1992).

> Manufacturing has been central to many of the great developments affecting society for over two hundred years. Despite the rise of services and the associated absolute and relative decline of manufacturing in the more developed economies of the world, its impact is unlikely to diminish if we take a global view of current trends. (Chapman and Walker, 1991, p. 17)

Though postindustrial societies are marked by the relative decline of manufacturing and the rise of services, it is the interplay of manufacturing and service sectors that is truly responsible for the growth of the tertiary sector.

The spatial distribution of manufacturing is illustrated in Figure 9.1. Note that manufacturing is largely associated with just a handful of nations. Note that the bulk of manufacturing "value added" remains highly concentrated in just three regions. These regions comprise the **global economic triad**—*the three leading production regions of the core*. Note that these three regions are located in the NORTH. All in all, the global economic triad is composed of the world's top-seven industrialized core nations; generally referred to as the **Group of Seven** or **G7**. The G7 is composed of the United States, Canada, Germany, Italy, the United Kingdom, France, and Japan. Since the mid-1970s, the leaders of the G7 have met on a regular basis to discuss trade and economic development matters. The mature market-based economies of the G7 control the large majority of the world's wealth and production capacity (Figure 9.1).

Though excluding the former Soviet Union and China, formidable production regions in their own right, the G7 accounts for roughly 70% of the world's total manufacturing value added during the 1980s and 1990s. The dominance of the G7 continues despite the growth of export manufacturing in South and East Asia and Latin America.

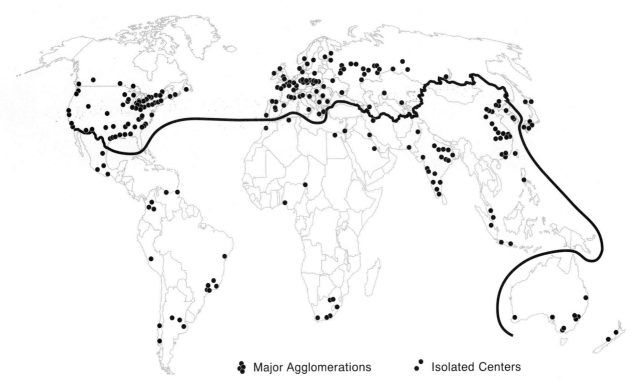

Figure 9.1 Global manufacturing: major agglomerations and isolated centers.

An historical breakdown of manufacturing value added is given in Table 9.1. The dollar figures provide a rough idea of the manufacturing clout of economies and regions listed. Note also the regional shifts that have occurred over the past several decades in the growth of manufacturing, particularly the growth of manufacturing value added in Asia and Latin America.

Note, however, that the G7 had roughly four times the manufacturing value added of all low- and middle-income economies combined in the late 1980s and early 1990s. In 1989, the U.S. economy alone recorded a higher total manufacturing output than the sum total of all the economies of Latin America and the Caribbean, South Asia, sub-Saharan Africa, and East Asia combined! In 1992, the manufacturing value added of the G7 was ten times that of all the economies of Latin America and the Caribbean and ninety times that of the economies of sub-Saharan Africa. Currently, the U.S. economy produces roughly ten times the amount of manufactured goods of China in terms of value added.

Concentrations of manufacturing are highly associated with the presence of production and transportation infrastructure. Investment and distribution of transportation infrastructure is highly correlated with economic development status and wealth. Locationally, manufacturing and transportation infrastructure are mutually associated with large urban areas or corridors which provide easy access to major markets. The locational tendencies of manufacturing activities are, in general, highly regionalized and located in or near the very regions where demand for manufactured products is large; that is, within core economies.

The advantage of direct market access and the availability of production and transportation infrastructure ensure that the G7 will maintain a leadership role in the world economy for some time to come. The influence of the United States especially, as the leading member of this elite group, is most impressive. Note that the U.S. economy would generate more GNP per capita than most of the less-industrialized economies of the world if the United States relied only on the revenue (value added) of General Motors (GM) alone (Figure 9.2).

To illustrate the economic clout of GM and the top-20 U.S. companies (as measured by revenue), a list of companies and nations are provided in Table 9.2. Note that the output of GM alone, when measured in per capita terms, produced more revenue or output per capita than the economies of China, India, Nigeria, or Bangladesh in 1990. The top-20 U.S. companies generate more income than the economies of Taiwan, Singapore, Malaysia, South Korea, India, Indonesia, and Bangladesh combined. Imagine, then, the economic clout held by all firms within the G7.

The prominence of the G7 is a by-product of 200 years of economic and industrial growth and capital accumulation. It is a testimony to the global competitive advantage and clout of the G7 in the world economy.

TABLE 9.1	Regional Variations in Manufacturing Clout		
	Manufacturing Value Added (in millions of dollars $)		
Nation or Region	**1970**	**(1989$) 1989**	**(1992$) 1992**
Low-income economies	46,144	227,368	273,098
Low-income economies (excluding China and India)	7078	39,091	41,279
Sub-Saharan Africa	3595	18,131	45,698
East Asia and Pacific	37,466	250,728	343,418
South Asia	10,357	52,644	94,160
Latin America and Caribbean	34,769	211,640	351,563
China	30,465	141,583	147,302
Indonesia	994	15,574	27,854
Nigeria	543	2989	2021
Bangladesh	387	1390	2164
Low- and middle-income economies	114,957	727,781	836,500*
Thailand	1130	14,760	31,185
Turkey	1930	16,793	27,465
Mexico	8449	46,932	67,157
Argentina	5750	18,646	50,009
Brazil	10,429	98,880	90,062
South Korea	1880	54,212	85,454
Singapore	379	7406	13,568
Hong Kong	1013	10,781	12,020
Austria	4873	33,723	46,739
Belgium	8226	33,809	48,432
Netherlands	8545	45,236	58,476
Finland	2588	22,370	20,785
Australia	9047	41,697	43,679
"Group of Seven" (G7)			
Canada	16,942	63,516	71,620
United Kingdom	35,739	140,879	201,859
France	68,201	202,734	271,131
Italy	29,061	192,884	250,345
Germany	70,888	377,173	565,603
Japan	73,339	831,779	1,023,048
United States	253,711	865,605	1,502,375
Totals for G7:	547,881	2,674,570	3,885,981

Source: *World Development Report 1991*, pp. 214–215, pp. 208–209; *World Development Report 1995;* pp. 166–167, pp. 172–173.

* Estimate.

Figure 9.2 A highly automated General Motors production facility. The use of robotics and high technology has greatly enhanced the performance and competitiveness of large corporations.
9.2: ©Photo Edit

across many economic sectors, regions, and boundaries, have led to shifts in the power base of the G7. Russia has recently been added to the group, which is now referred to as the **Group of Eight** or **G8**.

Note that the effects of physical distance and separation are less relevant today than they were forty years ago. As a result, the location and orientation of industry has changed. Firms no longer need to locate in or near the markets in which they sell their products. There are at least two reasons for the recent change in location preference: (a) even the most remote regions of the world economy have become highly accessible to all other places (that is, the world is becoming increasingly space-time converged) as a result of advanced transport and communications technologies; and (b) growth in demand has been high in the emerging markets of the SOUTH.

No industry or region is completely insulated by national boundaries or free from international competition. Large corporations have adapted by becoming multilocational and multijurisdictional in their operations. The increasing vulnerability of regions and industries to foreign rivals and intensified competition has brought about dramatic improvements in the production process and the quality of products produced. Global competition and the expansion of the global production network have helped to integrate regional economies of the NORTH and SOUTH. No longer are the economies of the SOUTH relegated (as they once were) to the subordinate role of raw material or low-cost labor suppliers for the production regimes of the NORTH. Instead, the industrializing nations of the SOUTH have become more intimately connected in the value-added sequence. These regions grow increasingly more global conscious and competitive through time.

Throughout the 1980s and 1990s, the core economies of the NORTH have experienced a marked contraction of

Recently, significant shifts have occurred within the G7. These changes are tied to the restructuring of manufacturing activities in the world economy brought about by the internationalization of business and commerce. Specifically, the globalization of production and the emergence of a global production network, one that cuts

TABLE 9.2	Economic Clout: A Comparison of U.S. Companies and Selected Nations (1990)		

Country or (#rank) Company	GNP or Revenues ($ billions)	GNP or Revenues per Capita ($ thousands)
United States	5400	21.7
Top-20 U.S. Companies	896	3.6
#1 General Motors	124	0.5
Japan	2100	16.9
Germany	1157	14.5
France	873	15.3
United Kingdom	858	15.6
Italy	844	14.6
Singapore	35	11.7
Spain	435	11.2
Taiwan	152	7.2
Portugal	57	5.7
Hungary	61	5.5
South Korea	238	5.5
Saudi Arabia	79	4.4
Turkey	178	3.0
Brazil	388	2.6
Malaysia	43	2.4
Argentina	70	2.1
Algeria	53	2.0
Iran	80	1.4
Egypt	37	0.7
Indonesia	94	0.5
Haiti	3	0.5
China	393	0.4
Kenya	9	0.4
India	287	0.3
Nigeria	28	0.3
Bangladesh	20	0.2

Companies with $30 billion or more in revenues:

#2 Exxon	103
#3 Ford	89
#4 IBM	65
#5 General Electric	60
#6 Mobil	57
#7 Philip Morris	48
#8 Dupont	38
#9 Texaco	38
#10 Chevron	37
#11 Boeing	30
#12 Chrysler	30

Source: United Nations, 1992; World Almanac, 1993.

many key production-for-export sectors. The birth of globally competitive manufacturing in the SOUTH has not only changed the nature of international business and commerce, but has lessened the control of the nations of the G7 as a political unit. International competition and globalization of production have made development planning and economic policy formation more difficult as governments become less able to effectively regulate economic change and industrial growth.

Two important shifts have occurred within the G7 since the early 1960s. First, there has been a notable decline of the United States as a world manufacturing leader. In 1963, the United States produced approximately 40% of the world's manufacturing output by value. By 1987, its share had diminished to slightly less than one-quarter of the global manufacturing output. Today, U.S. manufacturing is increasingly challenged by overseas competitors, and accounts for only about 20% of the world's manufacturing value added. Import penetration of manufactured goods from Asia and Latin America has contributed greatly to the demise of many U.S. manufacturing sectors. Heavy losses have been sustained in industries such as consumer electronics, appliances, vehicles, tools and machinery, textiles, and apparel. Despite rising labor productivity levels in most economic sectors, real wages in U.S. manufacturing have been steadily on the decline.

Second, and more strikingly, is the stellar performance of Japan as a manufacturing-for-export entity. In just 20 years, Japan has experienced phenomenal growth in its manufacturing base and has convincingly established itself once again as a world superpower. With an average annual growth rate of approximately 14% in manufacturing value added from 1960 to the late 1980s—a figure roughly two-and-one-half times greater than that of the United States—Japan has increased its economic clout within the G7 and the world economy. Japan's share of world production increased from roughly 5% in 1965 to 20% by the mid-1990s, putting them at par with the United States and in a co-leadership position. The intense rivalry between the United States and Japan continues.

Differences in trade and investment policies are largely responsible for Japan's success and America's decline, although Japan has yet to succeed in its attempt to "break the back" of the U.S. economy. While the United States has long been a supporter of free-market economy and free trade, the Japanese have a long tradition of closing their markets to outside competitors. As Japanese business flourished throughout the 1980s and 1990s, the lack of competition has hurt Japanese industry and consumers.

Though domestic investments in industry and infrastructure have been instrumental in the rise of Japan to superpower status, Japan's economic maturity has begun to threaten its economy. Overindustrialization, price inflation, increases in the cost of doing business, and a shaky banking system, have combined to weaken Japan's economy. Increasing production costs in Japan have forced Japanese producers to look elsewhere in Asia and North America for alternative production sites and investment possibilities.

U.S.-Japanese foreign relations have recently heated up as long-lasting trade imbalances (in electronics, automobiles, and automobile parts) continue to hurt the U.S. economy. Leaders from both nations continue to be at an impasse on trade matters as the Japanese continue their reluctance to open their markets to foreign competition and investors. The threat of retaliatory measures by the United States in response to Japan's unwillingness to cooperate on the trade front has been reduced, as policy makers view a trade war as counterproductive to the economies of both nations (Friedman and Lebard, 1991). Some trade analysts contend that the decade-old bilateral trade deficit with Japan is partly due to an overvalued U.S. dollar (in comparison to the Japanese yen). A "strong dollar" has helped to erode America's industrial base and stifle U.S. exports. U.S. policy makers have done little to reduce the unnaturally high value of the dollar in the world currency market. Japan's recent economic slowdown is likely to ensure that the value of the yen will remain low relative to the U.S. dollar. This will undoubtedly help boost Japanese exports to the U.S. market and reduce the competitiveness of U.S. goods in Japan, further contributing to U.S. manufacturing decline and trade problems. But the Japanese are not solely to blame for this predicament. The relative decline in the competitiveness of the U.S. manufacturing base, subsequent job losses, and decreases in real wages in manufacturing sectors—the "deindustrialization" of America—has been attributed to nine factors:

1. the use of cost-cutting, labor-saving technology in production and increasing capital-to-labor ratios in modern industry;
2. relocation of various U.S. manufacturers to overseas locations in search of a production cost advantage;
3. a loss of consumer loyalty to U.S. owned and operated companies or American-made products;
4. the penetration of U.S. markets by foreign competitors;
5. an increasing presence of foreign competitors on U.S. soil (displacing jobs and industry) as a result of foreign direct investment;
6. irresponsible government and misguided macroeconomic policies and the general ineffectiveness to reverse manufacturing and merchandise trade deficits;
7. the increasing use of low-wage, temporary, and part-time labor by industry in attempt to remain cost competitive (resulting in a suppression of wages and increasing competition for skilled labor for scarce full-time employment opportunities);
8. the globalization of production and the decentralization of industry worldwide; and
9. the rise of high-technology sectors and service industries in core economies.

The increasing capital intensiveness of production and the movement of production facilities to overseas locations in the attempt to remain competitive have been greatly aided by technology and information access. The decision-making processes of firms and domestic consumption have become less nationalistic. The loss of loyalty of the U.S. consumer to American-made products is also a reflection of consumers' attitudes toward U.S. manufacturers' lack of commitment to quality control and the changing needs of the marketplace, not to mention corporate America's lack of loyalty to the U.S. economy as they continue to shift their production facilities to foreign soil.

Failure of industry to keep jobs at home and its inability to produce reliable, inexpensive, and price competitive consumer goods left Americans looking to foreign suppliers for low-cost alternatives, further adding to the decline of sales, revenues, and jobs. This was particularly true in the case of automobiles shortly after the 1970s oil crisis (despite the fact that there was no real energy crisis). As U.S. consumers switched to more compact and fuel-efficient imports, U.S. auto manufacturers economic clout began to erode rapidly. This all came at a time when American producers lagged behind the Japanese in terms of product design and production efficiency in the manufacturing of steel and passenger vehicles. Modernization of U.S. production facilities and product lines proceeded very slowly, as corporations were more interested in bottom-line profits than reinvesting to maintain competitiveness. Subsequent losses in efficiency and productivity followed, along with losses of market share. As many U.S. companies abandoned America in search of low-cost production environments elsewhere around the globe, consumers became increasingly disenchanted with U.S. industry.

The growth of international trade and the industrialization of Southeast Asia also contributed greatly to the loss of America's manufacturing base, as foreign competitors flooded the U.S. market with inexpensive consumer goods. Although low-cost consumer goods increased the purchasing power of U.S. consumers, the subsequent loss of domestic industry had a boomerang effect on the U.S. economy. While U.S. consumers enjoyed a flood of cheap consumer goods from abroad, consumers' ability to earn a living was reduced as American factories began to close their doors.

To promote competitiveness in U.S. manufacturing exports, policy makers supported the devaluation of the U.S. dollar against the Japanese yen and German (deutsche) mark. A weaker dollar was intended to help U.S. manufacturers as it made the price of U.S. goods less expensive to consumers abroad. This strategy backfired, however, as it led to a general lack of confidence in the U.S. economy. Many interpreted a weak dollar to mean a weakened economy. The dollar's faulty reputation led to volatility in foreign currency markets and prompted many American companies to move operations overseas instead of reinvesting at home. U.S. manufacturers gravitated toward business environments which offered greater exchange stability and less financial risk as the future of the dollar remained in question. This combined with government's failures to control trade deficits, "dumping" of high-value-added products in the U.S. market, and changing consumer behavior, hastened the decline of U.S. manufacturing and America's position within the G7.

Recent and draconian cuts in defense spending have further deepened the extent of this economic crisis, causing further contraction of the secondary sector. The impact of lost income and wages continues to ripple and echo through the U.S. economy. Although defense-related industries are generally highly capital intensive, they do support jobs in transportation, construction, health and financial services, retail and wholesale trade, and government. As the value of products from defense-related manufacturing sectors are high in comparison to most other sectors of the economy, wages are generally higher in these industries and income multipliers are large. Lost income and revenue associated with cuts in defense spending have multiplied many times over, resulting in the loss of hundreds of billions of dollars to the U.S. economy over time.

Economic restructuring, the rise of alternative industries, and the erosion of traditional manufacturing sectors, much of which has been precipitated by changes in global markets, has had adverse socioeconomic effects. In particular, the "polarization" of socioeconomic groups can be traced to uneven earnings potential, a direct outcome of the switch in emphasis from traditional manufacturing to high-technology sectors and services.

While high-technology sectors generate a fair amount of high-wage and salary employment for skilled segments of the U.S. labor force, service industries have created a vast number of low-paying jobs. Wages in the service sector are far below wages once paid in the more traditional manufacturing sectors that have now been eliminated or geographically displaced. The shrinking manufacturing sector has left America with fewer middle-income employment opportunities and a steadily disappearing middle-class. As the demands of employers in high-technology and services sectors have increased in terms of the minimum skill level required to secure employment, enlarging segments of the U.S. population, particularly the poor and educationally deficient, have been rendered unable to compete in the modern-day job market (Morris et al., 1994; Bradshaw, 1996).

Despite the general decline of America's manufacturing base and the increasing competition from foreign manufacturers in low-value manufacturing sectors, the U.S. economy has recently showed signs of rebuilding its reputation as a world economic superpower. Table 9.3 provides a comparison of the performance of G7 members, as represented by their contribution to world output for selected years. Note that both the United States, and the G7 as a whole, continue to enhance their position as production leaders in the world economy. The recent rebounding U.S. economy is associated with vigorous exports from high-value-added manufacturing like aircraft, transport, high tech, and telecommunications. A list of major exporting firms is given in Table 9.4.

The swift ascent of Japan to economic superpower status has been largely attributed to high-valued exports, not to mention Japan's protectionist philosophy. Japan has moved up the ranks of the G7 quickly—from a distant fifth place in 1965 to the second-leading producing nation by mid-1990s. Japan now ranks as the top steel-producing nation in the world, with roughly 14% of the more than 750 million metric tons of crude steel produced annually in the world economy (compared to the United

TABLE 9.3	Production and Export Clout of the G-7

G7 Member	Percentage of World Output			Annual Average Growth Rate (manufacturing) 1980–1994
	1965	1989	1994	
United States	34.8	25.8	27.1	–
Japan	4.5	14.1	18.2	5.6
Germany	5.7	5.9	8.3	–
Italy	3.3	4.3	4.3	2.8
France	4.9	4.8	5.4	0.9
United Kingdom	4.8	3.6	3.5	–
Canada	2.3	2.4	2.1	2.3
Total	60.3	60.9	68.9	

	Percentage of World Exports (merchandise trade)		Average Annual Growth Rate of Exports	
	1988	1994	1970–1980	1980–1994
United States	11.9	12.5	7.0	5.1
Japan	9.4	9.8	9.2	4.2
Germany	11.7	10.2	5.6	4.2
Italy	4.8	4.6	6.9	4.3
France	5.9	5.6	6.8	4.5
United Kingdom	5.2	4.9	4.3	4.0
Canada	3.9	3.9	4.5	5.6
Total	52.8	51.5		

Source: *World Development Report 1991*, p. 209; *World Development Report 1995*, p. 167 and p. 187.

Note: Data for Germany prior to unification.

States, which now accounts for approximately 12% of world steel production).

Currently, the G7 control well over 60% of total world output and more than 50% of the world's exports by value. Slowdowns in the annual growth rate of G7 exports over the last decade is a signal that the G7 is somewhat vulnerable to global competition. The G7, however, is unlikely to yield its position as its members continue to post export growth rates that are typically above the world average.

Two important regional economic trends have been observed in the developing SOUTH since the 1970s: (1) China's miraculuous ascent to a world manufacturing and export powerhouse; and (2) the steady and impressive surge of a handful of rapidly industrializing countries in Asia and Latin America.

Consider the fact that China was the thirtieth largest trading nation in the world in 1977, with a sum total of imports and exports that was less than $15 billion (in 1990 U.S. dollars). By 1990, China's total imports and exports exceeded $160 billion . By the year 2000, the value of Chinese imports and exports will reach the $250 billion mark, placing it among the top-ten world exporters. Note that Chinese manufacturing exports have increased at an average annual rate of well over 20% from 1978 to 1990. Most of these exports are of low value, however. Textiles, apparel, footwear, toys, electronic gadgets, and recreational equipment account for the majority of Chinese exports. Nevertheless, China's economy produced more than 6% of the world's output in 1990, as compared to 2% of the world's output in the early 1960s. Though producing mostly low-valued commodities for

TABLE 9.4 | **Top-20 U.S. Exporters**

Rank (1992)	Exporter	Major Exports	Value of Exports (in $ millions)
1	Boeing	commercial aircraft	17,486
2	GM	motor vehicles and parts	14,045
3	General Electric	jet engines, turbines, medical systems, plastics	8200
4	IBM	computers and peripherals	7524
5	Ford Motor Co.	motor vehicles and parts	7220
6	Chrysler	motor vehicles and parts	7051
7	McDonnell Douglas	aerospace products, missiles, electronic systems	4983
8	Philip Morris	tobacco, beverages, food products	3797
9	Hewlett-Packard	measurement and computational products, systems	3720
10	DuPont	specialty chemicals	3509
11	Motorola	communications equipment, semiconductors	3460
12	United Technologies	jet engines, helicopters, cooling equipment	3451
13	Caterpillar	heavy machinery, engines, turbines	3341
14	Eastman Kodak	imaging, chemicals, health products	3220
15	Archer Daniels Midland	protein meal, grain, flour, vegetable oils	2700
16	INTEL	microcomputer components, modules, systems	2339
17	Digital Equipment	computers and peripherals	1900
18	Allied-Signal	aircraft and automotive parts	1180
19	UNISYS	computers and peripherals	1795
20	SUN Microsystems	computers and peripherals	1783
		Total	102,722

Source: *Fortune* (June 14, 1993), p. 131.

export, China ranks third in the world in overall production behind the United States and Japan (Lardy, 1992, 1994).

As the "export machine" of the Pacific Rim, China has outperformed all nations of the world in terms of export and economic growth over the past 20 years. Since 1978, China's merchandise trade grew at a rate that was almost twice that of the world economy. From 1980 to 1993, China recorded an astonishing average annual GDP growth rate of 9.6%. This rate was considerably greater than the economic growth rates of the leading NICs who averaged just over 5% per year over the same period. Much of China's economic success is credited to industrialization policies that favor import substitution as a means to stimulate domestic production to satisfy population-driven increases in demand. This was part of a multilevel strategy in which China sought to stimulate export-led growth, attract foreign direct investment, protect its domestic industries, and pursue economic development and modernization. China's rapid transformation from communism to **market socialism**—*a mixed economic system which retains characteristics of both command and free-market economies.*

In China, state-controlled production has become increasingly replaced with production activities which are market- and consumer-driven. In many coastal manufacturing areas, production activities proceed without direct intervention from the state. State-run industry has given way to the formation of independent businesses and market-oriented production networks, industrial diversification, rapid urbanization, and the birth of export-driven supercities. Economic change has brought about noticeable changes in Chinese society, including a greater sense of individualism, a shift of human resources from labor-intensive industries to high-technology sectors, and the slow yet steady rise of women as an entrepreneurial force. Moreover, a growing middle class has allowed China's urban areas to evolve from an export-led system to a consumer-driven economy (Naisbitt, 1996) (Figure 9.3).

With the exception of China, manufacturing clusters outside of the core continue to undergo only modest growth. The majority of that growth has been limited to a small group of semiperipheral economies which are in "transition," and currently moving through a rapid take-off stage and drive to maturity. These newly industrializing countries, or "NICs" as they are sometimes referred to, have experienced two decades of double-digit annual average growth rates in manufacturing output and exports, not to mention healthy growth in services (see Table 9.5). Despite the impressive economic performance of the NICs, they have captured only a small percentage of the world's overall economic output. Despite manufacturing growth rates of 10% and higher from 1970 to 1993, South Korea, Malaysia, Thailand, and Indonesia managed

Figure 9.3   A modern-day retail shopping environment in a Chinese city.
9.3: ©Bill Bachmann/Photo Edit

to account for a total of only 2.8% of the world's output. Note that of the top-12 NICs listed accounted for less than 10% of world output in 1993, as they still do today.

Rapid industrialization has come at a cost. Many analysts contend that overindustrialization has occurred and has impaired the future growth potential of many NICs. In particular, overindustrialization has led to internal economic problems in the Pacific Rim. Specifically, the production capacities of many Asian nations have grown faster than the growth of world demand. Overindustrialization and fierce global competition among core and semiperipheral economies has led to reduced export potential for individual nations, low and suppressed commodity prices, lost revenues, and little or no after-production profit. Global competition, sluggish export growth, and underutilization of existing production capacity has forced many Asian companies to the edge of bankruptcy. As companies go out of business and/or default on their loans, they continue to place enormous pressure on the banking system and regional economy.

Many failing economies in Asia and elsewhere continue to look to core economies for assistance and government support of a bail out. Overindustrialization and export stagnation has resulted in higher-than-expected unemployment and social discord in nations like Indonesia, Malaysia, Thailand, and the Philippines. Rapid economic growth and industrialization in Latin America has also come at a substantial cost. Double- and triple-digit inflation has plagued many Latin American economies. Consider the case of Brazil, a nation where the benefits of two decades of economic growth were largely negated by an average annual rate of inflation of more than 300% from 1980 to 1995.

TABLE 9.5	Growth of Manufacturing and Output of Leading NICs

	Average Annual Growth Rates			Percent of World Output[*]	
	Manufacturing		Services		
NIC	1970–1980	1980–1994	1980–1994	1965	1994
South Korea	17.7	12.3	8.3	0.19	1.43
Singapore	9.7	7.2	7.4	0.01	0.23
Malaysia	11.7	10.3	5.5	0.15	0.28
Thailand	10.5	10.8	7.7	0.21	0.54
Indonesia	14.0	11.8	6.9	0.19	0.63
India	4.5	6.3	6.4	2.52	0.97
Pakistan	6.1	7.3	6.3	0.27	0.20
Turkey	6.1	7.0	4.6	0.38	0.67
Brazil	9.4	0.7	3.3	0.97	1.92
Mexico	7.0	2.1	1.6	1.08	1.48
Chile	–0.8	4.4	5.4	0.12	0.18
Argentina	1.3	0.4	1.0	1.00	1.10
			Total	7.09%	9.63%

Average annual growth rate of GDP for NICs listed above (1980–1994): 5.1%

Average annual growth rate of GDP for China (1980–1994): 9.6%

Source: *World Development Report 1995*, pp. 164–165; *World Development Report 1991*, pp. 208–209.

[*]Note that estimates for percentage of world output for 1965 are computed without including data for the Soviet Union, whereas estimates for 1993 include republics of the former Soviet Union. Hence, the increase in the percentage of world output from 1965 to 1993 is somewhat understated.

THE EUROPEAN UNION

While the United States and Japan remain as the leading production centers on the planet, the nations of the "European Union" (or EU) now form the largest single concentrated production and trading zone in the world economy. The EU (formerly the European Community) was officially established with the recent ratification of the Maastricht Treaty in 1993. The **European Union (EU)** *is an eco-political organization, with a combined GDP of approximately $8 trillion and a population of almost 400 million people, whose goal is to strengthen economic interaction and ties between its members through free trade.*

The EU consists of 15 members (in 1998): Austria, Belgium, Denmark, Finland, France, Germany, Great Britian, Greece, Ireland, Italy, Luxembourg, the Netherlands, Portugal, Spain, and Sweden. Its primary objectives of the union are to (1) create a single, unified, and common European market with no trade barriers; and (2) implement a single currency (**the Euro**) as the official legal tender of the region to facilitate business transactions and trade throughout the region. With the recent introduction of the Euro, the EU is well on its way to achieving full market unity within the region.

Nations of the EU have found it difficult to proceed with unification given that some members are reluctant to give up their economic sovereignty. The more prominent members of the union (Germany, France, and Great Britian) are most likely to undergo the greatest sacrifices in the economic transformation process. With concerns over unemployment, excessive tax burden, and inflation, the European Union faces potential price stabilization problems. Nevertheless, economic unification and future expansion of the EU will undoubtedly make it the world's largest single free trade zone.

Note that the economies of the EU maintain a high development status with a per capita income that well exceeds $18,000. The recent addition of Austria, Finland, and Sweden into the EU in 1995 has expanded trade possibilities for EU members and the G7. As a unit, the EU rivals Canada as the leading trading partner for the United States with well over $200 billion in total goods trade in 1994 (see Table 9.6). Strong political and economic ties between North America and the EU are likely to enhance trade possibilities within and between core economies.

The extent to which the prevailing development gap between the core and semiperipheral economies will be closed depends largely upon the ability of semiperipheral economies to maintain a global competitive advantage as core economies become increasingly integrated and unified over time. This may prove to be difficult in light of the core's improving production-and-trade possibilities under a unifying Europe and a world economy dominated by the G7.

THE WORLD MANUFACTURING EXPORT EXPLOSION

Increasingly large shares of imported goods in the NORTH and elsewhere are coming from overseas producers in the SOUTH. Yet as semiperipheral economies of the SOUTH become increasingly integrated and experience the economic benefits of growth, it is anticipated that exports from the NORTH will also increase over time. Consider that the share of U.S. manufacturing exports going to developing countries increased from 29% in 1978 to more than 32% by 1990, despite the relative decline of U.S. manufacturing during this period.

As new markets for consumers and producers are established worldwide, there will be fierce competition between production regions for a share of the world's **big emerging markets** (or **BEMs**)—*consumer markets associated with the leading high-growth, semiperipheral economies*. Sizeable exports gains are expected within the G7 as they intensify manufacturing and exports to meet a growing demand for a wide variety of high-valued products in Argentina, Brazil, the **Chinese Economic Area (CEA)**—*China and Taiwan*, India, Indonesia, Mexico, Poland, South Africa, South Korea, and Turkey.

Table 9.7 provides an overview of the increasing market potential for a selected number of emerging markets, using U.S. trade data. Many of the BEMs listed have posted unprecedented triple-digit growth rates in their imports from 1987 to 1994. As the worldwide boom in exports to peripheral and semiperipheral regions continues, there exists a tremendous potential for export-led growth of the United States, Japan, the G7, and the EU. The industrializing economies of the SOUTH are likely to continue to exhibit a healthy demand for high-value-added goods and services produced in the core, products such as heavy machinery, transportation and construction equipment, high-end electronic devices, as well as financial and information services. This will undoubtedly reinforce the dominance of the G7 in the world economy—leaders in high-value-added manufacturing. With the collapse of the Soviet command economy in 1991, and recent wide-sweeping economic reforms in Russia and Eastern Europe, the reorientation toward market-driven economies will continue to mean further growth in exports for the United States and its G7 counterparts.

Russia is now one of the ten fastest-growing markets for high-valued U.S. manufactured products (mostly infrastructure-related goods and services such as trans-

TABLE 9.6	Goods Trade with United States—Major Trading Partners		
Nation or Region	Goods Trade* with U.S. 1994 ($ billions)	Imports from U.S. 1994 ($ billions)	Percent change in Imports 1987–1994
European Union	214	102.8	70%
Selected EU Members			
Germany	51	19.2	62%
France	30	13.6	72%
Italy	22	7.2	31%
Spain	8	4.6	48%
Canada	243	114.4	91%
Japan	173	53.5	90%

Source: *U.S. Global Trade Outlook 1995–2000: Toward the 21st Century*, U.S. Government Printing Office, Washington, D.C., March 1995.

* Goods Trade = ($ imports plus $ exports).

portation and telecommunications equipment, and scientific and electronic devices). It is expected that annual U.S. exports to Russia and other republics of the former Soviet Union will easily exceed the $25 billion mark by the year 2005.

The world export explosion has brought about tremendous trade possibilities for developing countries. Note that semiperipheral economies became a much greater source of manufactured imports for the United States, accounting for more than 30% of U.S. manufacturing imports in 1990, up from 17% in 1978. The heavy flow of imported textiles, apparel, footwear, small appliances, and low-cost consumer goods is a trend that is not expected to reverse itself in the near future. Despite the low value of these imports, it is expected that developing nations of the SOUTH will account for well over one-third of all U.S. imports by the turn of the century. Developing nations in close proximity to the leading economies of the core and the BEMs have already begun to capture a significant share of the world's low-valued manufacturing market. Nations such as Romania, Ukraine, Pakistan, Oman, Indonesia, Thailand, India, and Bangladesh have begun to take advantage of cheap and abundant labor forces, new production technology, and increasing consumer demand worldwide. These nations now vigorously compete with China, Mexico, and South Korea for the rapidly expanding markets for clothing and fashion accessories, athletic shoes, toys, trinkets, recreational equipment, and consumer electronics (Figure 9.4).

Note that the economies of the SOUTH have increased their share of world exports in manufactured commodities from 4.3% in 1963 to well over 12% by the 1990. It is expected the SOUTH's share of world exports will increase to about 15% by the year 2005. Note that

TABLE 9.7	U.S. Trade Data for Selected Big Emerging Markets					
			Exports to U.S.		Imports from U.S.	
BEM/Nation	**Imports 1994***		**1987**	**1994** (in billions 1994$)	**1987**	**1994**
Argentina	20.2	(248%)	1.1	1.7	1.1	4.5
Brazil**	27.7	(83%)	7.9	8.7	4.0	8.1
Mexico	79.3	(496%)	20.2	49.5	14.6	50.8
China***	115.7	(167%)	6.3	38.8	3.5	9.3
Taiwan	83.5	(138%)	24.6	26.7	7.4	17.1
India	24.8	(46%)	2.5	5.3	1.5	2.3
Indonesia	31.4	(143%)	3.4	6.5	0.8	2.8
S. Africa	20.4	(44%)	1.3	2.0	1.3	2.2

Source: *U.S. Global Trade Outlook 1995–2000: Toward the 21st Century*, U.S. Government Printing Office, Washington, D.C., March 1995.

Note: Numbers in parentheses show percentage growth of imports from 1987 to 1994 rounded to the nearest percent.

* Total imports in billions of U.S. dollars.

** The largest market for U.S. exports in South America.

*** "a nation that actively maintains an intricate system of import controls to implement its industrial and trade policies" (p. 75).

manufacturing exports from economies of the SOUTH grew at an average annual rate of almost 20% from 1970 to 1990, compared with the 12% average growth rate of exports posted by the market economies of the NORTH over the same period.

The importance of manufacturing exports from developing countries as a proportion of their total exports increased dramatically from 1970 to the mid-1990s. In 1970, manufacturing exports accounted for only 25% of all exports leaving the SOUTH. By 1986, this figure rose to 55%; and by 1995, manufacturing exports from economies of the SOUTH accounted for two-thirds of their total exports. The slow and steady shift from primary sector exports to secondary sector exports is likely to help the modernization and development efforts of the economies of the SOUTH as they diversify their production and export bases.

Geographic variations in manufacturing export growth are summarized in Table 9.8. Note that manufacturing exports still comprise a relatively small proportion of total exports in Africa, Central and South America (with the exceptions of Brazil, Mexico, and Argentina), and West Asia. These three regions still lag behind the average developing nation in the SOUTH in terms of establishing a viable and globally competitive manufacturing-for-export base. This, combined with excessive population growth, will undoubtedly hamper economic development efforts in these regions.

THE OECD

While Asian NICs have had relative good success penetrating the markets of the core, exports from the SOUTH are still relatively small in comparison to the flow of goods and services to and from nations of the NORTH. Consider that the top-twelve NICs accounted for less than 15% of the world's total exports by value in 1990, while a large majority of the world's exports were associated with core economies. In general, world trade is and will continue to be dominated by the nations which make up the **Organization for Economic Cooperation and Development (OECD)**. The OECD is *an alliance of 25 nations which share common goals in development matters, and coordinate economic and trade policies which work in the best interest of the group*. The OECD is composed of Australia, Austria, Belgium, Canada, Denmark, Finland, France, Germany, Greece, Iceland, Ireland, Italy, Japan, Luxembourg, Mexico, the Netherlands, New Zealand, Norway, Portugal, Spain, Sweden, Switzerland, Turkey, United Kingdom, and the United States. All OECD members are developed market-based economies of the NORTH with the exception of Turkey—a semiperipheral economy of the SOUTH.

Table 9.9 highlights the OECD's dominance in world trade. Note that the OECD accounted for almost 80% of world exports of manufactured commodities and more than 70% of all world commodity exports in 1990. The vast majority of OECD exports are traded within the OECD itself. Note also that the export dominance of the OECD is something that has remained largely unchanged since the 1970s. In fact, the OECD has actually improved its trade position over time. In 1990, the OECD accounted for 73.2% of world trade, up from 71.4% in 1970. The OECD is expected to account for 75% of world trade by the year 2000. As the world export explosion continues, the OECD will continue to demonstrate its economic clout as it secures the lion's share of export revenues and the associated economic benefits of export income.

Figure 9.4 Exploited workers in Asia's booming export business. Here a worker painstakingly makes umbrellas in Thailand, while a Vietnamese woman sews garments in a clothing factory in Saigon.

9.4a: ©John Fik/Stock Boston; 9.4b: ©Owen Franken/Stock Boston

THE NORTH AMERICAN FREE TRADE AGREEMENT

The economic position and leverage of the United States and the EU has been strengthened due to their G7 and

TABLE 9.8	Geographic Variations in Manufacturing Export Growth by Region

Geographic Region	Manufacturing Exports as a Percentage of Total Exports*			
	1970	1980	1986	1998**
North				
Developed market economies (13%)	66	67	73	76
South				
Developing market economies (20%)	25	37	55	65
South and East Asia (21%)	42	51	69	75
Latin America (17%)	10	16	24	32
Africa (12%)	7	6	10	14
West Asia (21%)	4	4	12	15

Source: P. Dicken, *Global Shift* (1992), p. 34, originally adopted from UNCTAD (1989) *Handbook on International Trade and Development, 1988*, Table 3.2.

* Rounded to the nearest percent.

** Estimate based on current trends.

Note: Average annual growth rate of manufacturing exports from 1970–1986 shown in parentheses.

OECD affiliations. The lessening of trade barriers worldwide has helped industries in these core economies to gain entry to new markets, expand their export bases, and enlarge their production networks. In addition to increasing export income, many economies have the added benefit of being part of a free trade alliance. Increasingly, free trade alliances will include developing nations as in the case of the **North American Free Trade Agreement (NAFTA)**—*a trade accord (enacted on January 1, 1994) which helped to establish a free trade zone between the United States, Canada, and Mexico.* The goal of the NAFTA is to facilitate cross-border movements of industrial and agricultural goods and services through the gradual elimination of tariffs and nontariff barriers in the tri-nation area. This will increase export possibilities throughout the region and stimulate economic growth and industrialization. As goods and investment dollars flow freely across international borders, it will promote competition and enhance employment opportunities throughout the free trade zone. Moreover, the agreement will provide adequate and effective protection and enforcement of intellectual property rights in each members territory and create useful procedures for the speedy resolution of disputes.

The NAFTA has served not only as a framework for further trilateral, regional, and multilateral cooperation, but has created one of the world's largest free trade zones with a combined regional Gross Domestic Product that exceeds $7 trillion. As roughly 87% of regional output is

TABLE 9.9	International Trade Flows and the Dominance of the OECD

Manufactured Products

Year	Value of World Manufacturing Exports (in $U.S. millions)	The OECD's Share of World Manufacturing Exports (in percentage terms)
1970	216,296	81.6%
1990	2,499,106	80.0%
2000*	3,800,000	80.5%

All Products

Year	Value of World Manufacturing Exports (in $U.S. millions)	The OECD's share of World Manufacturing Exports (in percentage terms)
1970	321,057	71.4%
1990	3,316,672	73.2%
2000*	5,400,000	75.0%

Source: *World Resources 1994–95* (1994), p. 263, Table 15.5.

* Forecast based on current trends.

accounted for by U.S. production, the NAFTA has solidified the position of the U.S. economy as a production leader in a unified and expanding regional market with more than 370 million prospective consumers. It has also made North America an increasingly competitive production region in the world economy. As a result of the NAFTA, the North American contingent of the G7 is likely to experience another take-off period (similar to what is expected in the EU) as they experience further export-led growth.

The United States has much to gain from the NAFTA and the subsequent vitalization of the Mexican economy. Growth will undoubtedly help increase Mexico's already healthy appetite for imported goods. Currently, more than two-third's of Mexico's imports come from suppliers in the United States. Note also that Mexico and Canada accounted for half of the U.S. export gain in 1994 according to the *U.S. Global Trade Outlook* (1995, p. 92). Although the United States has experienced a temporary trade imbalance with Mexico as a result of the peso devaluation of 1995 and Mexico's sluggish growth in the late 1990s, exports within the region are on the rise, and it may be just a matter of time before the multiplier effects of increasing export income ripple through the region.

The NAFTA is expected to bring about many benefits as the ailing Mexican economy recovers and grows and wages and income rise. Increasing income in Mexico will bring about increasing consumer and producer demand for high-valued products (much of which will be satisfied by U.S. industry). It is likely that U.S. industry

will also supply the majority of materials and production inputs needed to build up the infrastructure that is necessary to support the growth of Mexico's manufacturing, transportation, and telecommunications industries.

Growth of BEMs, the formation and proliferation of free trade zones, and the ongoing integration of economies such as Russia in the world economy will enhance the export opportunities of the G7 and the OECD. This is particularly true for high-value-added manufacturing, producer services, and trade-related sectors. Hence, the overall performance and competitiveness of the G7 and the OECD will remain strong as export opportunities increase across sectors and geographic regions. Unfortunately, it is also likely that the U.S. trade deficit will remain large due to the increasing demand for inexpensive imports and as a result of the growing pains of the NAFTA. Although some jobs in the United States will be lost as producers search for production-cost advantages within North America and abroad, the competitive advantage of the G7 and the OECD will help ensure that the most important jobs will remain "at home." Corporate downsizing trends and the use of cost-cutting measures such as "decentralization" of the production process (that is, moving portions of the production process to overseas locations), have allowed firms to remain viable and affiliated regions to reap the rewards of the boom in international trade.

Trade liberalization under the NAFTA has not caused a mass exodus of high-value-added jobs from the United States to Mexico as predicted by many experts. Although production shifts have been observed within the region, relocation of production facilities has been mostly limited to low-skilled, labor-intensive industries. Firms in these industries would most likely have relocated elsewhere with or without the NAFTA. Consider the Fender Musical Instrument company (originally based in Fullerton, California). This well-recognized leader and exporter of world-famous electric guitars and amplifiers had already moved some of its production and assembly facilities to Mexico and various Asian nations before the signing of the NAFTA. In some ways, the NAFTA has encouraged a greater production emphasis in Mexico than Asia given Mexico's proximity to the United States—the world's leading market of electronic musical equipment. Note that most of Fender's high-end product lines, those with the highest value added, continue to be produced in the United States. This is a production trend that is common across most industries with international production facilities.

The very fact that a large share of high-value-added producers has remained within the United States and that a large share of low-value-added production has remained within the tri-nation region is a benefit to the entire region. As wages and income rise in Mexico and Canada, and as the purchasing power of Mexican and Canadian consumers increase, it will create new export opportunities for U.S. companies (and their high-end product lines). In general, the industrialization and modernization of Mexico will mean an increase in demand for U.S. products, more trade, more growth, and inevitably more U.S. jobs.

The most visible impact of the NAFTA, thus far, has been in the transportation sector. As a business or producer service, the number of two-way border crossings between Mexico and the United States rose to 1.8 million per year in 1993, more than double the number in 1989. Border crossings by truck now exceed 2 million per year. It is a volume of traffic that is expected to rise even further as import and export restrictions are gradually lifted and routes and linkages between suppliers and consumers are established. The NAFTA has provided virtually unlimited access for long-distance truckers from Canada to the border states of Mexico. The NAFTA has also brought about new opportunities to financial sectors and the North American banking industry. It has given U.S. and Canadian banks and brokerage firms unlimited access to the Mexican market as the flow of money and investment capital become less restricted. Free trade and economic growth has led to new investment possibilities within the tri-nation area and an increased demand for financial services.

Opponents of the NAFTA have argued that the elimination of trade barriers between the United States and Mexico will accelerate the movement of U.S. firms southward. They contend that this movement will result in a substantial loss of revenue and jobs in the states as cheap imported products allow companies based in Mexico to capture a considerable share of the U.S. market. These arguments are not compelling, however, given that the U.S. market is still growing and that free trade (manufacturing-for-export) zones already existed prior to the NAFTA. In addition, Mexico is somewhat far removed (in terms of its location) from the major U.S. markets along the Eastern Seaboard. This separation adds a transportation cost premium to the price-competition equation, making Mexico less attractive to producers marketing their goods in New England and the Middle Atlantic states. Moreover, the net benefits of firm relocation are not entirely clear from a production-cost standpoint. U.S. labor offers both a wide range of skill levels and a significant productivity advantage in comparison to Mexican labor (despite its inability to compete in terms of wages). While many firms in core economies have aggressively moved their factories to overseas locations during the 1980s and early 1990s to take advantage of cheap labor in the SOUTH, many more have not relocated their operations.

Critics of the NAFTA claimed that the United States would lose as many as 500,000 jobs in the first year after the agreement was signed, although this was something that could not really be substantiated (as most job losses were related to defense budget cuts and the ongoing U.S.

trade deficit). As few people are interested in losing their job for the prospect of lower priced merchandise, the opposition to the NAFTA continues to be fierce. On one hand, scare tactics such as job-loss predictions are frequently used to try to sabotage trade accords. On the other hand, staunch supporters of free trade often exaggerate or overstate the potential benefits to the parties involved, expounding highly inflated export and job-creation estimates. A recent government report from the U.S. Labor Department claims that no more than 16,000 jobs were actually lost due to the NAFTA in 1994. By contrast, the U.S. Commerce Department estimates that the NAFTA created somewhere in the vicinity of 100,000 new jobs out of a total of 1.7 million American jobs created in 1994 alone, as trade between the United States and Mexico increased by 19% during the first year of the agreement. Naturally, NAFTA's critics claim that the government is covering up the real losses, in light of the recent U.S. trade deficit with Mexico. Meanwhile, exports to Mexico are up, and though the growth in exports to Mexico did not keep pace with the growth in imports, export earnings to U.S. firms are up. As a result, supporters of the NAFTA claim that job losses have been at least offset by job gains as a result of increased export earnings and production. Rising exports to Mexico continue to provide a great stimulus to the U.S. economy (Figure 9.5).

NAFTA-related job losses have been overstated as many U.S. industries had already become increasingly vulnerable to foreign competition prior to the passage of this agreement. Nonetheless, the NAFTA has led to job losses in low-technology sectors (mostly affecting rural areas. In particular, agriculture and the clothing industry have experienced the greatest losses. The overall effect of the NAFTA on various economic sectors will unavoidably translate into an uneven distribution of both costs and benefits over space. It is uncertain just how severe the

Figure 9.5 An anti-NAFTA protest from workers in Austin, Texas.

9.5: ©Bob Daemmrich/Stock Boston

losses will be in any particular region as linked industries in the North American production network slowly respond to the changes brought about by this agreement. As a result of the uncertainties, opposition and support (before and after the fact) will vary tremendously over space and the industries in question.

NAFTA's "rules of origin" provision will ensure that value added in production will be retained within the tri-nation region. This will allow multiplier effects to stimulate regionally linked industries in ways that have yet to be realized.

The flow of investment capital to Mexico would probably have occurred with or without the NAFTA. Mexico, with its modest earnings in the secondary sector, offers a production cost advantage not only in terms of wages but in terms of its location (proximity to the world's premier consumption region). In addition, productivity levels in Mexico grew at an annual rate of about 6% per year from 1988 to 1994, twice that of the United States. The growth of real wages in Mexico has not kept pace with the rise in worker productivity, making Mexico a very attractive work and production environment. This will continue for some time as real wages in Mexico are not expected to reach U.S. levels for at least several decades. Consequently, Mexico is likely to remain a viable destination for FDI and U.S. industry well into the next century. Table 9.10 highlights wages in the United States and Mexico in various manufacturing sectors.

Labor cost differences worldwide are largely responsible for the exodus of firms from the core to overseas locations. In fact, production cost differentials have hastened the pace of market penetration as imports continue to capture an increasing share of consumer markets in the core. Labor costs and international production shifts are two factors that are largely responsible for the decline of the U.S. textile and apparel industry. In 1980, imports only accounted for about 18% of all textiles and apparel consumed in the United States. By 1995, imports captured roughly 50% of the U.S. textile and apparel market.

REGIONAL INVESTMENT TRENDS

Relocation and decentralization of production is an outcome of globalization and the changing patterns of **foreign direct investment (FDI)**—*investments made abroad*. It is all part of a new philosophy amongst globally conscious firms to geographically diversify their operations and investment portfolios to increase revenue, maximize profit, and ensure their competitiveness in the ever-changing world of international business. As the hub of the G7, the United States has been the number-one foreign direct investment sink in the world economy.

A large portion of FDI to the United States comes in the form of money flows to **subsidiaries**—*branch plant*

operations of foreign firms. Money flows are categorized as either retained earnings or direct transfers of funds from parent firms abroad. Note that the majority of FDI is in the form of stocks, bonds, and private and government financial securities. These types of assets accounted for approximately 80% of the total FDI to the United States during the 1990s. The remaining 20% of FDI is attributed to investments made in real estate or industrial infrastructure (Ondrich and Wasylenko, 1993).

The inward flow of foreign direct investment to the United States grew at an annual rate of roughly 14% from 1982 to the mid-1990s, despite the effects of a lingering global recession and slowdowns in the growth of real income. While the boom in international trade is likely to bring new prospects for investment opportunities worldwide, it is expected that FDI flows to the United States will remain strong well into the next century. Investors from four countries continue to account for the majority of FDI in the United States. In 1990, roughly three-quarters of all foreign investments in the United States originated from Canada, the United Kingdom, the Netherlands, and Japan (see Table 9.11). Note Japan's quick ascent as an investing nation. Japan ranked fourth in terms of FDI in the United States in 1982. By 1996,

TABLE 9.10	**Mexican Wage Rates: A Comparison**

Mexico versus the Asian Giants (1991), for Workers in Industrial Sectors

Nation	Industrial Sector Wages U.S. $/hr. 1991
4 Tigers	
Hong Kong	$ 3.58
S. Korea	4.32
Singapore	4.38
Taiwan	4.42
Mexico	2.17

Mexico versus the United States (1991), for Workers in Selected Industrial Sectors

Sector	U.S. $/hr. Mexico	United States
Textiles	1.94	10.31
Food and tobacco	1.67	13.42
Electronic equipment	1.54	13.96
Industrial and office machinery	2.02	15.97
Chemicals	2.61	18.96
Automotive parts	2.75	21.93
Iron and steel	3.20	24.26

Source: *Business Week*, (April 19, 1993), pp. 84–85.

Data above are average wages and benefits of industrial workers.

TABLE 9.11	Major Investing Nations in the United States			
		Top Investing Nations		
	Total in millions	Canada	U.K.	Netherlands	Japan
Year	(1990 U.S.$)	**Percent of Total FDI to U.S.**		
1982	124,677	9.4	22.8	21.0	7.8
1990	396,702	7.6	25.9	16.1	20.6
1998	785,000	6.8	26.2	15.3	26.5

Source: Ondrich and Wasylenko (1993), p. 9., with estimated values for 1998.

Japan claimed the title as top foreign investing nation in the United States with approximately $150 billion of FDI.

The United States remains a favored destination for investment dollars in the world economy. As such, many argue that investment flows from abroad will enable the U.S. economy to reap the rewards of income transfers and associated multiplier effects. They argue that FDI will ensure that the U.S. economy will expand and create a multitude of employment opportunities in a wide variety of sectors and regions. Currently, there are seven states absorbing the bulk of FDI: California, Georgia, New York, Texas, Tennessee, North Carolina, and Illinois. This FDI trend reveals several important characteristics of international investment flows to the United States:

1. a large share of FDI gravitates toward destinations that are readily accessible to the major markets or cities along the coast or interior;
2. FDI is also absorbed by high growth areas, such as the case in the southeastern United States, where an increasing amount of investment flows are finding their way into the **new southern manufacturing belt**—*a rapidly industrializing region spanning the states of Tennessee, the Carolinas, and Georgia*;
3. foreign investment in the form of incoming firms is predominantly locating in areas that are away from the high-cost and established production regions; and
4. FDI from a given country of origin shows a tendency to form regional investment clusters.

The last two points suggest that physical and human capital requirements do not drive the location decisions of firms. Instead, other factors such as agglomeration economies, market access, and production cost considerations seem to be more important. Nonetheless, there are trade-offs to choosing a production location. For instance, while the established industrial corridor of the northeastern United States offers adequate infrastructure, skilled labor, and direct access to market, the southeastern United States offers a low-cost (low-wage) production environment in a high growth area. Not only does the southeastern United States offer new market opportunities, but it is far removed from the industrial heartland where organized labor is king, land is expensive, and taxes are high. Nations like the United Kingdom and Japan have had remarkable success in transplanting subsidiaries to American soil. Government at all levels—national, state, county, and city—has encouraged foreign investors to locate in the United States as part of an aggressive development strategy aimed at expanding regional employment opportunities and augmenting existing production networks (Figure 9.6).

Distinct locational patterns have emerged on the landscape. For example, Japanese subsidiaries competing in traditional manufacturing sectors such as automobiles and parts production have shown a strong location preference toward "right-to-work" environments. In other words, they have avoided regions which have a long history of supporting "organized labor" or labor unions (for example, the Northeast).

The rise of the new southern manufacturing belt has been attributed to more than just foreign firms trying to avoid organized labor. Many southern states have actively recruited foreign companies as a way to secure "good jobs" for their region or localities. By offering incentive packages which include grants, subsidies, tax breaks, land, and public financing of infrastructure to incoming corporations, many states in the South have had overwhelming success in attracting FDI. Once foreign-owned production facilities are established and prove to be profitable, the region may have the added benefit of attracting even more investment dollars having earned its reputation as a low-risk and low-cost production environment. In other words, FDI plays an important role in the creation of geographic clusters of firms from a given country of origin. There is empirical evidence to support the contention that foreign investors gravitate toward locations offering "agglomeration economies" (that is, savings which accrue from the spatial concentration of industry and proximity to users and suppliers) and that they generally exhibit a "follow-the-leader" mentality, thereby creating regional investment clusters.

Consider that Japanese investment in the United States has led to the formation of regionalized production networks in which Japanese firms show a distinct tendency to cluster in space. These affiliations often involve subcontracting of components and mutually beneficial ties between final assembly operations and suppliers (Head et al., 1995). For example, Japanese automotive parts suppliers tend to locate near Japanese-owned auto assembly operations. The agglomeration of suppliers and assemblers not only reduces total transport and interaction costs, but creates mutually beneficial linkages that allow suppliers to achieve "scale economies" in production. The attractiveness of a given location or region is dependent upon the economic advantages of clustering, yet it is also fortified by the development of a social network, as

Figure 9.6 Foreign manufacturers in the U.S. Here cars roll off the assembly line in Honda's production facility in Marysville, Ohio.

9.6: ©Barth Falkenbers/Stock Boston

TABLE 9.12	**Concentrations of U.S. and Japanese Manufacturing Firms by State: A Comparison**

	(%) Percentage of Total Firms		
State	**(900 firms) Japanese Manufacturing Firms**	**(229,292 firms) U.S. Manufacturing Firms**	**Difference between % Japanese and U.S.**
States with High Concentration of Japanese Firms in Comparison to U.S. Firms			
Georgia	7.8	2.0	5.8
Kentucky	5.6	0.9	4.7
Indiana	6.1	2.3	3.8
Ohio	9.0	5.4	3.6
Tennessee	5.0	1.7	3.3
Washington	3.3	1.5	1.8
South Carolina	2.4	1.0	1.4
North Carolina	3.9	2.6	1.3
Alaska	1.3	0.1	1.2
California	16.1	15.0	1.1
States with High Concentration of U.S. Firms in Comparison to Japanese Firms			
New York	2.3	9.4	−7.1
Pennsylvania	1.8	5.2	−3.4
Florida	0.4	3.6	−3.2
Massachusetts	0.7	3.4	−2.7
Texas	3.9	6.0	−2.1
Wisconsin	0.4	2.4	−2.0
New Jersey	2.8	4.7	−1.9
Minnesota	0.1	1.9	−1.8
Missouri	1.1	2.0	−0.9
Louisiana	0.1	0.9	−0.8

Source: Head et al. (1995), from map on p. 231.

Note: U.S. figures based on data from the Census of Manufacturing for the years 1982 and 1987. Japanese figures represent new manufacturing investments made after 1979 (excluding acquisitions).

people from like cultures begin to cluster in a given region. Social and economic networks have been equally important to the Japanese labor and management that accompany firm transplants to the United States and elsewhere.

It is interesting to note that FDI in manufacturing sectors may not actually create additional employment opportunities. As much of these investments come in the form of firm acquisitions or take-overs and not new production facilities, the overall impact of FDI may be overstated. In fact, many of the jobs attributed to FDI are not really additional manufacturing jobs, but "may be substitutes for, rather than complements of, domestically generated jobs" (Warf, 1990, p. 423). In other words, foreign direct investment may only serve to displace jobs from one region to another.

Despite expanding employment opportunities to the regions successfully absorbing FDI, the overall net benefits are questionable.

Agglomeration effects imply that any benefit received from attracting a single investment will be magnified by an increased probability of attracting subsequent similar investments. [Although] subsequent investment is likely to be small for states that Japanese investors perceive as relatively unattractive. Furthermore, the winners in bidding wars between state governments may find that the price paid in terms of subsidies and added infrastructure may offset any gain derived from attracting a foreign manufacturer. (Head et al., 1995, p. 243)

The differences in locational tendencies of U.S. and Japanese firms are highlighted in Table 9.12. Noticeably, the states which have relatively high concentrations of Japanese subsidiaries are states with relatively low concentrations of U.S. firms. The agglomeration and clustering effects described previously are highly visible, as well as the tendency of the Japanese to avoid the high-cost

labor areas in the Northeast. Note also that states which are absorbing the largest share of Japanese investment tend to be states that have not experienced large losses of industry or jobs over the past several decades (see Table 9.13).

The location of Japanese firms in the states exhibits a general avoidance of U.S. competitors. These non-confrontational locational patterns are also evident in **high-tech** industries—*activities in technologically sophisticated sectors which support a large percentage of highly skilled labor (for example, scientists and technicians) and have a heavy reliance on research and development expenditures targeted toward product innovation and improving production processes* (Malecki, 1985, 1991). High tech include software and computer systems, biotechnology,

TABLE 9.13	States with Significant Losses in Percentage of Total U.S. Manufacturing Employment (1970–1990)

States Showing Greatest Relative Decline in Total Manufacturing Employment (descending order)

1. New Jersey
2. Connecticut
3. Pennsylvania
4. Maryland
5. New York
6. Illinois
7. Massachusetts
8. New Hampshire
9. Maine
10. West Virginia

Source: Summary of data presented in Wheeler et al. (1998).

semiconductors, medical systems, telecommunications, pharmaceuticals, and photonics equipment.

Note that three nations account for roughly two-thirds of all foreign-owned high-tech employment in the United States—Germany, Japan, and the United Kingdom; with investments concentrated in states that are generally outside of areas containing the majority of U.S. competitors in these sectors (Warf, 1990). Exponential growth in the demand for high-tech products worldwide and the emergence of high-tech clusters within the core have hastened the globalization of production. In fact, the evolution of interregional production networks and changing investment flow patterns have been labeled as "techno-economic-driven" processes (Henderson, 1987).

FORCES OF CHANGE

Production trends and regional economic development have been greatly influenced by seven forces of change:

1. the **transnationalization process**—*the propensity for capital to flow across international borders in search of production cost advantage and the integration of regional production networks*;
2. **multinationals**—*transnationalized firms which operate in two or more nations*;
3. the emergence of **global capitalism** as a prevailing economic system;
4. the proliferation of **flexible production**;
5. **established FDI channels**—prevailing patterns of foreign direct investment within and between core economies;
6. **trade liberalization**—*the tendency of policy makers to support free trade and the deregulation of commodity and service flows across international boundaries*; and
7. the reemergence of **regionalism** at a larger scale—*the formation of alliances and policy measures to protect trading partners, zones, or blocs within the larger geographic region.*

The Transnationalization Process

Recent increases in the mobility of capital and the integration of regional production networks are changes attributed to transnationalization. The ever-increasing tendencies for capital to move across international boundaries has increased the mobility of industry, labor, and technology. It is a process which reflects the growing importance of managing risk by diversifying assets over the economic landscape. Subsequently, the economic activities of regions and nations are becoming more interdependent through time as virtually all globally competitive production networks now have **transboundary linkages**—*production ties and dependencies which span several political jurisdictions or geographic regions.*

As firms are finding it increasingly necessary to do business in external geographic markets, they continue to segregate production activities and the value-added sequence. Final goods and services produced in any one geographic region are likely to be composed of production inputs from several or many geographic regions. Consider the history of the U.S. automobile industry—an industry which is known to have a large and extensive production network. The transnationalization of the motor vehicle industry can be viewed as a five-stage process (Hartshorn and Alexander, 1988, pp. 242–254).

Stage 1: Domestic Production—characterized by domestic production (exclusively) to meet domestic demand. In the early years of the U.S. automobile industry, emphasis was on "mass production" and assembly line operations which relied on "scale economies" (that is, lowering the average cost of production by increasing the volume of output). Firms were initially located near places where product research and development occurred and where successful prototypes appeared. Later, as the industry matured, firms began to cluster near sites where production and transport cost advantages were realized or near places which had become dominant production hubs for the industry. In general, the location of firms and production was highly influenced by **historical inertia**—*the tendency to locate and form a production advantage in places in which a particular industry had an early start.* In the case of the U.S. automobile industry, major producers tended to locate within the confines of the so-called "automotive triangle"—a region centered around southern Michigan. Access to steel (a major production input), proximity to the major markets along the northeastern seaboard, and Detroit's early start in the business are the major reasons why the automobile industry flourished in this area. As domestic markets became saturated from the sale and resale of automobiles, the growth of sales slowed. Firms began to look to outside markets to help pick up the slack in domestic production.

Stage 2: Direct Exportation—characterized by the transition from domestic to international marketing in an attempt to sustain sales in the domestic market while capturing an increasing share of the demand in external markets. While the primary focus of stage two is still on sales in the domestic market, there is a growing secondary focus on sales abroad. As a result, international competition intensifies. Sometimes protectionist measures are employed to shield domestic producers as marketing within the industry becomes international in scope. Foreign competitors are restricted to have only limited access to the domestic market as regulations are placed on content, quality, etc., and governments impose quotas and import tariffs. Locational tendencies of firms do not change as they remain clustered in areas where the industry has established its roots and linkages. The primary locations of automotive parts and final auto-assembly operations in the United States remained concentrated in the automotive triangle (a production zone which spanned from Buffalo, New York, to Cincinnati, Ohio, to Janesville, Wisconsin).

Stage 3: Local Assembly of Products in Foreign Countries—characterized by the production of finished products on foreign soil. During this stage, subsidiaries or branch plants are set up in at least one "host nation" for the purpose(s) of establishing direct market access, sidestepping import restrictions, and/or reducing the cost of transporting a finished product to market. Consider the Japanese motor vehicle production facilities that began to appear in the United States during the early 1980s. Japanese firms were eager to locate in the states to take advantage of the large and growing U.S. market for "compact cars," while sidestepping restrictions and taxes imposed on "imports." The influx of Japanese automobile producers in the United States was not dissimilar to the movement of American-owned motor vehicle production facilities abroad (to Latin America and Europe) during the 1960s and 1970s.

Locations of foreign-owned motor vehicle production facilities in the United States during the 1980s were numerous, including Honda in Ohio, Toyota in Kentucky, and Nissan in Tennessee. The production landscape of the automobile industry began to change rapidly as foreign-owned firms opted to locate in areas that were outside traditional production areas. The locational preferences of foreign investors were in direct response to rising labor

and production costs in the automotive triangle. Domestic producers also tended to locate new production facilities in low-labor cost regions. For example, consider the location of GM's Saturn Division in Tennessee or the Ford Taurus plant in Atlanta, Georgia.

By the late 1980s, GM and Ford Motor Company had already established production plants throughout the world (GM in Brazil, Venezuela, Mexico, the United Kingdom, West Germany, Australia, the Philippines, Canada, Spain, New Zealand, and South Africa; Ford Motor Company in Argentina, Brazil, Columbia, Venezuela, Mexico, the United Kingdom, West Germany, Australia, Canada, Spain, Taiwan, New Zealand, and South Africa). In fact, during the 1980s, the European divisions of GM and Ford accounted for approximately one-quarter of sales in the European market, a figure that is comparable with the sale of Japanese cars in America over the last two decades.

Although U.S. companies have diversified their operations and sales geographically, the majority of market revenues are associated with either domestic sales or sales in the old established markets within the core. Consider the fact that more than 60% of overseas profits of all U.S. companies in 1991 came from the sale of products in Europe, from subsidiary operations, numbering in the tens of billions of dollars (roughly twice the percentage earned in all of Asia and Latin America). Though the pattern of overseas profits reflects how U.S. firms have diversified their investment and production portfolios in the world economy, like most highly industrialized nations, the bulk of foreign direct investment dollars have migrated to core nations in the NORTH.

Stage 4: The Integration of Regional Production Networks—when a large percentage of parts and purchases of a branch plant or subsidiary originate in the host country rather than from abroad. In stage four of the transnationalization process, foreign-owned firms begin to use "domestically" produced production inputs and parts in the assembly of their final products. Essentially, there is an assimilation of the foreign-owned facility into the production network of the host region. This accomplishes two objectives: (a) it strengthens network linkages and relations between domestic and "invading" industries; and (b) from a psychological standpoint, it fosters pride and prestige in making domestically produced foreign products.

To secure the benefits of network integration, local content requirements may be imposed (agreements or restrictions that limit the use of foreign-produced parts in the manufacturing process). Recent declines in the stockpiling of parts by manufacturing firms have contributed to an increased reliance on local suppliers, yet a very large percentage of parts still originate from production facilities outside of the host nation. Despite the "assembled-in-America" label, for example, it is well known that

Japanese producers show a clear preference for using their own supply network or imported parts from Japan rather than increase their reliance on U.S. parts producers.

Stage 5: Exportation of Products from Host Nations and the Emergence of a Foreign-Owned Production Network. The last stage of the transnationalization process involves the exportation of products from a host country. Consider that Japanese production facilities in the United States have now begun to export their automobiles to markets in Canada, Europe, and Latin America. This final stage not only allows firms to take advantage of low-cost production environments in multiple geographic regions, but also allows firms to increase their competitiveness in foreign markets by disguising their products as exports from another country. Foreign producers are now able to claim "made-elsewhere" status when exporting manufactured goods as a response to import restrictions and content requirements. The "American made" Japanese car can essentially be portrayed as a U.S. product when exported to Europe or Canada. This has allowed Japanese firms to further their penetration of markets within the core.

U.S. producers continue to take advantage of geographically diversified production and the NAFTA as they import parts and finished products from low-cost production facilities in Canada and Mexico. Ford has recently expanded its operations in Ontario, Canada, to increase production of the Windstar minivan for the U.S. market. Currently, more than one-half of Chrysler, Dodge, and Plymouth minivans sold in the U.S. market are built in Canada.

Stage five has also brought about the emergence of foreign-owned production networks in host nations as companies continue to diversify their holdings and operations across regions and economic sectors. Consider that almost 300 transplanted Japanese firms now reside on U.S. soil. In fact, Japanese subsidiaries operating in the United States are now the largest source of American-based exports to Japan. The overwhelming majority of these exports involve shipments to Japanese-owned parent or linked companies. Note that Japanese subsidiaries in America now export more goods to Japan than do all other U.S.-based exporters combined (Encarnation, 1992).

Transnationalization has not only changed the way companies do business, but it has changed the way in which companies and regions compete.

Multinationals

Economic growth in world and regional economies has been driven largely by exports, investment, integration of regional production networks, and the transnationalization of capital. The transnationalization process has given rise to multinational corporations or **multinationals (MNs)**—*private firms or companies that have legitimately (and with the consent of host governments) estab-lished operations which cross international boundaries and span several geographic markets, countries, or political jurisdictions.* As multinationals establish production linkages across national boundaries, they have become important to the development of the SOUTH. Despite their reputations as "exploiters," MNs are in big demand throughout most developing regions.

Consider the fact that global competition has led many large multinationals to internalize more of the value-added sequence, as opposed to "contracting out" at various stages of production. Hence, MNs have become more adept at activities that were once jobbed out (ranging from accounting services to marketing and inventory control, and the transport, distribution, and storage of production inputs and outputs). Large MNs are taking advantage of their size and potential scale economies, internally supporting every aspect of production from the acquisition of raw materials to the processing, fabrication, assembly, advertising, marketing, and distribution of the finished products. This transformation has led to the emergence of the **multinational enterprise (MNE)**—*a highly vertically integrated and globally competitive corporation or conglomerate that operates or does business in most countries and is involved in production of goods and services over multiple sectors.*

The Japanese version of the MNE, known as the "keiretsu" (or enterprise group), is typified by the joining together of three functionally distinct types of firms—those involved in manufacturing, marketing, and finance. These firms are linked by mutual ownership, interlocking directorships, and operational ties based upon mutual understandings. They are firms that act for the common good of the enterprise. Japan's success in the world economy has been traced to deliberate attempts to retain the majority of value added within the keiretsu. Japanese firms have also created market opportunities for the keiretsu by taking a "long-term" view of investment possibilities. Ironically, while General Motors and Ford laid-off tens of thousands of American workers in the 1970s and 1980s, Japanese automobile companies were in the process of erecting factories in the United States to build cars within a deindustrializing economy. Japanese firms intended to compete with American producers head on (on their own turf) to secure and possibly increase their share of an industry with hidden and unlimited growth potential. This would inevitably bring about future investment opportunities for the keiretsu on American soil as production linkages and networks materialized, not to mention export possibilities. Japanese production capacity in the United States now exceeds 1.5 million new vehicles per year with recent investments in assembly facilities in the Midwest and in the southeastern states. In all likelihood, Japanese vehicle production will continue to displace domestic production as Japanese exports from the United States are likely to rise over time. While exportation of Japanese products from the United States to markets abroad will help to reduce America's "on-paper" trade deficit, displacement of market share and exports will make matters worse for America's Big-Three automakers and the U.S. economy.

Multinational enterprises continue to steal the limelight as stars of a globalizing world economy. One of the best-known MNEs is the Mitsubishi Group—a world renowned electronics and automobile producer, and leader in financial and marketing services, not to mention a top-ten lending institution. Another well-known MNE is General Electric (GE)—a globally competitive conglomerate that produces or does business in most geopolitical jurisdictions around the world. GE continues to extend the breadth and depth of its empire with plans to expand operations in China, India and Mexico, further diversifying its holdings over many industries and sectors. GE operates as producer of household appliances, jet engines, plastics, and heavy machinery, and is a leading supplier of power and medical systems equipment and an international commercial lending and investing agent. Note that by the turn of the century, GE is expected to make more money from sales of its products outside the U.S. market than inside.

Other variations of the multinational enterprise are known to exist across cultures. Consider, for example, South Korea's powerful family-owned corporate conglomerates known as "chaebol." These MNEs control a multitude of firms that produce a wide variety of commodities within the global production and marketing network. Conglomerates such as Samsung, Hyundai, and Daewoo, for instance, have production interests internally or through acquired companies in electronics, telecommunications, ship-building, automobiles, and heavy machinery. Pressure from competitors and government, however, have prompted many of these conglomerates to reduce their holdings and acquisitions and lessen family control by making more of their corporate holdings accessible to the public.

Location decisions of multinationals are based on considerations to minimize production costs and maximize revenue and/or profit. This typically involves finding suitable locations in regions with easy acquisition to raw materials and other production inputs, low-cost and productive labor, and low fixed and/or variable operating costs in terms of expenditures on land, buildings, and infrastructure, relaxed or tolerable environmental laws and regulations, and political jurisdictions that are relatively stable, cooperative, and largely void of corruption, with governments that are nonintrusive. Multinationals also seek to avoid paying unnecessarily high taxes on income by gravitating toward low-tax or tax-exempt regions. In addition, multinationals seek to maximize their access to regional and world markets, taking into account existing tax codes and trade rules, regulations, and barriers that might cut into profits or inhibit the flow of raw material inputs or finished products.

Hence, a careful inspection of the business or production climate, as well as close scrutiny of the protectionist policies of a potential host region are necessary before a multinational decides to set up shop. In general, multinationals are sensitized to geographic variations in tax burden and barriers including tariffs, quotas, licensing agreements, customs and product documentation procedures, local content requirements, standards and product guidelines, credit restrictions, environmental regulations, etc. The location decisions of multinationals are also influenced by the relative locations of production sites with respect to major markets and the geography of income and wealth. Hence, industrialized regions of the core and semiperiphery are viewed as more appealing than remote or peripheral locations, production costs, taxes, and trade barriers permitting.

The two most common types of barriers to trade are "tariffs" and "quotas." A **tariff** *is a tax or duty imposed against a particular category of merchandise entering or leaving a nation.* Import and export tariffs are levied to: (1) protect domestic producers from foreign competition and minimize leakage; and (2) generate revenue for the purposes of support development initiatives or subsidizing domestic industries. The United States generates revenues by imposing import tariffs on a variety of products including shoes from Italy and Brazil and export tariffs on foreign-owned made-in-the-U.S. products exported to markets outside of the United States. Governments of underdeveloped nations ordinarily rely upon such revenues to fund social programs, industry, or development efforts. In many developing regions, export tariffs are used to counter the effects of the opportunity costs of exporting raw materials (especially strategic minerals) or other physical and human resources.

The general effects of tariffs imposed on imports include: (a) the increase in domestic prices (in comparison to price levels in the absence of a tariff; and (b) less import penetration and competition from outside suppliers or regions. By design, import tariffs and government subsidies are used to protect industries that are not globally competitive. Ultimately, consumers lose as they must contend with higher and noncompetitive pricing of commodities. Furthermore, less competition means less incentive to promote, sustain, and improve standards, quality, and quality control in production, resulting in a less-than-efficient use of resources and production factors.

While tariffs are popular ways in which to generate revenue and protect domestic industries, quotas are also effective as protectionist measures. A **quota** *is a specific limitation on the quantity of exports or (more usually) imports that a country will permit over a certain period of time.* Quotas are especially favored as a means of protection against foreign competitors, and have been imposed with a fair amount of regularity in cases involving agricultural commodities (for example, quotas on imported sugar to protect domestic sugar producers in the United States) and standardized manufactured products such as finished textiles and fabrics. Governments tend to favor quotas when the supply of a product is very great and the product is available at a low price. The imposition of a quota restricts supply and keeps prices artificially high. Administrators favor quotas because they produce sudden, drastic, and certain results, and are easy to impose, remove, and/or adjust depending upon the situation and policy goals. Note that economic sanctions imposed against a given nation, particularly an all-out embargo (banning all interaction and trade with that region or partially restricting the flow of commodities to that region from one or more supplying regions), represent the most severe type of quota.

Important nontariff barriers include government involvement in industry. This may take on many forms. For instance, several leading European nations provide government price supports and subsidies to guarantee prices for agricultural products in an attempt to protect European farmers from foreign competition. Discriminatory practices such as local or domestic content requirements (as imposed in U.S. steel and automobile industries) are still another form of a nontariff barrier. Governments may also establish strict customs and entry procedures, such as requirements concerning product standards, documentation, health and safety restrictions, and classification and packaging requirements. For example, recent health-related problems with imported fruits and vegetables from Mexico has made the U.S. government ready to impose tougher quality and handling standards for imported agricultural products (despite their commitment to free trade).

European beef producers continue to insist on restrictions on beef imported to European markets. U.S. beef was considered to be of unacceptable quality by European nations who had imposed tough standards to protect European producers from the less-expensive imports, as European producers faced a loss of sales revenue from outside competitors. The Europeans claimed that U.S. beef was tainted by the chemicals and steroids fed to the animals during the fattening and finished feeding stages of production. Although arguing that the growth-hormone-fed beef was unfit for human consumption, the intervention was a way to stop low-cost and superior-quality American beef from reaching European consumers. Health and safety requirements and other imposed restrictions can easily be used as a way to protect domestic industries.

Consider the existence of government agencies like the U.S. Food and Drug Administration (FDA)—an organization that imposes strict quality control, test and performance standards on foods and drugs (in prescription or nonprescription form) which reach the marketplace. The law requires that FDA testing and approval is necessary before any new food or pharmaceutical is marketed and sold to the public. Essentially the FDA acts as a gate-keeper, keeping foreign drug manufacturers out of the U.S. market in the name of consumer safety, thereby limiting access to the U.S. market.

The greater the array of trade barriers and restrictions that producers must face in the world economy, the more difficult it is has become for those producers to compete for a share of protected markets. The vast array of government restrictions and the increasing levels of government involvement in trade matters is somewhat of a contradiction. While governments have long realized the economic benefits of promoting free trade and competition amongst regions and producers, they continue to legislate from a protectionalist mindset. At times, it is not clear whether their decisions are in the best interest of consumers or industry. Despite the challenges posed by protectionalist measures, multinationals have continued their steady rise to dominance in a world economy that seems less affected by local conditions and increasingly more responsive to the dynamics of interregional, international, and global change.

Multinationals are both friends and foes of development. On one hand they offer nations a way to buy time as they search for a cure for underdevelment. By furnishing regions with jobs and a sense of economic empowerment, multinationals offer hope in the struggle to gain economic sovereignty. On the other hand, multinationals provide little in the way of helping regions promote self-reliance.

Integration into the world economy means much more than just attracting foreign direct investment, or promoting manufacturing exports, or joining a trade alliance. As such, underdeveloped regions have found it increasingly difficult to secure or exploit their own production advantage as they must accommodate or compete with powerful multinationals. The "if-you-can't-beat-'em-join-'em" philosophy has led many regions of the SOUTH to concentrate their efforts on attracting multinational investment as a least-resistance path to development. The influx of FDI in the form of multinational subsidiaries is one approach that is favored by many prospective host nations in their attempt to institute broad-sweeping economic reforms to promote integration and as a way to supplement economic growth. The continuing challenge of baiting and hooking multinationals has been made more difficult as host governments walk the fine line between creating conducive business and production climates and imposing restrictions, tariffs, and "performance requirements."

As a way of securing jobs for the skilled portions of the indigenous workforce, multinationals may be required to use a specified portion of local personnel in production and management, or set up a timetable to incorporate portions of the local labor force directly into company operations. Host governments must be careful, nevertheless, not to impose too many restrictions or place too many demands upon multinationals, as they run the risk of scaring them away. This is especially true when it comes to taxes on exports. Cognizant of this reality, many developing nations have set up **special export processing zones (SEPZs)**—*geographic areas set aside by government to use as production-for-export sites in which incoming firms are to be given preferential treatment.*

It is common for governments to permit low-tariff or tariff-free exporting of finished goods from final assembly operations located in SEPZs. The use of consigned materials and equipment in SEPZs allows incoming firms to sidestep regulations, taxes, and the required paperwork associated with carrying out similar activities in other host countries. Incoming multinationals are persuaded to incorporate skilled and managerial segments of the domestic labor force in the day-to-day operations of the company. As infrastructural requirements are complex, SEPZs are typically in or near large established urban areas (although by no means are these zones excluded from rural settings). The "industrial recruiting" process in developing regions has been greatly aided by the creation of globally competitive SEPZs. These specialized production zones have been instrumental to China's rise to economic superpower status.

Urban and economic reforms initiated during the late 1970s under China's "open door policy," including provisions to accommodate foreign-owned firms in SEPZs, enabled the Chinese government to attract FDI from around the globe. The majority of FDI to China ended up in specialized production zones located in "open cities" and numerous "open coastal areas" in locations that spanned the entire coast of China from north to south. By establishing SEPZs in close proximity to thriving economic regions like Hong Kong and Taiwan and other booming coastal cities which engaged in free enterprise, the Chinese government was successful in securing large amounts of FDI from investors in free-market economies despite its command economy orientation. The recruitment of multinationals in and along coastal China was intended to maximize the benefits of incoming investments and technology transfers by directing flows to a handful of geographic locations that were largely outside the direct influence of Beijing (and for all intents and purposes acted as centers of free-market activity). In short, the Chinese government adopted a hands-off approach, granting greater flexibility to both indigenous and foreign producers in matters that had previously been handled by the state.

The highly concentrated nature of FDI surely added to the enormous success of China's market-oriented reforms. Consider that approximately three-quarters of China's exports in the 1990s came from just five provinces along the coast—a geographic area that accounts for less than 1% of China's total land area. The most successful exporting region is the Guangdong province, home to three of five SEPZs. It is an area that is expected to be 50% urbanized by the turn of the century.

Guangdong province alone accounts for well over 20% of all China's exports.

China, nonetheless, is an economy of contradictions. While the majority of China's economic activities are under the watchful eyes of a repressive communist regime, the modern production-for-export zones offer a taste of the free-market economies of the Western world with seemingly endless profit-making opportunities for outside investors. Throughout the SEPZs, taxes are low, import and export restrictions are minimal, contract labor is cheap and plentiful, and land leasing is made simple by a government that seems to be very responsive to the needs of multinationals (de Blij and Muller, 1994, pp. 544–545). Outside of these zones, however, heavy-handed government continues to be visible, dominating virtually every aspect of production and distribution on the economic landscape.

The coexistence of a historically dominant communist regime and a series of newly established market-oriented regional economies is testimony to the power of capitalism and to the limitations of command economy. Unlike leaders in the former Soviet Union, Chinese officials had the foresight to embrace capitalist ideals, if only in a limited geographic sense and scope. By minimizing government intervention in SEPZs, China was able to realize the benefits of export economy and investment transfers, while the Chinese government maintained control of its domestic economy. In addition, SEPZs helped to increase the interaction between mainland China and overseas communities via Hong Kong and Taiwan (areas with strong ties across the globe). Essentially, Hong Kong and Taiwan served as both stepping stones to overseas markets and important sources of investment and human capital. Industrial linkages and commodity flows between Hong Kong, Taiwan, and China were also strengthened by the growing market for producer and consumer goods in coastal and inland China. The rapidly growing Chinese economy provided investment opportunity and market clout, and the dynamic economies of Taiwan and Hong Kong supplied vital linkages to the capitalist global economy.

The ever increasing economic integration of Hong Kong, Taiwan, and the mainland, a tri-region entity that has been labeled as "Greater China," has allowed China to move ever closer toward a reunified Chinese state (Shambaugh, 1995). Certainly, the probability of such an outcome has been increased as (a) the colony of Hong Kong (formerly under British rule) has been returned to China as of July 1997; and (b) China continues to exert tremendous political pressure on Taiwan, flexing its strengthened military muscle, in an effort to push for economical and political unification. A unified Greater China is likely to enter into yet another take-off period should further industrial linkages bring about the complete integration of the tri-regional production network.

As a result of these and other reforms, China has not only diversified the distribution of its national industrial output and increased its ties with markets and investors around the globe, but it also reduced the economic burden on the Chinese government to finance growth and development. In 1978, state-owned operations accounted for 81% of China's industrial output. By 1987, the production share associated with state-owned industry declined to 60% as production associated with collective and individual investments accounted for 26% and 14% of total production, respectively.

By establishing basic and nonobstructive ground rules for incoming industry, the Chinese government was able to exploit its low-cost, low-wage production environment and secure (by agreement) many managerial and management-training positions for its workforce. Thus, the presence of multinationals in China's economy served not only to improve coastal China's position as a global investment sink and as a globally competitive economy, but it also served to help train and upskill its labor force. Moreover, interactions with multinationals have helped China improve its production technology, as many foreign investors are forced to sell or divulge their production know-how, which can be later adopted by Chinese competitors. The diffusion of technology has aided Chinese companies in their attempt to gain access to world markets and compete with firms in core economies. Unfortunately, Chinese producers have also engaged in the pirating of technology, the illegal and unauthorized production, sales, and distribution of products, and international copyright and patent violations, despite the outrage of governments in the free world and the international business community.

In light of China's success in securing FDI and attracting multinationals, other developing nations have intensified their efforts to create a globally competitive export base. Export processing zones have gained popularity throughout the SOUTH, in economies like Mexico, Brazil, Indonesia, the Philippines, and Malaysia. Notwithstanding local employment, export, and income gains associated with multinationals operations, there is an ongoing debate on the pros and cons of multinational investment in the regional economic development process in developing economies.

In general, there are four major benefits of multinationals. First, multinational investment is a source of FDI that would otherwise be unavailable to many underdeveloped regions of the SOUTH. Viewed in this way, multinationals are an alternative to industrial development loans. Second, multinationals transfer advanced production technology and know-how which can aid in upskilling segments of the labor force and help overall in the social transformation and modernization processes. Third, subsidiaries of multinationals create not only jobs for the local economy, but increase employment opportunities throughout the region by forming linkages to other domestic industries. Fourth, multinationals assist host nations in realizing the full potential of their human and physical resources. From a radical perspective, this last benefit could be interpreted as a "cost" as it may imply the exploitation of a host nation's labor force or physical resources.

While the list of benefits seems impressive, the list of nonbenefits or costs is rather extensive. Basically, there are at least nine costs or nonbenefits of multinational investment.

1. Multinationals are **footloose industries**—*industries with little or no loyalty to a host region or nation, that tend to be highly mobile and can relocate with little warning once a production-cost advantage erodes*. Once the production location or facility ceases to be advantageous or profitable, the multinational simply closes its doors and moves on to another region or nation.

2. Multinational investment provides little "long-term" benefits for a host regional economy as investments are typically earmarked for selected geographic areas—primarily large urban areas. These places are already experiencing rapid growth due to increasing demand and an ongoing population explosion. Hence, the overall effect of an incoming multinational is marginal or small (and in some cases undetectable). In some instances, the incoming multinational may actually exacerbate the hyper-urbanization problem.

3. Investment flows of multinationals are sector-specific in nature. In short, money flows only to those sectors or industries in which the multinational operates.

4. Natural resource extraction and use activities are geared only toward the production activities of the multinational and are not necessarily in line with what is best for the host nation. There is little regard to the sustainability of the extraction activity or the opportunity costs of resource utilization.

5. Capital and technology transfers to host countries are somewhat limited, restricted to highly specialized applications in the sectors or industries in which the multinational is active. Most production technology transfers tend to be one step behind current state-of-the-art production technologies.

6. Some branch plant operations of multinationals tend to be fairly capital intensive, while others require only a very specific type of labor input or skill (for example, hand labor for sewing garments or soldering circuitry boards). Hence, host governments run the risk of subsidizing a process that is not easily duplicated should the multinational close its doors. Furthermore, host regions run the risk of training a labor force with skills that may or may not be useful to the region once the multinational terminates its operation.

7. The movement of multinationals into peripheral and semiperipheral economies has had a negative impact on the employment situations in the regions from which investment dollars originate (typically core economies). The outflow of capital from the core, albeit minimal, has been blamed for employment losses in several industries including textiles and clothing manufacturing. As firms decide to close down domestically owned and operated facilities, and relocate to production sites abroad, they cause major disruptions to the local economies.

8. When multinationals flood regional and world markets with inexpensive goods produced abroad (in labor-saving production locations in the periphery and semiperiphery), they compete away the market share of firms or competitors in the core. This boomerang effect can hamper growth and reduce employment opportunities in the industry in question and those linked to it.

9. Profits of multinational operations are usually expropriated back to the investing nation. In short, the large majority of revenue and profits associated with sales from commodities produced by foreign-owned subsidiaries will leak out of the host nation, allowing a smaller-than-expected multiplier effect.

While there is no clear-cut consensus among industry analysts on the net-benefits of entertaining multinationals and their role as instruments of development, many do believe that the costs of multinationals generally outweigh the benefits, particularly for the world's more peripheral regions. The bottom line here is that investors and corporations benefit greatly, while labor, regions, and resources tend to be exploited for financial gain.

The influence and clout of multinationals will undoubtedly increase as the world's production regions succumb to the forces of globalization. From an economic perspective, it can be argued that multinationals provide the world economy with an overall efficiency gain as they maintain or advance their competitive positions and increase competition with rival multinationals. Hence, it can be argued that multinationals assist in efficient "global sourcing" and use of world resources and production factors.

Global Capitalism

The globalization of production has led to many changes in the structure of industry, corporations, investment, and production. These changes have brought about a new variant of advanced capitalism, namely "global capitalism." **Global capitalism** *is a system of international production and exchange in which the process of capital accumulation and the organizational aspects of regional and interregional production networks are controlled almost exclusively by multinationals.*

Under global capitalism, corporations based in the dominant core economies of the NORTH exert the greatest leverage over all production regions as they attempt to stabilize or improve their positions and compete for shares of the world's established and emerging markets. It is international commerce dominated not so much by regions or nations, but by the world's most influential companies and conglomerates.

The onset of global capitalism has been paralleled by the corresponding decline of "monopoly capitalism," a term which describes the dominant influence of a single producer or industry in a given region or market. Under monopoly capitalism, an individual producer or industry is said to exert sufficient control over a region or market as to determine the extent to which others may have access to that region or market. Under global capitalism, no one producer, company, or industry is said to exhibit a dominant influence in a region as most geographic markets are now served by a wide array of suppliers and international rivals. The decline of monopoly capitalism is largely due to the inability of any one firm or industry to maintain dominance or sustain control of any one given geographic market.

Unlike the power base of organized labor which flourished in the era of monopoly capitalism (due to the regional concentration of industry), global capitalism has greatly reduced the bargaining power of labor. Collective bargaining with corporations has become increasingly difficult in the era of footloose industry. Workers throughout the world have lost considerable amounts of clout as highly mobile multinationals are apt to pack up and move elsewhere should the production climate become "hostile" or when the demands of labor become excessive from the standpoint of management or the owners. Labor organization has been made increasingly difficult in a production environment where multinationals have geographically decentralized production and the value-added process.

> Global capitalism eliminates the familiar meeting ground between labor and capital. Under the monopoly variant the primary workforce confronted management through the medium of stylized negotiations, job actions, and long-term contractual agreements. With the emergence of global capitalism these long-standing rules of the game no longer apply. The ability of capital to shift from one production branch to another, or from one factory production to informal subcontracting, and hence from one labor market to another, renders labor more dependent on . . . management. Workers are apt, as a consequence, to be more pliant or quixotic. (Graham et al., 1988, p. 477)

Multinationals and the globalization of production have also led many to question the relevance of the "nation-state" as a policy-making body. The **nation-state** is *a geopolitical unit comprised of production regions and human agents, as unified under a single political* *jurisdiction and possessing a certain amount of power over its own social, economic, and political destinies* (de Blij and Muller, 1994, p. 60). The nation-state represents a geopolitical unit in which the territorial state both coincides with the area settled by a group or culture and includes those human activities that are vital to the interest and security of the nation. Just as labor has become less powerful, so too are regions and nations becoming less in control of their economic destinies. Increasing interdependence and vulnerability of most regional economies to external forces have left the higher authority of the nation-state without much economic or political clout.

There is an ongoing debate on the role and the effectiveness of the nation-state in matters of national economic policy as production goes global and as multinationals dominate the economic landscape.

> As industry repositions itself with respect to the nation—ceding national markets to foreign competition and producing goods for domestic consumption in foreign locations with foreign labor—the traditional identification of the economy and state is undermined by radical contradictions. (Graham et al., 1988, p. 477)

Concerns over the ineffectiveness of the nation-state have given rise to discussions of forming a **world-state**—*a universal or global geopolitical unit.* Yet, government and policymaking at the global level may be impractical as the "internationalization" of humankind has been largely challenged by a world in which unity and division of cultures often go hand in hand. Critics also argue that the concept of a "powerless state" in the world economy is nothing more than a myth, as national governments continue to exert tremendous regulatory influence over industry in all corners of the world.

Flexible Production

Parallel to the emergence of large and geographical diversified multinationals and the internalization of many production activities, there has been an increasing tendency for smaller and more specialized firms to appear on the economic landscape in distinct regional pockets. This trend continues to transform the geography of manufacturing in Europe and North America. Intensified competition in regional and global markets has led many firms to reorganize the manufacturing process in ways which allow them to respond to "very particular and frequently changing demands of niche markets" (Knudsen, 1994, p. 137). In other words, highly specialized firms and production zones are appearing in response to a segment of market demand which is not generally satisfied by product lines offered by multinationals. Producers in these areas are said to be "flexible" in terms of their ability to meet the diverse and changing needs of niche markets.

The rise of "flexible" production has most certainly been initiated and aided by new information and commu-nications technologies and the increasing coordination of intra- and inter-firm activities in modern-day production networks. Flexibility can be achieved through the "vertical disintegration" of production, as firms externalize many of the tasks and functions previously performed within the firm, or through output specialization.

In general, **flexible production** *is a highly specialized production process in a network where the use of information and communications technologies permits the rapid switching of production processes to meet the diverse and changing demands in the marketplace, resulting in a diverse array of end-products that are tailored specifically to the requirements of individual customers.*

Examples include high-technology firms specializing in the production of prostheses, computer-aided manufacturing, custom tool and die operations, hand-made apparel, ceramics, and musical instruments, design-intensive manufacturing such as custom-made furniture, electronics, and computer systems and software, and financial and personal services. The ability of firms to remain competitive rests on external **economies of scope**—*cost advantages arising from an ability to rapidly switch from one production process or product and manufacture or create highly differentiated product lines.*

Flexible production networks have emerged in a wide variety of places including industrial districts in the northern and central regions of Italy (that is, Third Italy), the Jura region of Switzerland, central Portugal, and the high-tech production centers found in France and the United States. Moreover, less-visible flexible production networks have been formed in the labor-intensive sweatshops of urban clothing makers in the garment and high-fashion industries of global cities such as New York, Los Angeles, Toronto, and Paris (Scott, 1988).

Flexible production is most likely to be observed in localities which are largely insulated from organized labor activity. Typically, flexible production tends to be found in areas in which wages and skill levels are low and worker benefits are virtually nonexistent. Flexible firms not only prefer low-cost labor, but have an uncanny ability to survive and adapt to their surroundings, including

> an artisanal capacity to respond to new designs and market signals, as well as other factors such as self-exploitation and the use of family labour, the evasion of tax and social security contributions, low overhead costs, and the use of female and young workers. (Amin, 1989, p. 30)

Flexible production management techniques are now being adopted by large firms. The use of computer-integrated manufacturing systems, robotics, and production teams (which give greater responsibility to individual workers who must be able to complete a diverse set of work assignments) has led to dramatic changes in the high-value-added industries in France, Germany, the United States and Japan. This has placed increasing demands on labor, as workers are now required to do multiple tasks and have many skills. The changing needs of industry have increased the need for further investment in human capital and education. This will require a renewed commitment on the part of both the public and private sectors to help retrain workers if they are to meet the demands of the flexible workplace and ensure that affected regions make a smooth transition from conventional to flexible production. Increasing flexibility of the workforce has allowed many firms to survive the onslaught of deindustrialization and globalization as they take advantage of cost-saving production technologies and greater worker productivity (Figure 9.7).

Malecki and Veldhoen (1993, pp. 134–135) distinguish between three types of flexibility in production:

1. *intra-firm flexibility*—achieved through the adoption of advanced computer-integrated technology and systems (for example, robotics, best-practice machinery, and specialized computer programming applications) or other highly individualized production processes which are carried out within the firm;

2. *inter-firm flexibility*—achieved by improving the quality and efficiency of inter-firm interaction, network relations, information processing, and inventory management systems which control the flow production inputs and outputs (as tailored to meet the needs of producers in the network); and

3. *labor adaptability*—requiring that labor be flexible rather than rigid in carrying out various production tasks for which they are called upon to perform.

Adaptability of labor has meant that workers must move away from traditional task fragmentation and specialized Fordist assembly-line activities. Labor must now be able to work in teams, rotate jobs, learn by doing, and coordinate and integrate various production tasks and stages. This requires a working knowledge and understanding of social and physical organization of the firm and the entire production process. Flexibility of this variety has allowed small and highly specialized firms to stay competitive as they serve niche markets or act as suppliers within a regional production network.

The Geography of Foreign Direct Investment

Foreign direct investment has three major impacts on a region. First, FDI stimulates regional employment growth or at least has a stabilizing effect on a regional economy. Second, FDI promotes interaction between firms and industry, leading to the exchange of goods and services and the formation of linkages. Third, FDI increases competition and efficiency in linked and rival firms and the entire production network.

Figure 9.7   Flexible and computer-aided manufacturing continues to dominate the production landscape. Here a computer-controlled machine produces dough at the Nabisco plant in Richmond, Virginia.
9.7: ©Bob Crandall/Stock Boston

Data on foreign direct investment are usually reported as either "transactions" or "stocks." FDI **transactions** *are total investment flows for a given year*, whereas **stocks** *are accumulations of FDI flows over time (that is, the sum of all FDI transactions up to a given year)*. Note that the bulk of foreign direct investment in the world economy is heavily biased in favor of nations with high economic development status. In other words, most FDI stock resides in the core economies of the NORTH.

The growth of FDI has been truly remarkable over the last 20 years. From the late 1980s to the mid-1990s, the rate of growth in FDI was roughly three times that of the growth rate of international trade. Nonetheless, FDI transactions in 1990 were still only a small fraction of the value of commodities traded worldwide, with transactions amounting to less than 10% of the total value of all merchandise trade. The United States remains the favorite destination of global investors. This trend is unlikely to change as recent investment flows to Asia and Latin

America have been but a trickle of total investment. Note that FDI-associated production now accounts for roughly 10% of the U.S. manufacturing labor force, up from just 2% in the early 1970s. While prospective investors continue to explore the option of internationalizing their investment portfolios, recent financial woes in Asia and lackluster economic growth and unstable currencies in Latin America and Russia have reminded investors of the risks of international investment. The economic stability of core economies makes the core an intervening opportunity for conservative and risk-averse investors.

A breakdown of investment flows by income group shows the tendency for flows to be largely restricted to economies of high economic development status (see Table 9.14). The lion's share of FDI flows between the high-income nations of the core and secondly to middle-income or semiperipheral nations. The inability of the periphery to attract large amounts of foreign direct investment is something that will ultimately result in stunted economic growth and a persistent lack of competitiveness. In addition, the reputation of the periphery as a risky investment sink will further reduce its ability to attract investment capital in the future. This is a dangerous spiral which traps many low-income economies.

As noted by Hanink (1994), the distribution of FDI by sector for the U.S. economy does not follow a pattern that is consistent with the breakdown of labor force participation by sector or the contribution of sectors to gross output (see Table 9.15). The nominal shares of FDI do not correspond very closely to sectoral employment or national income shares for several reasons. First, the pri-

mary sector seems to be overrepresented in both outgoing and incoming investment flows. This is attributed to the importance of petroleum and mineral industries to core economies and the fact that the world's largest oil companies operate in and across many international jurisdictions and boundaries (being multinational corporations). Second, the share of FDI in the secondary sector, both incoming and outgoing, seems to be disproportionately large in comparison to the sector employment and national income shares. By contrast, the share of FDI in the tertiary sector seems disproportionately small in comparison to the tertiary sector's contribution to GDP. This implies that foreign manufacturers think they can do well by locating in America and American manufacturers think they can do well by producing abroad. Hence, the amount of FDI held by foreign interests in the United States (particularly in the secondary sector) indicates that exports and foreign direct investment are not direct substitutes; otherwise foreign producers would simply produce at home and/or in other low-cost production environments and export to the U.S. market (Hanink, 1994). This provides strong evidence that proximity to major markets of the core is an important consideration in the multinational investment decision and the industrial location process.

Two established investment flow channels have been identified in the world economy:

1. the primary investment flows between core economies (mainly FDI amongst the G7, the EU, and Japan); and
2. an increasingly important series of secondary flows from the core to semiperipheral areas. The former category represents an astonishing 75% of all investment capital transfers in the world economy, with the United States accounting for well over half of those transfers. Figure 9.8 provides an illustration of intra-core investment flows and the gravitational pull of the U.S. economy.

TABLE 9.14	Mean Foreign Direct Investment: Transactions and Stocks (1990)

	Transactions (in millions of U.S. $)	
Income Group	Outgoing	Incoming
All	3482.9	2771.5
Low	38.0	291.4
Low-Middle	12.1	414.4
High-Middle	174.7	625.6
High	12,517.5	9402.9

	Stocks (in millions of U.S. $)	
Income Group	Outgoing	Incoming
All	79,911.2	74,508.4
Low	nr	nr
Low-Middle	266.0	2510.0
High-Middle	3685.0	5986.0
High	115,181.1	106,286.9

"nr" means none reported.

Source: Hanink, 1994, p. 226.

TABLE 9.15	U.S. FDI, Labor Force Participation, and GDP by Sector: A Comparison

	Percentage of Total (by sector)			
	U.S. FDI (1990)		Percent of U.S. Labor Force	U.S. GDP
Sectors	Outgoing	Incoming	(1990–1991)	1990
Primary	16	12	3	2
Secondary	40	40	26	29
Tertiary	44	48	72	69

Source: Based on data from Hanink (1994), p. 229; *World Resources 1994–95*, p. 287; and *World Development Report 1991*, p. 209.

Note that investment flows from the core to semiperipheral regions exhibit a very noticeable and regular geographic pattern. In fact, the outflow of investment capital from the core tends to form several distinct regional investment clusters, with the flows going to regions that are proximate to the source (as illustrated by Figure 9.9). The majority of U.S. investments made outside the core tends to flow to nations in Latin America (with the notable exception of a minor investment cluster in the Indian Ocean Rim). Japan's investment outside the core seems to be largely confined to a handful of nations in the Pacific Rim. Noncore investments from the European Union show a clear preference toward investing in eastern Europe and republics of the former Soviet Union (Figure 9.9).

In response to inflationary pressures and the dramatic increases in the cost of both living and doing business in Japan throughout 1980s and early 1990s, Japanese investors looked for investment opportunities abroad. Transnational investment allowed Japanese investors to internationally diversify their portfolios and Japanese firms and businessmen the luxury to escape the high-cost urban environments. As Japan's cities became congested and expensive places in which to live and work, investment dollars flowed outward and firms exited to the low-cost and neighboring Pacific Rim nations like Malaysia, Taiwan, Indonesia, Singapore, Thailand, the Philippines, and South Korea. Japanese investors also looked to the United States for potential real estate bargains. In particular, Hawaii became a preferred off-shore location for the affluent Japanese corporate investor seeking a relatively low-cost, high-amenity environment for a primary or sec-

ondary place of residence or as a get-away retreat from the fast-paced and stressful urban lifestyle.

In general, FDI flows outside of the core economies tend to be highly regionalized, absorbed by nations that are in close proximity to the investing nation or source. As core economies extend their influence to the larger and proximate semiperipheral economies of the SOUTH, they will limit international investment and export opportunities amongst the world's most peripheral and more remote regions. These patterns reveal that the friction of distance and intervening opportunities play a dramatic role in impeding investment flows to peripheral or remote regions in the world economy. Core economies, via multinationals, now possess the ability not only to exploit a global competitive advantage in production, but also to exploit the comparative and competitive advantages of the entire developing world. This situation presents a rather formidable hurdle to overcome in a seemingly endless series of development obstacles which now face the more peripheral nations of the SOUTH.

CAPITAL MOBILITY AND INTERNATIONAL ECONOMICS

It has been well established that the growth of "international capital markets" has allowed investors to geographically diversify their assets and trade with agents outside their region. The rise of international asset holdings has been unprecedented since the early 1970s. As in the case of international trade of goods and services, it is expected that asset trade will continue to expand as

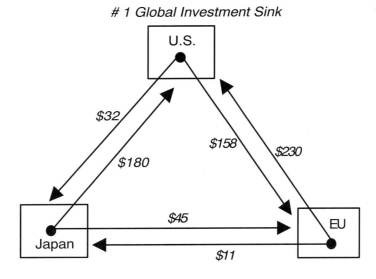

Note: FDI stock shown in billions of U.S. dollars (1995) and rounded to nearest billion.

Figure 9.8 Major Inter-Core Investments Flows.
Source: United Nations World Investments Report 1995.

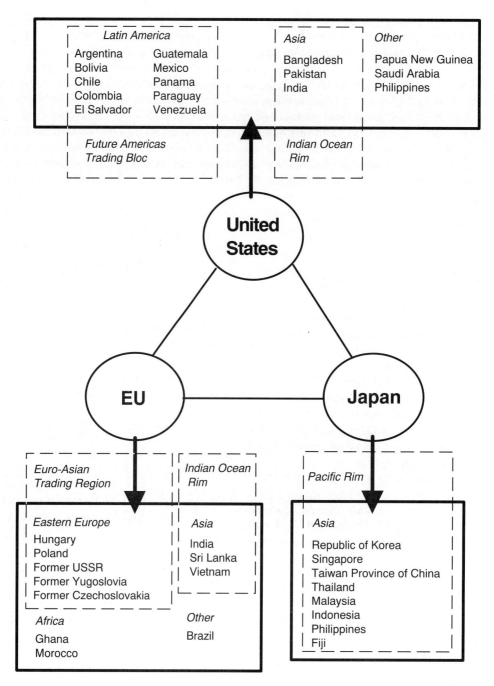

Figure 9.9 Major Investment Flows Outside the Core.
Source: United Nations World Investment Report (1992), p.33.

barriers to movement and investment are gradually dismantled or eliminated. Although the mobility of capital has increased over the past several decades with international diversification of investment holdings, capital mobility in the global economy has been impeded by several factors: (a) government and institutional restrictions on the flow of capital across international borders; (b) opportunity and transaction costs associated with the lag which exists between the recognition of a capital deficit in one region and flows from capital surplus regions in re-

sponse to that deficit; (c) segmentation, separation, and heterogeneity of geographic markets; (d) regionalization and the formation of free-trade zones which facilitate capital transfers within the region, while discouraging the transfer of capital outside the region; (e) differences in the evaluation of risk as based upon the perceptions of investors with imperfect information about markets and/or investment opportunities outside their own region; and (f) the regionally confined patterns of investment (as discussed previously).

Countervailing forces are hard at work at the regional and global scales to both encourage and discourage the flow of capital between regions. On one hand, capital mobility has been enhanced by increasing opportunities in big emerging markets geographically. Diversification of holdings across BEMs is becoming part of a strategy to reduce the risk of overseas investment. The more risk-averse investors are, the greater tendency they will have to diversify both the mix of assets in their portfolio and the regional and/or sectoral affiliations of those assets. On the other hand, low capital mobility is attributed to the fact that capital markets (including international equity markets), although segmented, are highly integrated and interdependent.

Consider the recent adjustments in the U.S. stock market, during the winter of 1997 and the late summer and early fall of 1998. Both events are inherently tied to downturns in Asian economies, the ongoing financial crisis in the Pacific Rim, and the declining value of stocks in Asian markets. As the value of stocks plummeted in reaction to regional banking crises in Japan, China, Malaysia, and elsewhere, international investors were quick to pull their holdings out of the region, further aggravating the decline. The impact was far reaching in that it caused a significant correction in markets around the globe, leaving little refuge for global investors.

Faulty investment loans to Asian industrialists and 20 years of overindustrialization were largely to blame. The Asian financial crisis has been blamed for a tidal wave of bankruptcies of Asian firms, slowdowns in spending and investment in the region, and the loss of millions of jobs from July of 1997 through January 1999. The loss of income and consumer purchasing power in Asia has led to reduced earnings for multinationals, many of which rely on revenue from sales in Asian markets. Expectations over reduced earnings in overseas markets and lower profits have led to downward adjustments in regional stock markets and a general decline in the share prices of many international corporations.

Stock market corrections were no surprise to financial analysts who cautioned investors throughout the 1990s that Asian stocks were overvalued and that expansion and industrialization policies in the region created a situation in which the total production capacity far exceeded regional and global demand for the products produced. Moreover, rising labor and production costs continued to erode the region's production advantage and reduced profit potential. In some cases, government-subsidized industry showed an increased vulnerability to outside competition. There was much skepticism over the prospects of sustaining noncompetitive industries in a region that relentlessly pursued free trade, privatization, and free-market development strategies.

A post-market-correction period pause, coupled with deflation, could bring about future downturns in the U.S. and Asian markets, although most analysts agree that future corrections will be relatively small and short-term. Low nominal interest rates (the sum of its expected real interest rate and its expected inflation rate) in the United States is likely to lead to further currency appreciation (that is, a stronger U.S. dollar). All other things being equal, the appreciation of U.S. currency and a strong dollar makes U.S. goods and services more expensive to foreign consumers and foreign products (imports) less expensive to domestic consumers. This imbalance will undoubtedly exacerbate the already high U.S. trade deficit and diminish the extent to which export-related growth contributes to overall economic growth. Problems in the U.S. economy are likely to be overlooked as long as consumer confidence remains high and interest rates remain low.

The late 1990s continue to bring about a flurry of consumer activity in housing markets, increased borrowing, and the refinancing of consumer debt. As a result, short-term growth rates are expected to be modest. By contrast, lackluster performances are expected amongst most economies in Asia and Latin America given their current debt structures and banking crises.

The growth of international trade, the integration of production networks, and the increase in capital mobility have hastened the growth of off-shore banking and currency trading. Many off-shore banking activities are not protected by the same safeguards imposed by national governments on their on-shore banks to deter bank failures. The internationalization of banking, nonetheless, has been attributed to the banking communities' desire to escape the highly restrictive regulatory environments established by governments while allowing depositors the options of holding currencies outside their issuing jurisdictions. By shifting portions of their operations abroad or to off-shore financial centers such as the Cayman Islands or the Isle of Man, banks can side-step some of the more prohibitive and costly regulatory aspects of carrying out financial transactions and in some cases reduce the "red tape" or tax burden. Many financial analysts and governments see increased regulation of the international banking community as inevitable, as they believe that global regulation of banking activity would lessen the chance of a massive worldwide financial system failure (despite the obvious difficulties associated with international regulatory cooperation).

Interest rates also play a pivotal role in the worldwide foreign-exchange market and the flow of money as deposits traded will ultimately earn interest for the investor. The foreign-exchange market is said to be in equilibrium when the expected returns on deposits of all currencies are equal—a condition known as "interest parity." This condition arises when there exists no potential for spatial arbitrage. In the market for money, the forces of supply and demand dictate that the market always moves toward an interest rate at which the supply of money equals the aggregate demand for money. Note that an excess supply of money generally causes interest rates to fall, while an excess demand for money causes interest rates to rise. Expectations over inflation will also affect interest rates, which in turn affect the appreciation (or depreciation) of a nation's currency. Moreover, expectations over inflation tomorrow will affect the pace of inflation today. For example, if consumers anticipate price levels to rise in the future, their expectations are likely to lead to a higher rate of inflation today with short-term increases in spending. Thus, changes in inflation expectations in any one given region can have a direct impact not only on interest rates in that region but also on deposits traded in the global foreign-exchange market. With the world boom in trade and investment and the increasing interdependence of production networks, regions are becoming increasingly susceptible to international economic trends and crises.

TRADE LIBERALIZATION AND REGIONALISM

Although most nations of the world favor trade policies and support the elimination of trade restrictions and barriers, many are hesitant to abandon policies which help protect various key exporting industries. Free trade may be in the best interest of the world economy, but it is not necessarily in the best interest of every region or industry. Nonetheless, the recent formation of various free-trade zones and regionalized FDI flows suggests the likely emergence of three major **continental trading blocs**:

1. The Americas Trading Bloc—a free-trade zone that is likely to include most, if not all, nations of North, Central and South American, and the Caribbean as part of a future Americas Free Trade Agreement;
2. A Pacific-Rim Trading Bloc—which would most likely include the most prominent economies in East and Southeast Asia (including Japan, South Korea, China, Taiwan, Thailand, Malaysia, Indonesia, Vietnam, etc.) with ties that stretch south to Australia and New Zealand and to the West coast of the United States; and
3. A Euro-Asian Trading Bloc which would most likely include most of Europe, republics of the former Soviet Union, some West Asian economies (for example, Turkey) and nations in the Indian Ocean Rim.

With the formation of continental trading blocs or alliances, we are likely to see a resurgence of protectionalism and a reversal of initial attempts to promote global free trade. Trade restrictions and barriers are likely to be lifted amongst intracontinental trading partners, yet there may be a tendency to impose restrictions on trade with outside members. In the end, free-trade areas are likely to

exist within continental trading blocs, but there is less of a chance of observing free trade between blocs.

The unification of Europe and the signing of NAFTA are just the tip of the iceberg. The rise of semiperipheral economies in South and East Asia, a zone that now accounts for more than 20% of world trade, the formation of a Pacific Rim trade alliance would surely exert tremendous clout in the world economy with a geographic market of more than 3 billion people! The sheer size of this market alone would almost guarantee that it would remain insulated from the influence of the West. Nevertheless, there is little that could be done should this region decide to impose trade measures (tariffs and quotas) to protect Pacific Rim industries from outside competition given the size of its internal market.

As markets of the world economy are opening up, trade regions are becoming compartmentalized as divisions are drawn on a larger geographic scale. It is possible that the formation of continental trading blocs will diminish the overall impact of the **General Agreement on Tariffs and Trade (GATT)**—*a worldwide trade alliance which supports the push toward global free trade.* The GATT was composed of 105 contracting nations in 1993, with an additional 27 countries included as so-called "de facto" members (as they agreed to the principles of the GATT, though not officially signed on). A seventh and most recent "round" of GATT negotiations (since its origination in 1947) was held in Uruguay during December of 1994. The Uruguay round boosted membership to 116 signees. Note that the nations under the GATT currently account for more than 90% of world merchandise trade.

Proponents of the GATT outline four major objectives. First, the GATT is to open markets by the removal of trade impediments worldwide. Second, the GATT supports nondiscriminating free and fair trade, a feature which would intensify competition, further the globalization of production, and improve the efficiency of interregional production networks. Third, the GATT will promote regional development through the free exchange of exportable commodities and greater investment flows to less-developed regions (as the perceived risks of investing in the periphery decline over time). Fourth, the GATT will promote responsible management of regional trading blocs (meaning that GATT members will discourage protectionalism).

The GATT is to liberalize world trade and intensify global competition by opening up all markets and eliminating all barriers which impede the movement of commodities or capital. Some feel that the emergence of regional trade agreements (like the NAFTA) are a sign that the world is already heading toward one global free trade zone, with or without the GATT. Others see problems looming on the horizon, as most regional trade accords already discriminate against producers outside the region. Skeptics see the emergence of "economic regionalism" as detrimental to the GATT. As regional trade alliances with the help of governments fortify production and trade linkages between countries that share a common border, many are likely to enact policies which exclude "outsiders" from capturing a share of the internal market. Import penetration from suppliers outside the region would weaken linkages between members and cause income to leak out of the region. Hence, it is logical to assume that trade alliances will protect their members. Should continental trading blocs become protective, it could result in widening development gaps between nations within the blocs and those outside. Multinationals have the least to lose as most already have subsidiaries located within each of the three dominant emerging continental trading blocs.

The removal of trade barriers and the movement toward global free-trade is expected to promote the optimal and efficient use of resources and production factors. It is expected to reduce the costs of goods and services worldwide and, as a result, increase standards of living and promote economic development throughout the world. Inefficient producers will be forced out of the picture as globalizing industry restructures and reduces wasteful allocations of human, physical, and financial resources. The underlying benefits of trade have long been realized by those nations seeking **most-favored-nation (MFN) status**—*a favorable position amongst an enlarging inner-circle of members that prosper exclusively from government-regulated bilateral trade agreements.* Recent controversy has surfaced over granting China MFN status as the Chinese government continues to be arrogant in matters of international security, noncooperative in international trade discussions, defiant of international laws, and notorious for human rights violations. The Chinese government persists in its mistreatment of the Tibetan people and the intimidation of neighbor Taiwan, intermittently flexing its military muscle. In addition, Chinese officials continue to support limiting access to its growing consumer markets through the use of tariffs and other import restrictions. China's recent sales of military weapons and nuclear technology to Pakistan and renegade nations in the Middle East has outraged members of the international community. Despite all this, China's MFN status is not likely to be revoked given its growing number of allies, its newfound superpower status, and the global-wide push to promote free trade.

Although regional trade alliances and the formation of continental trade blocs seem to be counter to the GATT, as they could possibly undermine global trade liberalization efforts, the GATT has a much larger problem. It is well known that the benefits are not something which would most likely be distributed equally across all participating nations, sectors, or industries. Table 9.16 provides a representative sample of the impact of the GATT on the U.S. economy. Surely the unequal distribution of benefits across industrial sectors and production regions is likely

TABLE 9.16	The GATT: Winners and Losers
Sector	**Impact of the GATT on the U.S. Economy**
Agriculture	Reductions of farm subsidies by nations of the EU will open up new opportunities for U.S. farm exports (grain and meats). Japan and South Korea will begin to import rice. Some U.S. agricultural sectors (citrus, sugar, and peanuts) will lose their production cost advantage as government subsidies are trimmed. Many big agricultural exporting nations in the SOUTH will benefit greatly (with hefty losses in various noncompetitive agricultural sectors in the NORTH).
Automotive Parts	As the United States will be permitted to protect only one industry with a "voluntary" restraint agreement that limits imports, the Big-Three automakers fear that the automobile industry may be abandoned by the government to protect high-technology sectors (not allowing restrictions to be placed on Japan's share of U.S. car sales).
Entertainment, Pharmaceuticals, and Software	New rules and regulations governing patents, copyrights, and trademarks to protect various industries of the core from pirating nations in the developing world will not be phased in for up to a decade. Meanwhile, losses to core economies could be staggering as NICs borrow innovations and products without consent or compensation. The outright refusal of France to liberalize market access for the U.S. entertainment industry continues to be a very controversial matter.
Financial and Legal Services	Free trade in services sectors has created a vast array of opportunities for U.S. industries, although specific terms must still be worked out with developing nations.
Textiles and Apparel	Strict quotas limiting imports from developing nations of the SOUTH will be phased out over a ten year period, further aggravating job losses in these low-value-added industries throughout the NORTH. U.S. consumers, however, are expected to benefit from lower clothing prices.

Source: *Fortune* (January 10, 1994), p. 66, with modifications.

to bring about a resurgence in protectionism. Economic regionalism in the form of regional trade accords, continental trading blocs, and bloc-specific trade policies will promote free trade, but on a restricted geographical basis (Gibb, 1994).

Multinationals continue to decentralize production activities over the economic landscape. This has resulted in an increasing amount of intraenterprise and intercontinental trade. The presence of many multinationals in each of the three continental trade areas is a hedge against possible trade wars between trading blocs. Intercontinental trade policies are likely to greatly influence the location strategies of globally minded firms as the balance of power in the world economy shifts away from the G7 and toward multinationals and nations unified under regional trade accords.

The formation of region-specific free trade areas is nothing new, however. Consider the case of the United States and Mexico, nations which established border production areas and corresponding free-trade zones long before the signing of the NAFTA. Foreign-owned and operated assembly plants along the U.S.-Mexican border have been in existence for almost two decades, stretching from California to Texas. These production-for-export zones continue to flourish as a result of Mexico's low-cost production environment and minimal regulation on industry. Mexican assembly plants or "maquiladora" can be found throughout this border region, in towns such as Tijuana, Nogales, Ciudad Juarez, and Nuevo Laredo. The maquiladora produce finished and semifinished goods and parts almost exclusively for the U.S. market. Many U.S. firms have relocated to Mexico during the 1980s and early 1990s to take advantage of an abundant, hard-working, and low-wage labor force, not to mention the lack of import tariffs and restrictions placed on commodities produced there and brought directly back to the United States. The maquiladora assembly plants are now part and parcel of the well-integrated production networks of Ford, General Electric, IBM, Nissan, Sony, and others.

The maquiladoras are primarily driven by external forces, mostly U.S. demand, proximity to the U.S. market, and Mexico's ultra-low-cost production environment. The virtual lack of Mexican-made inputs in the maquilas and absence of an internalized supply network, however, have limited regional expansion possibilities. It is unlikely, therefore, that many border production zones will ever become integrated with the larger Mexican economy. As such, the economic benefits of multinational involvement in border-town production areas of Mexico have been limited to the immediate areas in which production occurs. The major beneficiaries of the maquilas are U.S. multinationals (South, 1990) (Figure 9.10).

Corporate relocation strategies often involve moving to regions with known production cost advantages. The decision to relocate is viewed as a survival strategy by firms as they search for innovative ways in which to compete and improve profit margins. While relocating production facilities allows some operations to stay in business, plant closures in established production regions of the core have a devastating impact on the local economies where industry and jobs are lost. Relocation also presents a series of problems for firms as they attempted to manage an international and geographically extensive production network. In many instances, relocation results in unforeseen inefficiencies and hidden costs that can offset the advantages of relocation. This is particularly true for firms in textiles, clothing, and consumer electronics industries, where a move is commonly in response to labor cost differentials.

A low-labor-cost offshore manufacturing base increases a company's transport and tariff costs. It may also generate hidden costs, such as increased problems in quality control, a long pipeline of costly in-process inventory, and slow response time for coordination between designers and the factory or the factory and the field. (Magaziner and Reich, 1985)

The importance of some hidden costs has been reduced with the flexible practice of **just-in-time (JIT) manufacturing**—*a production and inventory control process where inputs, supplies, and/or products reach suppliers and consumers at their intended destinations just when they are needed.* JIT manufacturing has not only reduced the need to carry costly or excessive inventories, but has virtually eliminated lags in production and distribution. Moreover, JIT production has minimized downtime in manufacturing systems and waste associated with idle capital. This approach has changed the nature of international competition as suppliers and producers become increasingly more time and delivery conscious in a global market where both space and time are money.

The JIT model has been successfully adopted by the Japanese in production networks where operations require

Figure 9.10 Mexican workers repair telephones for AT&T in a Maquiladora plant near Nuevo Laredo, Mexico.

9.10: ©Bob Daemmrich/Stock Boston

flexible and punctual deliveries of parts from suppliers. In many ways, JIT production supports the move toward "re-internalizing" of core firms. Note, however, that functions are not brought back inside the firm, but are inside of a rapidly responsive and vertically integrated production network. This makes individual producers or plants vulnerable to problems within the network. For instance, should a labor dispute cause an unforseen strike at a facility which is a major parts supplier in the region, the shutdown could lead to problems for the entire production network. Wildcat strikes by United Auto Workers (UAW) locals at auto-part manufacturing plants in the United States commonly lead to lay-offs in assembly operations in other facilities once the flow of parts is disrupted.

JIT manufacturing can be viewed as an inventory and flow control system that relies on trust and reliability of suppliers in the production network. As network suppliers have become more geographically diverse and as production networks go global, firms have been made increasingly vulnerable to disruptions in the production and delivery system or from labor disputes. The increased risk that goes along with relying on outside suppliers has led many firms to internalize some parts production or "job out" to reliable sources. The Japanese, for example, have opted to transplant entire production networks to foreign

soils, in addition to locating in regions where organized labor activity is virtually nonexistent. This location strategy helps to guarantee the effectiveness of the JIT system by utilizing suppliers within the enterprise group.

Regional Trade Alliances

Free trade has been openly embraced by the majority of the world's nations. The past 40 years have brought about a wide variety of regional trade accords and alliances in assorted geographic and economic settings. A list of selected major trading alliances and trade areas is given in Table 9.17.

Notice that several regional trade alliances/areas are overlapping. For example, virtually all nations of the Association of Southeastern Asian Nations (or ASEAN) are also members of the Asia-Pacific Economic Cooperation group (APEC)—an organization which also includes the United States (the largest economy in the NAFTA), Chile (part of the Organization of American States and NAFTA bound), Ecuador (a member of OPEC), Malaysia (part of the British Commonwealth), and Japan (one of many members of the Organization for Economic Cooperation and Development, of which the United States, Germany, France, and the United Kingdom are also a part).

TABLE 9.17	**Selected Major Trade Areas and Accords**
Acronyms **Trade Area/Alliance**	**Participating Members (1995)**
NAFTA—North American Free Trade Agreement	U.S., Canada, Mexico
LAFTA—Latin American Free Trade Area	Brazil, Argentina, Mexico, Chile, Paraguay, Uruguay, Bolivia, Peru, Ecuador, Colombia, Venezuela
CACM—Central American Common Market	Guatemala, Honduras, Nicaragua, El Salvador, Costa Rica
AG—the Andean Group	Peru, Venezuela, Colombia, Ecuador, Bolivia
MERCOSUR—Mercado Comun del Cono Sur or SCCM—Southern Cone Common Market	Brazil, Argentina, Paraguay, Uruguay
USAO—U.S.A-oriented trading partners or OAS—Organization of American States	Mexico and Chile
EC—European Community (in 1993)	the U.K., Germany, Ireland, Denmark, France, Belgium, Italy, Netherlands, Greece, Portugal, Spain, Luxembourg
EU—European Union (1995)	EC, plus Austria, Finland, Sweden
EFTA—European Free Trade Association	EU, plus Iceland, Norway, Switzerland
EE—Eastern Europe Region (by agreement) regionally integrated with CIS	Poland, Slovakia, Czechlands, Hungary, Romania, Bulgaria, former Yugoslavia, Albania
CIS—Commonwealth of Independent States	Former republics of Soviet Union
ASEAN—Association of Southeast Asian Nations	Thailand, Vietnam, Malaysia, Singapore, Philippines, Indonesia, Brunei
CEA— Chinese Economic Area	China, Hong Kong, Taiwan
Northeast Asia	CEA, plus Japan, S. Korea, N. Korea
SADC—Southern African Development Community	Tanzania, Mozambique, Malawi, Angola, Zambia, Zimbabwe, Namibia, Botswana
SACU—Southern African Customs Union	Namibia, Botswana, Republic of South Africa, Lesotho, Swaziland

Source: Gibb and Michalak (1994); and *U.S. Global Trade Outlook 1995–2000* (1995).

Although most nations of the world are already connected to each other either directly or indirectly through a trade agreement of one kind or another, the benefits of increasing international trade continue to be unevenly distributed over participating members. Much of the unevenness is explained by differences in regional development goals and variations in global competitive advantage, not to mention the degree to which regions encourage global integration and foreign direct investment.

While it is true that regions cannot remain isolated in a globalizing world economy, careful thought must be given to the manner by which global integration is achieved (and how it ultimately affects the economy of a given region). While free and fair trade can be promoted across all regions and sectors, noncooperative nations face the real consequences of foregoing the benefits of an ongoing boom in world trade.

Consider the data in Table 9.18. Note that with the exception of the interwar period 1929–1937, the growth in world trade has exceeded the growth in world output. It is expected that world trade will continue to grow at a pace that exceeds world production for some time to come in light of the many trade alliances that have been established. Although trade liberalization policies help to promote economic growth and development around the globe, these policies may also work in reverse of their intended goal.

Kitson and Michie (1995) argue that the push toward global free trade and the GATT can backfire should the countervailing force of economic regionalism prevail.

> The 1994 GATT (General Agreement on Tariffs and Trade) agreement—the signing of the Uruguay Round—ended seven years of negotiations and created the new World Trade Organization to police international trade, to be established on 1 January 1995. It is claimed that the freer trade in manufactures, services, textiles, and agriculture, which will follow what will be the biggest tariff cuts in history, will add $250 billion a year to world output in ten years time. . . . The likely impact of the GATT agreement, a key determinant of its effect on output will be the extent to which it impacts differentially on countries' trade balances, since if one or more countries or blocs are affected adversely, the resulting trade deficits may prove unsustainable, leading to deflationary policies being pursued to slow down the growth of import demand. The resulting slowdown in economic growth will also reduce the growth of trade . . . provoking trade wars and protectionism. Free trade is quite capable of creating the conditions for its own demise. (Kitson and Michie, 1995, pp. 31–32)

The benefits of free trade may be viewed as a "positive-sum game" for some regions, but a zero-sum game for the nations of the world economy. In the long run, as the initial and secondary benefits of trade are realized and as growth of various regions or industries slow, intensified competition between the most-efficient production regions (especially core economies) will ultimately mean losses to regions supporting less-efficient producers and industries. Regions unable to compete or retain value added will find themselves increasingly segmented from the remainder of the world's production regions. Multinational involvement in these regions is also likely to diminish once their competitive advantage begins to erode.

> The genuinely footloose [multinational], willing to locate anywhere to ensure the highest returns, appears to define the ideal agent of resource allocation. . . . The presences of these firms, in itself, is testament to discontinuities in the economic environment and there can be no a priori reason for assuming that the activities of [multinationals], any more than international trade, will by themselves remove inequalities between and within regions. In fact, and recalling the experience of many countries caught up in an early period of international production and liberalization during the period before the First World War, weak states are more likely to establish weak, and ultimately unsustainable, growth paths and reinforce global segmentation. (Kozul-Wright, 1995, pp. 159–160)

Globalizing firms continue to seek locations with labor and production conditions which will allow them to turn a profit. Finding a stable and suitable pro-multinational environment in the periphery with adequate infrastructure and suppliers and accessibility to multiple markets is becoming increasingly difficult. Furthermore, the increasing mobility of capital has not diminished the importance of agglomeration economies nor has it made it easier for firms to find desirable locations for "networked" production. The fact that many industries must rely on a network of suppliers makes them less mobile once a network materializes on the landscape. Geography and history continue to play significant roles in influencing the direction of economic change and development potential. As regional economies in the core and semiperiphery are constantly being transformed by multinational

TABLE 9.18	Growth of World Output and Trade (1870–1990)	
	(Annual Percentage Growth)	
Period	Output	Trade
Pre-WWI: 1870–1913	2.7	3.5
Interwar: 1929–1937 ("the go-it-alone period")	0.8	–0.4
Post-WWII: 1950–1990	3.9	5.8
1950–1973	4.7	7.2
1973–1990	2.8	3.9
1990–2010	2.0*	4.5*

Source: Kitson and Michie (1995), p. 7.

Note that annual percent growth rates are calculated from peak to peak over period in question.

* Forecast.

activity, a large number of peripheral economies still remain out of the world's investment and trade loops.

> [Multinational corporations] have accelerated the pace and changed the nature of international production. The steady growth of FDI and the production activities of foreign affiliates attest to this change. Deeper integration between economies during the past one or two decades reflects the intricate ways in which the various components of the world economy are being organized and geographically arranged by multinational corporations. The emergence of a new integrated international production system under the governance of [multinationals] in many economic sectors is now visible. But its incidence is uneven, is quite vulnerable to adverse shocks, and does not rule out exclusion. (Kozul-Wright, 1995, pp. 170–171)

We are entering an era in which the traditional regional economic base model will soon be replaced by a global economic base model which recognizes the increasing importance of FDI stocks, interregional production linkages, and the production and revenue generating capabilities of branch plants or subsidiaries. The new model will recognize the way in which integrated production now spans multiple geographic areas. The data in Table 9.19 reveal three important trends that support this argument.

First, FDI continues to become an increasingly important component in the world economy, as it has been growing faster than world output. Second, the growth of FDI and the emergence of regionally concentrated foreign-owned production networks have led to increasing sales of production outputs from affiliates. In short, an increasing share of multinational outputs are being produced in the host regions. Third, world sales revenue of foreign affiliates has recently exceeded sales revenue from exports (Kozul-Wright, 1995). This implies that sales by foreign affiliates are preferred over the direct exportation of products from an investing nation, as consistent with the previously discussed findings of Hanink (1994).

The data also suggest that the traditional tools of managed trade (that is, tariffs, quotas, and other restrictions) will be increasingly less effective. New approaches must be found to manage transplants should regions decide to protect domestic industries as multinationals continue to shift their entire production networks over host countries. Local content and input requirements (by ownership) and corporate income taxation remain as viable options. Yet, truly little can be done to recover the economic damages associated with transplants and affiliates which have permanently replaced significant portions of a region's domestic production network.

The increasing importance of foreign affiliates has changed the way in which multinationals have organized their operations on the global economic landscape. Knox and Agnew (1994, p. 225) discuss the locational hierarchies of multinationals, noting five specific outcomes of the globalization process:

1. the spatial concentration of high-level management in world cities;
2. secondary concentrations of midlevel management and administrative activities in the larger metropolitan areas of the core and leading centers of NICs (mostly in the semiperiphery);
3. R&D clusters in regions supporting high-technology sectors and the innovative milieux within core economies;
4. high-tech industrial and specialized production industrialized regions or zones mostly within the core; and
5. nonspecialized and routine production activities that are decentralized in branch plants found in either peripheral regions of the core or metropolitan areas in SEPZs of semiperiphery.

Further questions regarding the relevance of the nation-state have surfaced with the emergence of flexible industrial districts, territorial production complexes, transplanted production networks, and intercontinental trading alliances. This has created interest in exploring the possibilities of microregulatory and subnational institutions (governmental or otherwise) and region-states with macroregulatory authority. Despite the fanfare, Gertler (1992) reminds us that there is no evidence to suggest that the nation-state has become less important as an agent or mediator of regional economic change.

> Most . . . nation-states have produced (and continue to produce rather distinct national systems of innovation which create particular possibilities for economic change while precluding others. . . . So too do nation-states define distinct regimes of macroeconomic policy, which have the potential to: (a) produce differentiation in economic conditions (interest rates, unemployment, currency exchange, etc.) between countries; (b) influence the basic conditions for production and the terms of competition for firms in each country; and (c) produce incentives for

TABLE 9.19	Foreign Direct Investment (FDI) in the World Economy for Selected Years					
	1960	**1980**	**1985**	**1991**	**1995**	**1999***
World FDI stock as a percent of world output	4.4	4.7	5.4	7.2	9.4	10.2
World sales of foreign affiliates as a percent of world exports	84	97	99	122	140	150

Source: Kozul-Wright (1995), p. 158, with additions.
* Forecast.

capital mobility within and between countries, as the national conditions to support manufacturing are made more or less favorable. . . . Despite the common observation that the advent of a strengthened Europe [European Union] or North American Free Trade will reduce the powers of individual nation-states to regulate their own economic affairs, it should be pointed out that, in contrast to the *rhetoric* of free trade, countries will continue to retain many powers, and markets will not be completely and unequivocally 'opened up'. Although the more subtle powers to harmonize social and economic policies across nations will be strong, the intensified competition expected to prevail will . . . enhance the importance of fostering a supportive national system for innovation. Consequently, it is difficult to see how the rise of supernational [continental] blocs will undermine these particular powers of the nation-state. (Gertler, 1992, pp. 270–271)

Note also that the absence of a "global growth strategy" has led many regions to independently forge expansionary economic policies. Noncoordinated national economic policies can prove to be self-defeating or negating, as the outcome or consequences of such endeavors are tempered by globally dominant forces (Kitson and Michie, 1995, p. 35). Many analysts view the rise of regionalist policy as a last-ditch attempt or protest to combat international economic superstructures and the globalization of economic activity. Some see globalization as a threat to national economies (and national security) as it is a process that has gradually reduced nations' control of their own economic destiny.

LABOR ISSUES

As multinationals continue to exert ultimate power and authority over labor within the globalizing production network, there has been a renewed interest in labor issues. The very fact that much of the world's labor force is still outside the global production and trade system, is all the more reason for initiating economic reforms to deal with a problem that some see as forthcoming. World Bank data show that wage employment is a relatively small share of the total employment in poorer countries (see Table 9.20). In the absence of clear-cut labor standards, it is the low-income and/or peripheral nations of the world that open themselves up for exploitation as multinationals become increasingly dominant. Proactive policies to prevent the exploitation of labor must gain international acceptance.

Clearly defined economic development policies must be implemented to (a) protect workers' rights; (b) establish minimum working ages and child labor laws; (c) insure equality of wages and equal access for men and women to adequate employment opportunities (in both number and quality); (d) assure safety of workers and decent working conditions; and (e) increase nonwage benefits and/or access to basic survival resources (Figure 9.11).

While much lip service has been paid to labor exploitation issues, particularly in regard to the developing SOUTH, the greatest challenge is enforcing international labor laws and workplace standards. While the intentions behind international labor laws are all well and good, monitoring workplace conditions, mediating disputes, settling grievances, and investigating worker abuse will be far too costly for most developing economies. Hence, supervision of labor and the work environment will require human and monetary resources from core economies.

PRODUCTION LIFE CYCLES

Trade and regional economic growth are largely affected by normal production cycles. The dynamic ebb and flow of production cycles, inherent to global capitalism, have

TABLE 9.20	Wage Employment as a Share of Total Employment by Sector and Income Group (1993)

| | Income Group/Nations | | |
Sector	Low	Middle	High
Wage Employment as Percent of Total Employment			
Industry	29.8	76.7	89.1
Services	46.4	68.2	85.6
Overall	17.1	57.4	84.4

Source: *World Development Report 1995*, p. 72.

Figure 9.11 The exploitation of woman and children in textile manufacturing in developing nations is commonplace. Here a young Thai girl works hard to sew garments in a typical "sweatshop" setting.

9.ll: ©Mark Richards/Photo Edit

had profound impacts on the geography of production and development. Regularities have been observed in the locational tendencies of industry in relation to the life cycles of products. Changes in production technology, consumer preference, and product development continue to influence industrial location patterns worldwide. As competition among regions intensifies, and as new geographic markets form for new products and services, the determinants of regional production advantage are continuously altered. Production efficiency has emerged as an important ingredient in the survival strategies of both firms and production regions. Yet sustainable economic growth requires either the expansion of existing geographic markets, or the creation of new products and markets (Casson, 1983).

Competitive production regions require the recruitment, attraction, and retention of industry, a feat that becomes increasingly difficult with rapidly changing production technology, the increasing mobility of capital, and the locational tendencies of firms in response to product and production cycles. Recent advancements in information processing, communications, and transportation have had profound effects on the development of new products and production techniques. The impact of technology and product innovation continue to influence the production and location decisions of firms, as well as the competitiveness of industries and regions. Considerations must, therefore, be made for the degree to which "product life cycles" affect the geography of production.

A **product life cycle** describes *a series of changes in total sales revenue or units sold in relation to various stages of product development and marketing*. It is not uncommon to see the sales of a product run a predictable course as the markets expand, contract, and later thin-out over time and space. In general, we may identify five stages of a product life cycle (as illustrated in Figure 9.12). These stages have many implications for the geography of production, investment, and development.

Stage one of the product life cycle—the **product development stage**—describes *the birth of a new product and its introduction to the market*. In this stage, producers/competitors (usually the innovators in the field) are few, as are the number of potential consumers. Both the demand and supply of the product are low and per unit price is high, as production costs are high, revenues low, and profit margins slim or nonexistent. The relatively small number of prospective buyers is attributable not only to cost, but also from the fact that consumers may have a limited knowledge of the use, capability, and reliability of the product. Only a limited number of units may be available or test marketed, and the diffusion of information about the product will be slow. Price levels of prototypes or initial product lines are usually high and beyond the reach of the average consumer. Hence, the greatest potential markets exist in high-income regions. Stage one is also characterized by high R&D expenditures in relation to product development. With costly R&D and specialized production and a limited customer

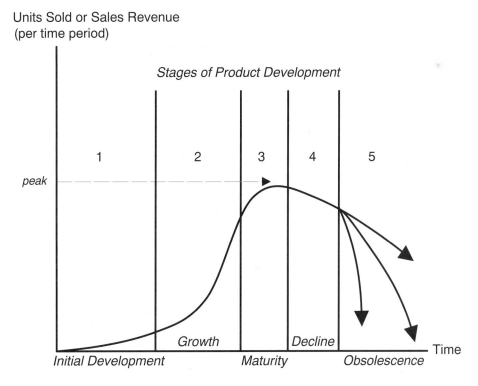

Figure 9.12 The product life cycle.

Source: After P. Dicken (1992) p. 111

base, the geographic location of producers is largely restricted to core economies.

Should a product prove to be marketable, sales revenues will ultimately increase and a second stage, called the "growth stage" will begin. The **product growth stage** represents *that period of time in which sales of a given product and sales revenue explode as a growing number of buyers emerge*. The majority of sales are associated with consumers in regions where household income levels are high (for example, core economies). During this stage, competition will intensify as new producers (rivals) enter the industry and vie for a share of the growing markets. Price levels start to decline as competition heats up and market supply increases as a result of increased production and firm entry. Marketing and advertising efforts also intensify and sales revenues skyrocket as the product gains general acceptance and popularity amongst mainstream consumers. The production technology becomes commonplace and in some cases production becomes standardized as competitors seek to achieve scale economies. In most instances, large producers within the industry become regional, concentrated in one or a few key locations within a supporting production network. Hence, agglomeration economies and market access considerations become increasingly important as the industry develops.

Historically, the U.S. automotive and steel industries (in the early-to-mid-1900s) exhibited a tendency to regionalize their operations. The locational patterns of steel and auto producers brought about the formation of America's Industrial Heartland, centered about the Great Lakes region and covering a large area that included roughly a dozen or so states. A regional network of heavy industry gave rise to a horde of big manufacturing cities such as Minneapolis, Chicago, Gary, Detroit, Pittsburgh, Cleveland, Cincinnati, Buffalo, and Baltimore. This premier manufacturing region was largely driven by classic location factors such as access to large and expanding consumer markets for automobiles, an increasing demand for steel to support urbanization and the demand for infrastructure, access to raw materials and production inputs, the availability of labor (skilled and unskilled), scale, agglomeration, and network economies, and relatively low material acquisition and distribution costs due to the advancement of interstate roadways and the proliferation of a dense railroad network and an efficient barge transport system.

Once growth in demand begins to slow and per period sales revenues begin to peak or plateau, the product is said to be in the "product maturity stage," the third stage of the product life cycle. The **product maturity stage** represents *that period of time in which a consumer market becomes satiated with a given product and the growth in sales revenues levels off*. During this stage, price levels generally start to decrease. Declining price levels and revenues will mean that only the more-efficient producers will be able to turn a profit and remain in production. Generally, only larger firms that are able to take advantage of scale economies and lower average production costs per unit output or flexible firms will be able to survive.

By contrast, many smaller, less-efficient producers, or inflexible firms will be forced to exit the industry. In an attempt to remain competitive, some firms will move some or all operations to low-cost overseas locations or compensate losses in one sector or region with gains from the sales of products in other regions or sectors. Competition amongst rival firms (both foreign and domestic) will intensify over time as sales level off. Only the strongest and savviest of firms will survive the shake out. Corporate downsizing and lay-offs are not uncommon as companies struggle to survive during a period of intensified regional and global competition.

Once per-period sales revenues begin to drop off and the exodus of noncompetitive firms spirals, a fourth stage, that of product decline, sets in. **Product decline** represents *that period of time in which the sales potential of a product declines as a direct result of decreasing demand*. Decreases in market demand for a product are generally related to any of several factors including oversatiation, product obsolescence, changing tastes or preferences, and the emergence of a substitute or superior. Firms that have survived up to this point will further intensify efforts to compete for a shrinking market share as sales decline. These firms may turn their attention toward developing and marketing new product lines or making improvements to old product lines to increase their marketability and sales potential. In the case of low-end products like consumer electronics or apparel, production facilities are more likely to gravitate toward production regions outside of the core and places that are, typically, far removed from the locations where the innovation occurred.

Eventually, the final stage of the product life cycle brings about the termination of the marketability of a product. This fifth and last stage, known as **product obsolescence**, *occurs when the demand for a product falls precipitously and revenues decline to zero*. This stage is marked by the abandonment of a given product by existing firms or the death of an industry.

Consider the once-thriving market for the eight-track tapedecks—a music reproduction technology that boomed in the 1960s and early 1970s. The eight-track tape format was quickly driven to obsolescence and extinction by the introduction of the more durable and better sound-reproducing cassette tape. The cassette tape industry is now entering into a product decline stage and rapidly entering into the product obsolescence stage as its existence is being threatened by compact disc (CD) technology. Today, the cassette tape industry is in jeopardy due to the ever-increasing popularity of compact discs and digital tape formats. These product innovations have certainly rendered the LP record obsolete. Although the death of nondigital tapes and record albums has been predicted for many years, there continues to be a growing and nostalgic consumer market for those that crave "vinyl" and the warmth of analog music reproduction technology.

Note that several variations in the product life cycle have been observed. Product life cycles may be altered or prolonged due to (a) substitution of a more advanced or superior product line; (b) finding new uses for an existing product; (c) change in production technology (in particular, the introduction of cost-saving production techniques that allow producers to sell products at lower prices; thereby allowing demand to rebound and market sales to increase); and (d) incremental innovation (which inspires consumers to replace an older variety product with a more modern version with all the latest features. Subsequently, the classic product life cycle curve as shown in Figure 9.12 may take on any of a number of forms as illustrated by Figures 9.13a–c.

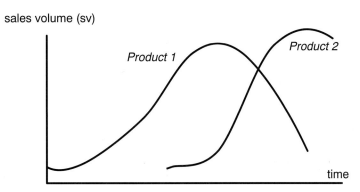

a. Interdependence of product life cycles: the introduction of a new product or substitute

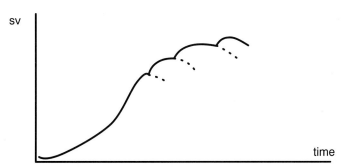

b. Extension·of product life cycle: the effects of improved design or expanded uses

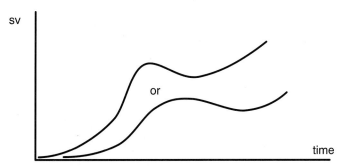

c. Extensions of product life cycles: the impact of changing production technology, innovation, or declining production costs

Figure 9.13 (a) Substitution of a new product; (b) extension of life cycle by updating or finding new uses for product; and (c) change in production technology or incremental innovation.

Analysts have observed that product life cycles are getting shorter over time as a result of rapid advancements in technology and the increasing awareness and sophistication of consumers. This has placed enormous pressure on firms to develop new product lines to keep up with what's "hot" and "in demand," and to make changes and improvements on existing lines. Flexibility in production and the growth of niche markets has also shortened product life cycles as an increasing number of highly specialized products are developed and sold to a small yet important number of high-end consumers or markets around the globe. These activities require a strong commitment on the part of firms to expand their research and development (R&D) activities and/or find new cost-cutting procedures to extend the commercial viability of products (and hence their life cycles) and maintain an acceptable margin of profit to allow them to stay in operation.

As firms try to adapt and respond to the changing needs of consumers and increasing competition, the need for information, producer services, and aggressive marketing, investment, and corporate restructuring strategies have become increasingly vital to the production process. Technological change, the increasing demand for producer services in advanced economies of the core, the need for information, the globalization of production, flexible production networks, and compression of product life cycles have spawned the growth of **multinational service industries (MSIs)**—*globally oriented service providers* (Bryson and Daniels, 1998). In short, the internationalization of the secondary sector and the emergence of multinationals and flexible firms (which serve global markets) have been instrumental to the internationalization of tertiary sector activities and the rise of MSIs.

Despite these trends, it is difficult for producers in remote areas or peripheral economies to respond quickly to changing market conditions or signals. Although information can be transferred virtually anywhere in the world at the blink of an eye, the production, packaging, shipment, and delivery of commodities still take time. The tyranny of space and separation are hard at work to negatively affect the competitiveness of some regions. Even in JIT systems, questions arise as to just how much of any one given product or input should be produced in advance and/or kept on hand at any one given time to assure a smooth flow of that commodity from production centers to markets. As a result, it is also important to consider to what extent the product life cycle will influence production and sales in various regions or markets.

Historically, product life cycles have been known to influence the location of production and patterns of trade (see Figure 9.14). The introduction of a new product generally takes place in the core, with production technology eventually diffusing throughout the core during the first two stages of the product life cycle. Core economies compete for a share of the global market, and are generally net exporters of that commodity until the maturity stage is reached. As production costs begin to rise, only the most efficient producers and producing regions within the core will survive. Although core economies continue to be net exporters during stage three of the product life cycle, relocation strategies are employed to maintain competitiveness and secure profits. Facing product maturity and declining sales, production generally shifts entirely out of the core and moves to the low-cost production environments of the semiperiphery or periphery (simultaneously benefiting from the growth of markets outside the core). Subsequently, core economies become net importers of the products that they originally introduced. This generally occurs by the fourth stage of the product life cycle. During this time, production in the core is displaced as imports are substituted for domestically produced goods (Figure 9.14).

Consider the shift in the production of color television sets from the core to the semiperiphery during the 1980s. This phenomenon was largely associated with an industry that has run its course in the product life cycle. Originally an activity dominated by the United States and Japan, a mature world market for color televisions is now predominantly supplied by producers in a handful of semiperipheral nations like China, Taiwan, South Korea, and Malaysia. U.S. consumers would be hard pressed to find a color television set produced in the states.

Changing production technology, product design, and innovation, have caused the product life cycle to shorten in terms of duration, especially stages four and five. New and improved product lines and upgrades constantly flood the global market, prematurely reducing sales or bringing about the obsolescence of an old product line (sometimes before its time and possibly even before it even reaches the shelf or showroom). Consider how VHS-format video cassettes are now being gradually replaced by laser disc technology or how the introduction of the Pentium processors (IBM trademark) in the mid-1990s lowered the price levels and shortened the sales potential and lifespan of computers supporting the obsolete 486 chip technology.

As the world's geographic markets for goods and services become more numerous and diversified, and as industries face shorter and shortening product cycles in light of the rate of technological change and product innovation, flexible production systems will become increasingly more important to regional economies. Some important questions remain, however, regarding economic growth and development potential in relation to shortening product cycles. Can and will flexible production networks be established and sustained in developing regions of the SOUTH when producers and production regions in the core are most likely to benefit from shorter product cycles (given that they will secure the greatest

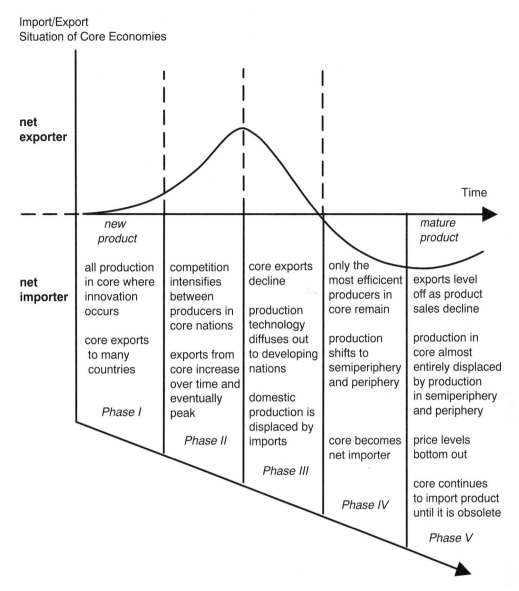

Figure 9.14 Product life cycle and locational effects on production and trade.
Source: Vernon (1996) as presented in Knox and Agnew (1994, p. 93), with modificarions.

export-related benefits of end life-cycle compression)? Will peripheral economies be able to show enough flexibility to secure adequate shares of the rapidly changing world markets to support enough of a manufacturing-for-export base to sustain industrialization and a viable flexible production network? If so, will this be enough to foster further export diversification to sustain economic growth and development?

One could argue that the dominance of the multinational will mean a greater retention of value added by core economies as they secure positions as dominant net exporters of products during the growth and maturity stages of the life cycle. This is certainly true in the cases of high-value-added and/or high-tech products and multinational service industries. These industries are less likely to offer developing regions of the SOUTH the same benefits that industries like textiles, apparel, and consumer electronics and appliances did (and continue to do), given that high-value-added industries tend to be more capital intensive, more inclined to adopt a flexible production system, and require a highly skilled labor force. These attributes alone are likely to keep the locations of high-value-added industries predominantly in core economies of the NORTH.

KEY WORDS AND PHRASES

big emerging markets (BEMs) 232

Chinese Economic Area (CEA) 232

continental trading blocs 254

economies of scope 248

established FDI channels 240

European Union (EU) 231

(the) Euro 231

flexible production 240/248

footloose industries 246

Foreign Direct Investment (FDI) 237

General Agreement on Tariffs and Trade (GATT) 254

global capitalism 240/247

global economic triad 223

Group of Seven (G7) 223

Group of Eight (G8) 225

high tech 239

historical inertia 240

just-in-time (JIT) manufacturing 256

market socialism 230

MFN status 255

multinationals (MNs) 240/242

multinational enterprise (MNE) 242

multinational service industries (MSIs) 264

nation-state 247

new southern manufacturing belt 238

North American Free Trade Agreement (NAFTA) 234

Organization for Economic Cooperation and Development (OECD) 233

product decline 262

product development stage 261

product growth stage 262

product life cycle 261

product maturity stage 262

product obsolescence 262

quota 243

regionalism 240

special export processing zones (SEPZs) 245

stocks (in reference to FDI) 249

subsidiaries 237

tariff 243

trade liberalization 240

transactions (in reference to FDI) 249

transboundary linkages 240

transnationalization process 240

world-state 248

REFERENCES

Adelman, I. "Beyond Export-Led Growth." *World Development* 12 (1984), pp. 937–949.

Amin, A. "Flexible Specialization and Small Firms in Italy: Myths and Realities." *Antipode* 21 (1989), pp. 13–34.

Blackburn, P., R. Coombs, and K. Green. *Technology, Economic Growth, and the Labour Process.* London: MacMillan (1985).

Bluestone, B., and B. Harrison. *The Deindustrialization of America.* New York: Basic Books (1990).

Bonker, D. *America's Trade Crisis: The Making of the U.S. Trade Deficit.* Boston: Houghton Mifflin (1988).

Bradshaw, T.H. "The Changing U.S. Labor Market: An Analysis of Employment, Earnings, and Education." M.A. Thesis, Department of Geography, University of Florida, Gainesville, Florida (May, 1996).

Braverman, H. *Labor and Monopoly Capital: The Degradation of Work in the Twentieth Century.* New York: Monthly Review Press (1974).

Bryson, J.R., and P.W. Daniels (eds.). *Service Industries in the Global Economy,* Williston, VT: Edward Elgar Publishing (1998).

Business Week. "Where America's Bottom Line May Be Squeezed Overseas," by G. Koeretz (September 20, 1995), p. 22.

Business Week. "A Flying Leap toward the 21st Century." by L. Nakarmi (March 20, 1995) pp. 78–80.

Business Week. "GE's Brave New World." (November 8, 1993), pp. 64–70.

Business Week. "Borderless Management." (May 23, 1994), pp. 24–26.

Business Week. "A Chaebol Charges toward the Top." (June 20, 1994), pp. 60–61.

Business Week. "Financing World Growth." (October 1994), pp. 100–103.

Business Week. "What Has NAFTA Wrought? Plenty of Trade." (November 21, 1994), pp. 48–49.

Business Week. "What to Do about Asia?" by B. Bremner (January 26, 1998), pp. 27–30.

Casson, M. (ed.). *The Growth of International Business.* New York: Allen & Unwin (1983).

Chapman, K., and D.F. Walker. *Industrial Location* (2nd edition). Cambridge, MA: Basil Blackwell (1991).

Cohen, S.B. "Global Geopolitical Change in the Post-Cold War Era." *Annals of the Association of American Geographers* 81 (1991), pp. 551–580.

de Blij, H.J., and P.O. Muller. *Geography: Realms, Regions, and Concepts* (7th edition). New York: J. Wiley & Sons (1994).

Dicken, P. *Global Shift: The Internationalization of Economic Activity.* New York: Guilford Press (1992).

Dicken, P., and P.E. Lloyd. *Location in Space: Theoretical Perspectives in Economic Geography.* New York: Harper & Row (1990).

Dunning, J.H. *International Production and the Multinational Enterprise.* London: Allen & Unwin (1981).

Dunning, J.H. *Explaining International Production.* London: Unwin Hyman (1988).

Dunning, J.H. *The Globalization of Business: The Challenge of the 1990s*. New York: Routledge (1993).

Durkheim, E. *The Division of Labor in Society* (translated by W.D. Halls). New York: Free Press (1984).

Ellison C., and G. Gereffi. "Explaining Strategies and Patterns of Industrial Development." In *Manufacturing Miracles: Paths of Industrialization in Latin America and East Asia*. G. Gereffi and D.L. Wyman (eds.). Princeton, NJ: Princeton University Press (1990), Chapter 14, pp. 368–403.

Encarnation, D.J. *Rivals Beyond Trade: America versus Japan in Global Competition*. Ithaca, NY: Cornell University Press (1992).

Fan, C. "Foreign Trade and Regional Development in China." *Geographical Analysis* 24 (1992), pp. 240–256.

Fishlow, A. "Latin American Failure against the Backdrop of Asian Success." *Annals of the American Academy of Political and Social Science* 505 (1988), pp. 117–128.

Forbes. "The Myth of United States Manufacturing's Decline." by J. Flint (January 18, 1993), pp. 40–42.

Fortune. "What's Next after GATT's Victory?" by L.S. Richman (January 10, 1994), pp. 66–70.

Friedman, G., and M. Lebard. *The Coming War with Japan*. New York: St. Martin's Press (1991).

Gertler, M.S. "Flexibility Revisited: Districts, Nation-States, and the Forces of Production." *Transactions, Institute of British Geographers* 17 (1992), pp. 259–278.

Gibb, R. "Regionalism in the World Economy." In *Continental Trading Blocs: The Growth of Regionalism in the World Economy*. Gibb, R., and W. Michalak (eds.). New York: Wiley (1994), pp. 1–36.

Gibb, R., and W. Michalak. *Continental Trading Blocs: The Growth of Regionalism in the World Economy*. New York: Wiley (1994).

Glick, L.A. *Understanding the North American Free Trade Agreement*. Boston: Kluwer (1994).

Glickman, N., and D. Woodard. "The Location of Foreign Direct Investment in the United States: Patterns and Determinants." *International Regional Science Review* 11 (1988), pp. 137–154.

Gluck, F. "Global Competition in the 1980s," In *Strategic Management of Multinational Corporations: The Essentials*. H.V. Wortzel, and L.H. Wortzel (eds.). New York: Wiley (1985), pp. 9–15.

Graham, J. "Firm and State Strategy in a Multipolar World." In *Trading Industries, Trading Regions*. H. Noponen, J. Graham, and A.R. Markusen (eds.). New York: Guilford Press (1993), pp. 140–174.

Graham, J., K. Gibson, R. Horvath, and D.M. Shakow. "Restructuring in U.S. Manufacturing: The Decline of Monopoly Capitalism." *Annals of the Association of American Geographers* 78 (1988), pp. 473–490.

Griffin, K.B. *Globalization and the Developing World: An Essay on the International Dimensions of Development in the Post-Cold War Era*. Geneva: UNRISD (1992).

Griffiths, A. *European Community*. London: Longman Group (1992).

Grub, P. (ed.). *The Multinational Enterprise in Transition* (2nd edition). Princeton, NJ: Darwin Press (1984).

Hamilton, I. (ed.). *Resources and Industry*. New York: Oxford University Press (1992).

Hanink, D. *The International Economy: A Geographical Perspective*. New York: J. Wiley & Sons (1994).

Harris, C. "The Market as a Factor in the Location of Industry in the United States." *Annals of the Association of American Geographers* 44 (1954), pp. 315–348.

Hartshorn, T.A., and J.W. Alexander. *Economic Geography* (3rd edition). Englewood Cliffs, NJ: Prentice Hall (1988).

Head, K., J. Ries, and D. Swenson. "Agglomeration Benefits and Locations Choice: Evidence from Japanese Manufacturing Investments in the United States." *Journal of International Economics* 38 (1995), pp. 223–247.

Hefler, D.F. "Global Sourcing: Offshore Investment Strategy for the 1990s." In *Strategic Management of Multinational Corporations: The Essentials*. H.V. Wortzel, and L.H. Wortzel (eds.). New York: Wiley (1985), pp. 343–349.

Henderson, J.W. *Globalization of High Technology Production*. London: Routledge (1987).

Johnson, C. "Enter the Dragon: 10 Reasons to Worry about Asia's Economic Crisis." *U.S. News & World Report* 30 (March 1998), pp. 40–41.

Kaplinsky, R. (ed.). *Third World Industrialization in the 1980s: Open Economies in a Closing World*. London: Cass (1984).

Kitson, M., and J. Michie. "Trade and Growth: A Historical Perspective." In *Managing the Global Economy*. Michie, J., and J.G. Smith (eds.). Oxford: Oxford University Press (1995), Chapter 1, pp. 3–36.

Knox, J., and J. Agnew. *The Geography of the World Economy* (2nd edition). London: Edward Arnold (1994).

Knudsen, D.C. "Flexible Manufacturing." *Growth and Change* 25 (Spring 1994), pp. 135–143.

Kozul-Wright, R. "Transnational Corporations and the Nation State," in *Managing the Global Economy*. Michie, J., and J.G. Smith (eds.). Oxford: Oxford University Press (1995), Chapter 6, pp. 135–171.

Krugman, P.R. *Geography and Trade*. Cambridge, MA: MIT Press (1991).

Krugman, P.R., and A. Venables. *Globalization and the Inequality of Nations*. Cambridge, MA: National Bureau of Economic Research (1995).

Kurtzman, J. *Decline and Crash of the American Economy*. New York: Norton (1988).

Lardy, N.R. *Foreign Trade and Economic Reform in China 1978–1990*. New York: Cambridge University Press (1992).

Lardy, N.R. *China in the World Economy*. Washington, D.C.:

Institute for International Economics (1994).

Leontidades, J.C. *Multinational Corporate Strategy: Planning for World Markets*. Lexington, MA: Lexington Books (1985).

Magaziner, I.C., and R.B. Reich. "International Strategies." In *Strategic Management of Multinational Corporations: The Essentials*. H.V. Wortzel, and L.H. Wortzel (eds.). New York: Wiley (1985), pp. 4–8.

Mair, A., R. Florida, and M. Kenney. "The New Geography of Automotive Production: Japanese Transplants in North America." *Economic Geography* 64 (October 1988), pp. 352–373.

Malecki, E.J. "Industrial Location and Corporate Organization in High Technology Industries." *Economic Geography* 61 (1985), pp. 345–369.

Malecki, E.J. *Technology and Economic Development: The Dynamics of Local, Regional and National Change*. London: Longman (1991).

Malecki, E.J., and M. Veldhoen. "Network Activities, Information and Competitiveness in Small Firms." *Geografiska Annaler* 75 (1993), pp. 131–147.

Mansfield, E. (ed.). *Monopoly Power and Economic Performance: Problems of the Modern Economy*. New York: W.W. Norton (1978).

McGarrah, R.E. *Manufacturing for the Security of the United States*. New York: Quorum Books (1990).

Michalak, W. "The Political Economy of Trading Blocs." In *Continental Trading Blocs: The Growth of Regionalism in the World Economy*. Gibb, R., and W. Michalak (eds.). New York: Wiley (1994), pp. 37–72.

Michie, J., and J.G. Smith (eds.). *Managing the Global Economy*. Oxford: Oxford University Press (1995).

Mirza, H. (ed.). *Global Competitive Strategies in the New World Economy*. Williston, VT: Edward Elgar Publishing (1998).

Morris, M., A. Bernhardt, and M. Handcock. "Economic Inequality: New Methods for New Trends." *American Sociological Review* 59 (1994), pp. 205–219.

Naisbitt, J. *Megatrends Asia*. New York: Simon & Schuster (1996).

Nijkamp, P., and S.A. Rienstra. "Private Sector Involvement in Financing and Operating Transport Infrastructure." *The Annals of Regional Science* 29 (1995), pp. 221–236.

Ondrich, J., and M. Wasylenko. *Foreign Direct Investment in the United States*. Kalamazoo, MI: Upjohn Institute (1993).

Peck, J. "Labor and Agglomeration: Control and Flexibility in Local Labor Markets." *Economic Geography* 68 (1992), pp. 325–347.

Peet, R. (ed.). *International Capitalism and Industrial Restructuring*. Boston, MA: Allen & Unwin (1987).

Pinder, J. *European Community: The Building of a Union*. Oxford: Oxford University Press (1991).

Poniachek, H.A. *Direct Foreign Investment in the United States*. Lexington, MA: D.C. Heath (1986).

Redmond, J., and G. Rosenthal. *The Expanding European Union: Past, Present, and Future*. Boulder, CO: Lynne Reiner Publishers (1998).

Romm, J.J. *The Once and Future Superpower: How to Restore America's Economic, Energy, and Environmental Security*. New York: W. Morrow & Co. (1992).

Rourke, J.T. (ed.). *Taking Sides: Clashing Views on Controversial Issues in World Politics* (8th edition). Guilford, CT: Dushkin/McGraw-Hill (1998).

Sachar, A., and S. Oberg (eds.). *The World Economy and the Spatial Organization of Power*. Brookfield, VT: Gower (1990).

Sadler, P. *The Global Region*. Tarrytown, NY: Pergamon Press (1992).

Schoenberger, E. "Multinational Corporations and the New International Division of Labour." *International Regional Science Review* 11 (1988), pp. 105–120.

Schott, J.J. "Trading Blocs and the World Trading System." *The World Economy* 14 (1991), pp. 1–17.

Scott, A.J. *New Industrial Spaces*. London: Pion (1988).

Shambaugh, D. (ed.). *Greater China: The Next Superpower*. New York: Oxford University Press (1995).

Sklair, L. *Assembling for Development: The Maquila Industry in Mexico and the United States*. San Diego, CA: Center for U.S.-Mexican Studies, UCSD (1993).

Sklair, L. *Capitalism and Development*. London: Routledge (1994).

Smith, D., and R. Florida. "Agglomeration and Industry Location: An Econometric Analysis of Japanese-Affiliated Manufacturing Establishments in Automotive-Related Industries." *Journal of Urban Economics* 36 (1994), pp. 23–41.

South, R.B. "Transnational Maquiladora Location." *Annals of the Association of American Geographers* 80 (1990), pp. 549–570.

Stoddard, E.R. *Maquila: Assembly Plants in Northern Mexico*. El Paso, TX: Texas Western University Press, University of Texas at El Paso (1987).

Stopford, J.M., and J. Dunning. *Multinationals: Company Performance and Global Trends*. London: MacMillan (1983).

Storper, M., and A.J. Scott. *Pathways to Industrialization and Regional Development*. London: Routledge (1992).

Swartz, T.R., and F.J. Bonello (eds.). *Taking Sides: Clashing Views on Controversial Economic Issues*. Guilford, CN: Dushkin Publishing Group (1995).

Tolchin, M., and S. Tolchin, *Buying Into America*. New York: Berkley Books (1988).

U.S. Global Trade Outlook 1995–2000: Toward the 21st Century. Washington, D.C.: U.S. Government Printing Office, (March 1995).

Vernon, R. *Storm over the Multinationals*. New York: Harvard University Press (1978).

Warf, B. "U.S. Employment in Foreign-Owned High Technology Firms."

Professional Geographer 42 (1990), pp. 421–432.

Watts, H.D. *The Branch Plant Economy*. London: Longman (1981).

Watts, H.D. *Industrial Geography*. London: Longman (1987).

Wheeler, J.O., P.O. Muller, G.I. Thrall, and T.J. Fik. *Economic Geography* (3rd edition). New York: Wiley (1998).

Williams, A.M. *The European Community*. Oxford: Blackwell (1994).

World Development Report 1991: The Challenge of Development. (published for the World Bank). New York: Oxford University Press (1991).

World Development Report 1994: Infrastructure for Development. New York: Oxford University Press (1994), pp. 1–122.

World Development Report 1995: Workers in an Integrating World. (published for the World Bank). New York: Oxford University Press (1995).

Wortzel, H.V., and L.H. Wortzel (eds.). *Strategic Management of Multinational Corporations: The Essentials*. New York: J. Wiley & Sons (1985).

Xie, Y., and F.J. Costa. "The Impact of Economic Reforms on the Urban Economy of the People's Republic of China." *Professional Geographer* 43 (1991), pp. 318–335.

Yanarella, E.J., and W.C. Green. *The Politics of Industrial Recruitment*. New York: Greenwood Press (1990).

10

DEVELOPMENT PROSPECTS AND OBSTACLES

SYNOPSIS

The discussions throughout this book have highlighted regional variations in economic and demographic change. Those discussions have reminded us of the many challenges now facing the people, regions, and nations of an integrated yet divided world economy. While the global population explosion continues to hinder regional economic development efforts throughout the SOUTH, growing disparities between core and peripheral nations continue to limit the long-term development potential of many SOUTHern economies; particularly, those which are isolated from globalizing production, investment, and trade networks.

World population is expected to rise to approximately 11 billion people by the year 2100. It is estimated that world output will have to grow five-fold to guarantee a decent standard of living for all (that is, to ensure that each person's basic needs for food, clothing, shelter, health care, and personal security are met). Though world output will most likely fall short of this level, there is hope that science, technology, and the human spirit will bring about significant improvements in the quality of life of the average person in the world economy. In light of the advances of the Industrial Revolution and the more recent onset of the computer and information revolutions, there is a profound belief in **technocratic ideology**— *finding technologically based solutions to economic development problems* (Figure 10.1).

Disciples of technocracy contend that production levels and capacity of most systems will continue to expand at a rate that is faster than the world's population. As existing and new production technologies become more commonplace and cost competitive, they are likely to give birth to new and yet unforeseen technologies, further advancing the role of science in production economics. Disciples of technocratic ideology insist that history will repeat itself and the world will undergo yet another series of changes that will boost overall production capacity and levels as innovations are periodically introduced to agricultural, industrial, commercial, and transportation systems.

Certainly, the time lag between the birth of a technological innovation and its use or maturity will shorten with the compression of product cycles. Hence, the impact to regional economies will be more immediate. The unknown benefits of technological change and increasing accessibility to technology are things that will not be realized for some time to come.

As advanced production systems become more commonplace and cost competitive around the globe, it is likely that the world economy will realize **economies of synergy**—*efficiency and production gains and savings associated with the adoption and use of advanced production technologies and the collaborative action of producers, networks, or regions which combine to yield mutually beneficial interactions and spillover effects that improve the growth and development potential of all regions.* Although the NORTH-SOUTH distinction itself will probably remain intact, people in the developing SOUTH will enjoy a higher quality of life as the world economy redefines its sector-specific carrying capacities. The realization of economies of synergy in the core and semiperiphery will help improve access to "survival resources" worldwide. Many regions of the world will continue to support a population base that is growing faster than its regional economy, yet forthcoming economies of synergy will allow spillover effects that will improve regional development possibilities throughout the world economy. While excessive population growth today may counteract the benefits of economic growth today, the collective impacts of global economic growth and regional information and technology transfers will have many unknown and potentially desirable results, some of which are likely to improve the competitiveness of many, if not all, developing regions.

The economies of synergy argument is in opposition to the theory of "circular and cumulative causation," as introduced by Myrdal (1957) and later by Pred (1966). **Circular and cumulative causation** refers to *the process by which regional economic change and disparities are viewed as logical outcomes of a series of changes which improve the positions of favored or advanced economic regions, thereby creating greater inequalities between the development status of favored versus nonfavored regions.*

It is well documented that core economies have been favored destinations for capital, investment, and labor flows (at the expense of the periphery). As growth begets growth, favored regions tend to attract investment dollars and retain industry and value added. Thus, favored regions are allowed to expand their production networks and export possibilities. Corresponding increases in gross regional product and income allow these regions to accumulate wealth and enjoy economic prosperity. The concentration of growth in a limited number of geographic areas amplifies existing inequalities and regional variations.

The theory of circular and cumulative causation suggests that poor or nonfavored regions will experience very little of the benefits of economic growth over time. In fact, it is likely that many peripheral nations will become relatively poorer over time as production and value added becomes ever more concentrated in dominant and favored regions. As a result, the forces of free-market economy tend to create a natural polarization of economic regions. Although trickle-down effects (from the core to the periphery) can help lessen the unbalance, the negative effects of polarization are difficult to overcome or reverse once established. In essence, circular and cumulative causation suggests that peripheral economies will be trapped in a cycle of poverty and underdevelopment indefinitely.

The circular and cumulative causation argument suffers, however, from one minor flaw—the simple fact that core economies have much to gain from improving the development status of all nations in the SOUTH. The rapidly growing and emerging markets of the SOUTH represent a formidable source of future revenues and profits for core-based multinationals. High population growth areas provide large and ever-expanding markets for exportables. While increasing disparities in income between the nations of the NORTH and the SOUTH have been observed for some time (as highlighted in Table 10.1), the widening development gap has been largely underscored by the fact that economic conditions have steadily improved in many SOUTHern economies. Many analysts see increasing foreign investment and aid and technology transfers from the core to periphery as forces that will eventually reduce polarization. Over time, the prosperous core and semiperiphery are likely to boost up the fledgling economies of the periphery and encourage the participation of all regions in a global economy, something that is mutually beneficial to all parties (Figure 10.2).

Figure 10.1 Large multinational corporations use the latest tools and technology to stay competitive. Here CEO's take part in a televideo conference at NEC in Tokyo, Japan.

10.1: ©Charles Gupton/Stock Boston

TABLE 10.1	Income Ratios of Richest and Poorest Nations for Selected Years

Year	Income of Richest 20% Divided by Income of Poorest 20% (ratio)
1960	30:1
1970	32:1
1980	45:1
1990	59:1
2000*	70:1

Source: United Nations (1992).

* Forecast.

Figure 10.2   Hope remains within the world's poverty stricken regions. Here children play in the slums of Sao Luis, Brazil.

l0.2: ©Sean Sprague/Stock Boston

It is likely that existing disparities in consumption will be also reversed. Consider that the developed NORTH, with less than 30% of the world's population, now consumes well over 70% of the world's resources and energy. The expansion of regional and global production networks, the rise of BEMs, and explosive population growth in SOUTHern economies, will inevitably cause the SOUTH to increase its demand and production of commodities, resources, and energy. By end of the next millennium, it is likely that the economies of the SOUTH will consume roughly half of the world's resources and energy. This has brought about increasing concerns over resource depletion and energy consumption as SOUTHern economies develop and modernize.

Increasing resource use and energy demand in southern Asia, the Pacific Rim, and Latin America is likely to pose several development challenges. First, there will be increasing tensions and competition between production regions. Second, intensified competition for world resources and nonrenewable sources of energy will cause the price of development and production factors to rise beyond the means of many of the world's poorer nations. This outcome is likely to slow the economic development of African, West Asian, and Central American nations.

All in all, the economies of the SOUTH have been plagued by numerous development obstacles, including:

- the continued domination of the world economy by the G7, the OECD, and multinationals—entities which control production, trade, prices, and investment;

- relative shortages of investment capital and the inability to attract significant amounts of foreign direct investment;
- capital flight (that is, the mass exodus of investment capital);
- brain drain (that is, the mass exodus of the skilled portion of the regional workforce);
- inadequate R&D expenditures, inferior production technologies, and noncompetitive industry or firms;
- unstable business and political climates (not to mention a high incidence of graft and corruption amongst members of the government, military, and industry);
- a lack of adequate infrastructure and services to support business;
- the footloose and noncommittal nature of multinationals and the exploitation of labor by globalizing industry;
- increasing deficits in regional food production and a continuing dependency on external suppliers for food and agricultural inputs;
- increasing competition between the numerous and low-cost production areas found throughout the SOUTH;
- the physical remoteness of many peripheral areas and their general lack of integration in the expanding global production-and-trade networks;
- a worsening international debt crisis;
- unfavorable terms of trade as tied to the relatively low value of exports;
- rapid population growth;
- hyper-urbanization and the inability of the public sector to provide for an increasing number of urban dwellers;
- internal discord, conflicts, war, civil unrest, etc.;
- a growing refugee problem;
- production-led ecological decline and environmental degradation; and
- natural and man-made hazards.

TRADE-OFFS AND PROGRESS

Human economic activities continue to deplete the earth's natural resources and degrade soils, forests, wetlands, grasslands, and cropland. Each year more than 40 million acres of tropical forestland is destroyed, along with 6 million acres of farmland worldwide. Approximately 20 billion tons of topsoil are lost annually due to floods and wind and water erosion. It is estimated that more than 10% of the world's vegetated lands have experienced moderate to severe soil degradation, a figure that is likely to rise as increased pressures are placed on the world's farming systems. Declines in the growth of agricultural output have been linked to soil erosion. Overgrazing, intensive subsistence cultivation, shortening fallow periods, land mismanagement, and urbanization continue to contribute to the loss of farmland around the globe. Deforestation and desertification have been accelerated by nonsustainable human activities, transforming countless tracts of productive range and cropland into virtual wastelands.

Environmentalists have also been troubled by a series of other problems including ozone depletion, acid rain, a global warming trend associated with increasing concentrations of greenhouse gases, as well as industry-related damage to the atmosphere and oceans. Crude-oil spills, seepage of chemicals, and the dumping of waste and toxic materials are now poisoning many of the earth's fragile marine environments. Coral reef and coastal upwelling areas have been particularly hard hit in the assault, especially those in close proximity to major population concentrations. Urban population growth has also placed an enormous burden on landfills, sanitation, and waste management systems. Perhaps a much larger problem is the impending shortages of water. Freshwater sources continue to be threatened by increasing demand, chemical intrusion, and toxins introduced by urban and agricultural systems. Effective management of water resources is of great concern in West Asia, East Africa, parts of Latin America and most cities worldwide. Transboundary pollution and contamination of ground and surface water by industry and urban run-off will continue to pose a challenge and a risk to human agents in affected regions (Figure 10.3).

Environmental concerns include overexploitation of marine and coastal environments, atmospheric pollution and corresponding climate changes, deforestation, the loss of bioidioversity and natural habitats due to disruption of fragile ecosystems, the absence of alternatives to wood for fuel, and general decline of environmental conditions in and along urban and industrial corridors. In addition, the lack of sufficient long-term storage facilities for low-level nuclear waste has increased the threat of radioactive waste exposure in many regions (Figure 10.4). As human economic activities and growth of regional production systems continue to advance environmental degradation, many see "progress" as the end result of the inherent trade-off between economic growth on the one hand and environmental quality and sustainability on the other hand.

Surely, industrialization and urbanization have encouraged resource-intensive consumption, nonsustainable production, and a diminishing quality of life in many places. Yet, modernization and development have also led to a greater awareness on the part of consumers and producers as they attempt to find new ways in which to expand output while minimizing damage to the environment. Meanwhile, the economic costs of environmental and land degradation are staggering.

It is estimated that damage to land-based resources and the environment now costs many industrial countries the equivalent of 1% to 5% of their Gross National Prod-

Figure 10.3 An environmental clean-up crew works hard in the aftermath of a crude oil spill.
l0.3: ©Jonathan Nourok

Figure 10.4 A large and near capacity urban landfill.
l0.4: ©Robert Brenner/Photo Edit

uct (GNP) in terms of lost output and damages, not to mention health-related problems associated with industrial pollution. In eastern Europe, the lack of environmental and air-quality standards and the virtual nonregulation of heavy industry have caused extensive damage to agricultural lands and forested areas along highly polluted industrial corridors. The loss of crops, cropland, and timber, not to mention the loss of wildlife, has been estimated at $30 billion a year (on average) over the past two decades (Figure 10.5).

Figure 10.5 Pollution in eastern Europe. Heavy industry continues to pose a threat to the quality of life where environmental regulations are lax or nonexistent.

l0.5: ©A. Ramey/Photo Edit

Productivity losses are expected to continue unless regulations and laws are put into place to protect environmentally sensitive areas. These laws must be sufficiently enforced, something which may prove difficult as a large proportion of high-polluting industries have moved out of core economies to peripheral and semiperipheral regions, areas which are notably lax in terms of placing environmental restrictions on foreign firms. Most regions of the SOUTH have opted to sacrifice environmental quality for the immediate benefits of jobs and industrialization.

The core economies of the NORTH continue to push the industrializing nations of the SOUTH toward the adoption of a **world view**—*a belief that regions should support economic activities that are sustainable and non-detrimental to the environment, along with the conservation and preservation of world resources.* The world view may prove difficult to impart on a rapidly growing and developing SOUTH, as for centuries the developed NORTH has not practiced what they now preach.

Regional economic growth and developmental strategies have changed dramatically over the past several decades. At one time, a competitive and viable industrial and export base was all that was needed to stimulate a re-

gion's economy. Today, exports have taken a back seat to the production and sales activities of foreign affiliates. Large and powerful multinational enterprises now control production on the economic landscape. Global capitalism has made it difficult for many nations of the SOUTH to forge sure-footed development policies. Developing nations must now compete in a world economy whose destiny is increasingly being controlled by globalizing industry and corporate-owned production networks. Adopting a world view may be fine for core economies or firms with ties to global production networks; yet, developing regions are more interested in supporting more "nationalistic" industrialization and development policies.

THE GLOBAL DEVELOPMENT AGENDA

The world view has spawned a sustainable development "movement," with mixed support amongst the international community. Advocates of **green development**—*environmentally friendly or compatible economic change or activities which lead to a higher quality of life*—suggest six fundamental changes to enhance the development positions of the world's nations. First, is the adoption of environmental "green" production technologies; that is, technologies that have a minimal impact on the physical environment. Second, there must be a commitment to energy and resources conservation and the abandonment of nonsustainable practices in the production and marketing of products. For example, product designs may be introduced with less packaging, fewer materials or parts, and longer life cycles. Elimination of useless or unwanted features and options is one way in which producers can simplify production and minimize waste. The renunciation of planned obsolescence in product marketing is viewed as a desirable characteristic of a modern-day production system. Production and distribution must be specifically tailored toward the needs of humans and not just for the sake of creating new markets or products.

Third, there must be a concerted effort to build a framework for economic and social change that is sensitive to externalities (that is, environmental spillover effects). This will require regulatory and compensatory measures such as taxation schemes to recover cost and damages to the environment, market incentives and rebates for firms adopting environmentally sensitive production methods and technologies, and tax-breaks for low- or nonpolluting industries. The objective is to correct environmental problems by curtailing destructive actions. Fourth, the framework for economic and social change must include internationally coordinated programs for poverty alleviation in the world's poorest regions, as well as protection of world resources, global diffusion of technology and information, and policies which equalize access to investment capital. As part of this endeavor, free trade initiates must be pursued with vigor.

Fifth, nations must work together to help stabilize regional population growth, create economic incentives for having smaller families, and fund education and up-skilling of the labor force to boost productivity levels in all economic sectors. This, of course, would require tailoring job-training programs to meet the needs of local communities and the regional production network. Stabilization of regional population growth would help to speed up the demographic transition while bringing a higher overall standard of living.

Sixth, regions must restrain consumption to levels which will ensure the sustainability of both regional and global production systems. Reusing, recycling, and refurbishing consumer products, and reducing the amount initially consumed, are critical to this effort. Sustainable development would thus encourage lifestyles oriented toward conservation, preservation, and **green consumerism**—*a consciousness that favors products, production processes, and behaviors that are less harmful to the environment and less wasteful.*

To achieve these goals, there must be complete cooperation between nations, regions, consumers, labor, industry, and government. If changes are made now, and if world development strategies are effectively employed, it is unlikely that the world's people will prematurely deplete precious world resources.

Critics of green development are quick to point out that such strategies do little in terms of reversing uneven development and are costly to both industry and consumers in terms of jobs lost and the excessive tax burden which would accompany subsidized transformation and the "greening" of industry and regional economies. To some extent, the cost of ecological deterioration is nonrecoverable or nonquantifiable, making taxation of polluting industries problematic.

Atop the global development agenda is the eradication of hunger. Currently, there are more than 500 million people suffering from hunger and malnutrition in the world. More than 1 billion people are chronically undernourished, a situation brought about when the body receives less protein and nutrients than required to resist disease and work productively. Hunger continues to drain not only the physical strength of the world's underprivileged people, but also negatively affects their drive, motivation, and creative powers, suppressing their intellectual and productive capacities. Hunger and poverty are symptoms of powerlessness, conditions that are no longer tolerable in a world society that understands the underlying causes and simultaneously possesses the resources to eradicate these conditions. Access to food is a right and not a privilege. As such we must forego the desire to produce in response to market forces only. In addition, food must never be used as a policy tool or weapon; it should be made available for all that need it regardless of their background or political viewpoints or socioeconomic status.

Increasing food aid to people in food-starved regions is not the answer. Food aid provides economic disincentives as it discourages small farmers from increasing local and regional production and encourages governments to stall rural development reforms and population containment efforts. Empowering the poor and hungry is a much better approach. This must involve government support of agricultural sectors, land reform in subsistence regions, improved access to land and capital, and active involvement to promote long-term food and economic security initiatives and sustainable agricultural practices.

Moreover, long-term solutions must be found for the world refugee problem. The psychic and physical suffering and hardship associated with people caught up in the crossfire of political and social turmoil is far too great to dismiss this most urgent development problem. Government and people in all regions must work together to promote peace and ensure security in troubled regions around the globe. Minimizing the refugee problem and promoting peace will have the added benefit of decreasing the likelihood of famine and lessening the potential for disruption in food distribution.

A Strategy for Sustainable Living

In a joint publication by the International Conservation Union, the United Nations Environment Program (UNEP), and the World Wide Fund for Nature (WWF) entitled *Caring for the Earth* (1991), contributing authors define what they call "a strategy for sustainable living." This "how to" book unveils a two-fold development objective:

1. "secure a widespread and deeply-held commitment to a new ethic . . . the ethic for sustainable living" and translate its principles into practice in everyday life; and
2. integrate the concepts of conservation and development by keeping our collective actions within the sustainable limits, and by fostering a world view that enables people everywhere to enjoy long, healthy, and fulfilling lives, without concerns that they, their earth, or future generations are at risk.

A sustainable society is contingent upon human beings' abilities to

1. respect and care for the earth, its physical systems, and its many life-forms;
2. improve the quality of the human existence;
3. conserve the earth's vitality and diversity;
4. minimize the depletion of nonrenewable resources;
5. keep within the earth's carrying capacity;
6. change personal attitude and practices (such that they are compatible with a sustainable way of life);
7. enable communities and people to care for their own environments without imposing limitations on others;
8. provide national and international frameworks for integrating regional development and a system to reward those who comply to established standards or meet certain goals;

9. create an international alliance that recognizes the value of conserving world resources (*Caring for the Earth*, 1991, pp. 8–12).

Critical objectives for sustainable development policies and the requirements for sustainable development are listed in Tables 10.2 and 10.3, respectively. Although these are simple blueprints for policy makers, they help shed light on the many challenges that remain in the policy arena. Critics of sustainable development and the world view argue that global policy efforts are doomed to fail as most regions are unlikely to support initiatives that cost rather than benefit them. Funding global development will certainly be costly for core economies. Taxpayers in the developed world are leery of dispensing more tax dollars to support international development efforts in light of the many economic development problems in their own backyard (including homelessness, poverty, hunger, crime, and disease).

When all is said and done, there is yet another formidable constraint that will hinder economic development efforts in the SOUTH. It is a timing problem, one which can be traced to the normal waves and cycles of economic change which are inherent to production networks of the world economy. Let us briefly discuss this phenomenon, and speculate on why it may be one of the greater challenges for the economies of the SOUTH to overcome.

ECONOMIC WAVES AND CYCLES

Regional and national economies are greatly affected by changes in production technology, product cycles, investment patterns, and the balance of trade. These same economies are also affected by what has been labeled as "long waves" of economic change as first observed by the Russian economist Nikolai D. Kondratiev some 70 years ago. Kondratiev argued that the industrialized nations of the world, when viewed collectively, have experienced successive cycles of growth and decline since the beginning of the Industrial Revolution [c.1750 to c.1850 in Britain and c.1800 to c.1900 for the United States]. It is interesting to note that the cycle of economic growth and decline has had a regular periodicity or length. Specifically, long waves of global economic change seem to run in a cycle which lasts roughly 50 to 60 years. Hence, the so-called **Kondratiev wave** *describes a 50- to 60-year cycle by which industrialized nations of the world experience economic growth and then decline with corresponding price inflationary and deflationary periods.*

There have been four complete Kondratiev waves since the beginning of the Industrial Revolution. Each wave is marked by an initial acceleration in the rate of price increases—that is, a price inflationary period. Price spiraling lasts through roughly half of the wave or 25 to 30 years. As the wave peaks, price increases begin to decline in magnitude until price deceleration takes over.

During the next 25 to 30 years the rate of price increase tends to decline. In sum, the Kondratiev wave is characterized by a price-inflationary period followed by a price-

TABLE 10.2	Critical Objectives in Sustainable Development Policies

1. Reviving growth and creating new pathways to growth
2. Redefining growth and progress
3. Meeting essential needs for jobs, food, energy, education, water, sanitation, health care, and other survival resources
4. Stabilizing regional and global population bases
5. Conserving and enhancing natural resources
6. Reorientating technology and finding new ways in which to manage resources and risk
7. Merging environmental concerns and economics in the decision-making processes of firms and individuals
8. Balancing production and consumption in environmentally friendly and responsible ways
9. Revealing the interdependent nature of development, the interconnectiveness of all regions in a globalizing economy, by discussing the common economic futures of the NORTH and SOUTH

Source: After Brundtland, 1987, p. 65 (as summarized in Adams, 1990, p. 60), with modifications.

TABLE 10.3	Requirements for Sustainable Development

1. A political system that secures effective citizen participation in decision making
2. An economic system that is able to generate production surplus and technical knowledge on a self-reliant and self-sustained basis, while ensuring that demands for basic survival resources are met
3. A social system that provides immediate solutions for the tensions arising from disharmonious development and conflict
4. A production system that respects the obligation to preserve the environment and the ecological basis for present and future development
5. A technological system that can search continuously for new solutions and readily adapt to change
6. An international system that fosters sustainable patterns of production, trade, investment, and finance in ways that do not promote uneven development or increased polarization of the core and periphery
7. An administrative system that is flexible and has the capacity for self-correction, one that is not excessive in size or domineering
8. A global market system that is efficient, impartial, and not over-regulated by government
9. A consumption system that is designed to minimize waste and conserve world resources
10. An economic-policy system that offers a balance between the rational forces of market-driven efficiency and more extensive participation in the determination of both economic ends and means

Source: After Brundtland, 1987, p. 65 (as summarized in Adams, 1990, p. 61), with modifications.

deflationary period, where the annual percentage change in price levels increases, peaks, and then decreases over time (Figure 10.6).

Economist Simon Kuznets discovered that within each Kondratiev wave are two cycles of economic growth. These were later described as **Kuznets cycles**—*cycles which describe change in the rate of economic growth*. Kuznets cycles are characterized by the following sequence of events, each of which last roughly 10 to 15 years:

1. recovery/growth;
2. prosperity;
3. recession; and
4. depression.

Note that the recovery/growth and prosperity periods define a "phase-A" Kuznets cycle, whereas the recession and depression periods define a "phase-B" Kuznets cycle. Note that there is exactly one phase-A and one phase-B Kuznets cycle per Kondratiev wave (as shown in Figure 10.6). Note also that the industrialized nations of the world continue to pass through this sequence with an uncanny regularity. These events have direct implications as to the economic growth and development potential of all regional economies as they are greatly affected by what goes on in the developed world. Let us briefly discuss each of the four periods and their characteristics.

The **recovery/growth** period is marked by the introduction of new technologies and products and the expansion and growth of geographic markets and regional economies which produce goods and services for those markets. This period is characterized by increases in the demand for labor and increasing consumer spending, with corresponding wage and price increases. Consumer confidence is high, with consumers seeking new outlets for their increased purchasing power. Economic growth is strong and seems to accelerate over time, gradually reaching a peak midway during the phase-A Kuznets cycle.

Healthy and steady economic growth eventually brings about **prosperity.** As high economic growth rates cannot continue indefinitely, economies begin to show signs of slowing down. Growth rates begin to taper off

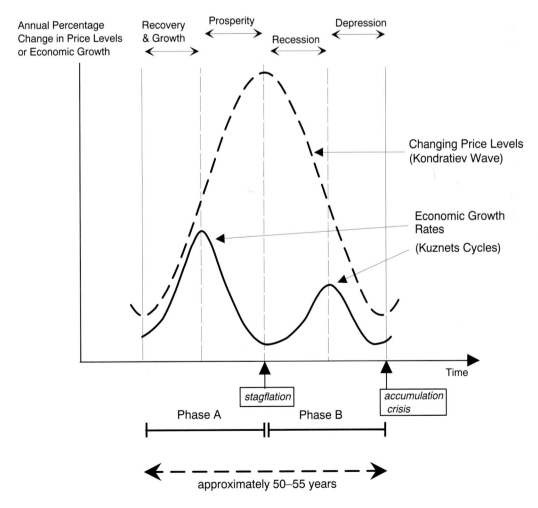

Figure 10.6 Relationship between rhythms of Kondratiev wave and Kuznets Cycles.
Source: After Berry (1991).

and a deceleration phase begins. Production technology diffuses throughout the industrial core and out to the semiperiphery. Markets expand and mature, and competition amongst rivals intensifies. Both demand and production output peak as regional production networks operate near capacity. Economic growth levels off despite the steady stream of investment dollars to industry. General price levels continue to rise and ultimately peak. Demand falls and economic growth slows even further.

The Kondratiev wave now reaches its peak or midpoint, and the phase-A Kuznets cycle is complete. The rate of economic growth is at its lowest point since the beginning of the cycle. This marks the end of a growth deceleration period, and a point where the price inflationary spiral has led to economic stagnation. Economies experience sluggish growth, as price levels remain high and demand is low and as "stagflation" sets in. **Stagflation** *is a crisis situation brought about by stagnant economic growth and price inflation*. Economic growth is also hindered by a shortage of financial/investment capital.

Stagflation marks the beginning of a period of **recession.** In this period, unemployment rises as less output is produced in relation to what can in fact be produced with most economies and sectors. Declining demand leads to declines in output and excess production capacity. As production continues to shift out of the core and move to the periphery, international competition intensifies and general price levels begin to decline as the world's markets are flooded with products from low-cost producers around the globe. Firms begin to downsize and streamline production. Many firms exit the industry and others move to overseas locations. Regional production networks begin to dismantle, restructure, and/or reorganize in space as firms attempt to survive the economic downturn and remain competitive. After the shake-out, a short economic recovery period begins as consumer demand responds to decreasing price levels and output levels rise. Economic growth rates in the core begin a slow ascent as the effects of recession linger. Recall that the United States experienced a long-lasting recession throughout the mid-1970s. It was a period marked by high rates of unemployment and inflation.

High rates of economic growth are not typically observed in the period following a recession. Economic output of industrialized economies is generally at a very low level. Reduced purchasing power from nagging unemployment and stagnant wages forces sales and revenue to decline, despite downward shifts in demand and falling prices. The lack of consumer confidence and purchasing power and reduced corporate profits continue to hinder economic growth. The decline of the general business climate (during this price deflationary period) forces economies to enter into a **depression.** A period of depression means that an economy is undergoing an economic slump. As a result, the price of stocks begin to fall as firms fall victims to shrinking markets, sales, and profits. Should the slump be severe or prolonged, an economy

could weaken and even collapse (as was the case during America's Great Depression). Hence, a depression typically brings about increasing pressure to implement economic policies to jumpstart economies and increase consumer demand. This can be done by lowering interest rates, increasing the money supply, or government bailout of failing industry. During a depression, economies must deal with an **overaccumulation crisis.** The overaccumulation crisis arises as a result of three conditions:

1. surplus labor;
2. surplus production capacity; and
3. surplus investment capital.

These conditions are brought about by low demand and correspondingly low levels of production. Low levels of output mean that an underutilized production capacity and labor force exist. Hence, a depressed economy is characterized by surplus production capacity, idle labor and/or unemployed and underemployed labor. In addition, investors earning profits, dividends, interest, and rents, etc. cannot find enough reasonably safe and/or profitable investment opportunities to satisfy their needs. Returns to investments are low and losses are high.

As economies near the end of the Kondratiev wave and the end of a phase-A Kuznets cycle, economic growth rates remain relatively low. In addition, price levels of consumer and producer goods bottom out. Hence, the overaccumulation crisis marks the end of a price deflationary period. An upward movement in price levels begins with the introduction of new products and new production technologies, and modernization. Growth of new product markets and new production methods signals an increased potential for higher rates of return on investments. This leads to renewal and growth. Rebounding economies experience increased production, expanding employment opportunities, as well as growth in income and earnings. Once again, economies move into the growth and recovery stage and begin another long wave of economic change.

Each of the four Kondratiev waves, observed since the onset of the Industrial Revolution, has been initiated by the development, use, and exploitation of new technologies and innovations by core economies (see Figures 10.7 and 10.8). In short, each long wave is technology driven. At the beginning of a wave, new products and technologies are introduced almost exclusively in core economies. This is not surprising as core economies are both centers of innovations and the leading markets for new products. It is not until those products and technologies mature that investment and production will diffuse to regions outside the core. Long-wave theory suggests that the flow of investment capital from the core to the semiperiphery, and later to the periphery, is most likely to occur in the period directly following stagflation (just beyond the midpoint of a Kondratiev wave). Note that the industrialized nations of the world are currently at a period which marks the start of a new Kondratiev wave, with the midpoint some 25 to 30 years away (Figures 10.7 and 10.8).

With the worldwide recession and depression recently behind us, many economic analysts are optimistic over the prospects of growth, acceleration, and expanding global markets. No surprise that the American stock market continues to boom despite the recent financial crisis in Asia. Though Asian economies are feeling the ill effects of the overaccumulation crisis, the more mature industrial economies in North America and Europe show signs of recovery and growth. In addition, the value of stocks of many multinationals continue to rise along with a growing optimism over the prospects for future earnings and profits associated with the proliferation of new global communications and information technologies. Technological innovation, new products, and new markets have given birth to a fifth Kondratiev wave.

Nonetheless, there are several economic development concerns. First, as patterns of economic change are ultimately linked to long-run waves, growth and development cycles have created a potential timing problem for the nations of the SOUTH. How can low-income, peripheral economies attempt to compete in the production of commodities for global markets when so much surplus capacity and surplus capital exists in the industrialized core economies and the semiperiphery? As the leading industrialized nations of the G7 and the Pacific Rim have recently experienced an overaccumulation crisis, it is unlikely that a flood of investment capital will reach peripheral economies. With underutilized production capacity in Europe, the United States, and Asia, it is unlikely that production will shift outside of established production regions.

Second, the industrialized world appears to be at the start of a price inflationary period. This does not bode well for developing regions as price spiraling will ultimately erode their already limited purchasing power. Expected price increases for products from the core and semiperiphery are likely to expand trade deficits in peripheral economies that rely on primary sector exports.

Overall, the economic benefits associated with a new growth and recovery period will be largely restricted to

Phase	Kondratiev Long Wave				
	I	II	III	IV	V
Recovery	1770–1786	1828–1842	1886–1897	1940–1954	1996–2010?
Prosperity	1878–1800	1843–1857	1898–1911	1955–1969	
Recession	1801–1813	1858–1869	1912–1924	1970–1980	
Depression	1814–1827	1870–1885	1925–1939	1981–1995	

Kuznets Cycles

Characteristic	Macroeconomic Characteristics			
	Recovery	Prosperity	Recession	Depression
Gross National Product	Increasing growth rates	Strong and Steady growth	Decreasing growth rates	Little or no growth
Investment Demand	Increase in replacement investment	Strong expansion of capital stock	Scale-increasing and overseas investment	Excess production capacity
Consumer Demand	Purchasing power seeks new outlets	Increases in demand in all sectors	Continued growth of new sectors	Growth at the expense of savings
Markets and Exports	Expanding regional and global markets	Explosive growth worldwide	Intensified competition and limited growth	Slowdowns in the expansion of markets and exports
Consumer Price Levels	Low rates of inflation	Moderate rates of inflation	Relatively high rates of inflation	Relatively low rates of inflation

Figure 10.7 Kondratiev waves and Kuznets cycles: a historical record.

Source: S. Kuznets, *Economic Change,* New York: W.W. Norton (1953); and Berry, Conkling, and Ray (1994, p. 329), with modifications and additions.

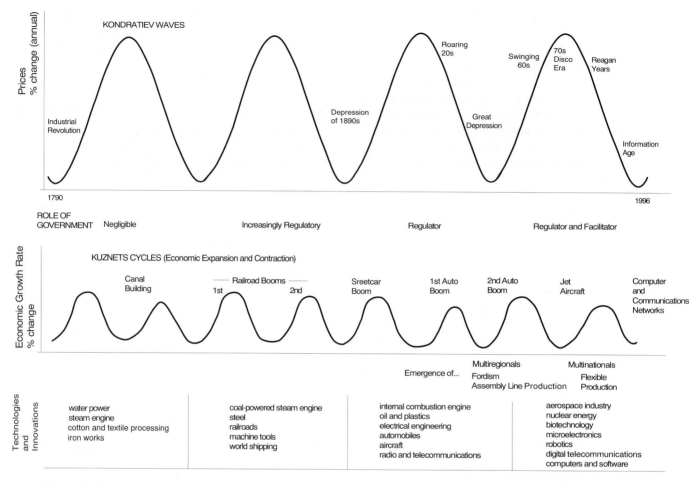

Figure 10.8 Economic change in the world's core economies, c. 1790 to present.
Source: Knox and Agnew, 1994, with modifications and additions.

core economies or regions with excess production capacity (those with the greatest ability to meet the demand in core and emerging markets). This situation can only serve to worsen the existing polarization of core and peripheral economies. This outcome is counterproductive from a development standpoint. As the world economy stands ready to embark on the next long wave, it is not difficult to envision greater tensions between the economies of the NORTH and the SOUTH for at least the next 20 to 25 years. While technology continues to promote a growing sense of interdependence and unity, bringing people and regions closer together, the regions of the world will become even more polarized and fractured in economic terms as waves and cycles run their logical course.

The global information and communications revolutions have many implications for regional development. One could argue that globalization of economic phenomena, increasing access to information, and the rapid diffusion of technology have the potential to shorten or compress economic cycles. Even so, it is not clear that the compression of Kondratiev waves and coinciding

Kuznets cycles will speed up economic development of all regions of the SOUTH. Globalization of economic activities has allowed the benefits of growth to be dispersed over a larger geographic area. Yet, the so-called globalization has been limited in terms of spatial coverage to a handful of emerging economies. As developing regions compete to attract foreign direct investment and attempt to form globally competitive industries, many will be unsuccessful given that they may simply be "out of sync" with the larger cycles of economic change.

In many ways, the stability and dominance of the core and the rapid emergence of the semiperiphery have thwarted efforts to elevate the economic development status of peripheral nations. Economic waves and cycles continue to work against the less-developed regions of the SOUTH. Nations of Central America and the Caribbean, West Asia, the Middle East, North and sub-Saharan Africa will remain continually at a competitive disadvantage. Peripheral regions may be forced (indefinitely) to adjust to change rather than be entities that are responsible for initiating change.

GLOBAL MARKETS, INFRASTRUCTURE, AND DEVELOPMENT ALTERNATIVES

The proliferation of global product markets has given rise to the transnationalization of capital and production. The rise of multinationals and the globalization of production are trends that have been encouraged by space-time convergence, reductions in transportation costs due to improvements in containerization and shipping, growth of overseas markets, advanced information and communications technologies, and the global trend toward Western-style consumerism. The big emerging markets in the SOUTH continue to increase demand for high-valued manufactured products such as turbines, aircraft, communications and high-voltage transmission equipment, computers, locomotives, and personal transport vehicles. Industries like robotics, fiber optics, lasers, advanced semiconductors, satellite networks, plastics and synthetic materials, energy technology, and precision machinery continue to expand globally, shaping and reshaping regional economic landscapes and international production networks. Modernization of the SOUTH has brought about a renewed dependence of the SOUTH on the production regimes of the NORTH as consumer and producer demand becomes development-driven. Hence, transformation of the world economy continues to find the bulk of value added in production networks which are dominated by firms of core economies.

While technology, international trade and investment, and transnationalized production networks have inspired both rivalry and cooperation amongst competitors and regions, these forces have also changed the way in which corporations do business. Firms from developing countries find it difficult to enter and compete with already established enterprises and large conglomerates, as the world's leading firms employ the latest production technologies and marketing approaches. The organizational and management structures of multinationals have become increasingly more globally conscious, coordinating each and every stage of the value-added sequence to best suit the needs of the company and its customers. Large-scale multinationals have become increasingly flexible in production and the development of product lines (as they respond to market signals and the diverse and changing needs of consumers in the global marketplace).

Global industries have increased the interconnectiveness of regions, but have reduced the likelihood that disadvantaged areas will ever rise up to the status of core economies. The increasing mobility of capital, the footloose nature of multinationals, the global integration of the value-added process, and the wide range of opportunities and locations for production and sales have combined to ensure the survival and success of multinationals. Developing regions that currently lack a global competitive advantage of their own are unlikely to develop one without external assistance.

This is not to say that multinationals do not face challenges of their own. As global competition heats up and as regions and nations deal with economic cycles, companies may find it increasingly difficult to expand their empires above and beyond the expected growth in demand associated with demographic change. Perhaps the only way to achieve further growth or increase market share is through corporate mergers, acquisitions, and take-overs. Instead of competing for a larger share of a market, many companies may simply buy or acquire it. Many companies could run the risk of becoming too large and/or too diverse (sectorally and spatially) and impossible to manage. As a result, **diseconomies of size and diversity** may set in—*economic losses associated with an operation being too large or geographically diverse.*

Industrial modernization and the development of social and economic networks are key survival strategies for firms and regions that must sustain their global competitiveness. Cooperation between government and industry is essential to this process. Many argue, however, that there is a trade-off between intervention and control on the one hand and growth and efficiency on the other. This is particularly true in the case of financing and operating infrastructure. As there has been increasing attention given to the role of infrastructure in the economic development process, there has also been much criticism levied toward government as an authority for identifying infrastructure needs and as overseer in the construction, maintenance, and management of infrastructure. Most analysts agree that industrial modernization must be accompanied by suitable investments in basic infrastructure such as roads, railways, power plants, water lines, dams, canals, bridges, ports, irrigation, drainage, and sanitation systems, and telecommunications networks. The growing importance of infrastructure to industry will hurt developing regions in their attempt to attract direct foreign investment and expand the size and depth of their regional economic bases. The social and economic benefits associated with infrastructure-led growth are many indeed, yet the distribution of these benefits may be geographically biased toward the core and semiperiphery.

Research findings from the World Bank show that there is a positive correlation between Gross Domestic Product (GDP) per capita and "infrastructure stock" per capita. The empirical evidence suggests that as a country's income grows so does the amount spent on infrastructure (see data in Table 10.4). This has led many policy makers to support **infrastructure-led development**—*the stimulation of regional economies through investments in infrastructure* (Figure 10.9).

While it is widely accepted that investment in infrastructure yields enormous returns to a regional economy, there is also evidence to suggest that the type of infrastructure put into place may determine the extent to which the benefits of growth will aid in poverty alleviation. Properly targeted investments in infrastructure and

TABLE 10.4	Infrastructure Coverage and Economic Performance		
	Reforming		
Indicator	Low-Income Economies	Middle-Income Economies	OECD Members
Coverage			
Main (tele-communications) lines per 1000 persons	3	73	475
Households with access to safe drinking water (%)	47	76	99
Households with electricity (%)	21	62	98
Performance			
Paved roads not in good condition (%)	59	63	15
GNP per capita 1991 (U.S. $ 1991)	293	1941	20,535
GNP per capita growth rate 1980–1991 (%)	–0.2	–0.6	2.0

Source: *World Development Report 1994*, p. 112.

Figure 10.9 Rapid urban development in Asia. Yet another skyscraper is erected in ever booming Hong Kong.
l0.9: ©Owen Franken/Stock Boston

satisfactory maintenance of existing infrastructure can help contribute not only to economic and environmental sustainability but can also lead to savings from efficiency gains, less waste, and increased reliability of service. The corresponding economic windfalls could then be targeted to human development projects.

Policy makers are concerned, however, with the speed at which infrastructure is being added to developing regions. Some analysts contend that inadequate attention is being paid to critical engineering details as structures are rapidly erected in order to keep pace with demand. To cut costs, many projects have been forced to use cheaper materials or inherently flawed designs. Concerns over structural integrity and subsequent shortened lifespans of many buildings and bridges have forced policy makers to reexamine the benefits of infrastructure-led development as defective or poorly designed structures continue to pose a threat to human life. The recent collapse of several large structures in Asia has brought worldwide attention to this growing problem.

Investment in infrastructure will undoubtedly yield returns that are not only beneficial to a regional economy but beneficial to the poor. Access to clean water and proper sanitation, the use of nonpolluting energy sources, safe disposal of solid waste, and better manage-

ment of urban traffic flows can greatly reduce the risk of accidents, exposure to disease, or the possibility of being subjected to dangerous or life-threatening situations. Priority must be given to regions where the demand for infrastructure exceeds the ability to supply that infrastructure. It is estimated that the need for infrastructure among the developing nations of the SOUTH is immense. It is estimated that more than $1 trillion must be spent on infrastructure over the next 10 years just to keep up with demand. Without a viable export base and export earnings, many developing regions may find themselves unable to divert or commit a sufficient amount of funds to much-needed infrastructure. The opportunity costs of not investing are high, as it will greatly affect the quality of life of billions of people throughout the SOUTH.

While many developing economies are doing well as a result of investments made during the 1980s and 1990s, there is still an enormous need for investment capital to fund present and future development projects, not to mention the general maintenance of now-aging infrastructure. Governments in developing countries are finding it increasingly difficult to fund the demands of an enlarging public sector, let alone maintain deteriorating industrial infrastructure. Many regions must turn to the private sector and multinationals to help finance, build, operate, and maintain infrastructure—particularly in transportation-related sectors. In the case of publicly managed transport authorities, it has been recognized that private sector involvement offers several distinct advantages, including (a) improved operational efficiency of existing transportation infrastructure; (b) reduced risk in the financing of new infrastructure as private companies carry some or all of the financial burden; (c) reductions in project completion time as private investors seek to expedite the process of putting infrastructure in place; and (d) the ability to better identify attractive investment projects and shift funds toward funding those projects with the highest benefits (Nijkamp and Rienstra, 1995).

A world economy in transition has certainly brought about a new set of rules and challenges. Globalization of production and international finance have altered the way in which regional economies view the nature of competition and investment. Instead of looking to governments of the developed world or agenda-constrained lending institutions for monetary assistance in matters of development, many regions continue to seek out help from the private sector. Global capitalism has brought about many new possibilities for funding growth and modernization. As the world's developing regions become increasingly aware of the benefits of privatization and private sector involvement, they are less likely to seek additional development loans from governments or institutions like the World Bank to meet their infrastructure requirements. Turning to private enterprise and foreign investors to fund economic growth will soon become a subtle yet popular form of indirect debt relief.

Transport and communications technologies will play an ever expanding role in the globalization process. Worldwide communications networks such as the Internet will intensify competition and increase consumer access to information on products, prices, and suppliers. Internet-related retail sales and consumer banking activities are likely to expand exponentially, bringing about new ways in which to market and sell products. As global communication systems become increasingly accessible to consumers in terms of cost and availability, they will redefine the notion of what constitutes a geographic market and/or a globally competitive firm or region.

Information systems, computers, and telecommunication technologies are transforming business and industry, changing the geography of production and commerce as well as the implied development potential of regions. While many industry analysts predict that location will become less important over time with the global-wide proliferation of advanced communications technologies and information service providers, not to mention increasing access to the Internet, there is evidence to suggest that the information service revolution may be geographically restricted in terms of the distribution of benefits. Recent research has suggested that the growth of information systems and telecommunications infrastructure has been largely confined to geograhic areas where money income and financial resources are plentiful and demand and profit potential are high (Gorman, 1999). Hence, large metropolitan areas in core or developed regions continue to benefit greatly from the growth of the world's information networks, having access to an increasing number of information networks. A distinct production advantage, therefore, is emerging in places boasting rapidly expanding agglomerations of information services and advanced telecommunications infrastructure. By contrast, more remote and/or less serviced locations continue to fall further behind. Inadequate communications infrastructure throughout the developing world has further expanded the development gap between core and peripheral economies and the gap between large metropolitan areas and places located in the hinterland. While advanced telecommunications, information technologies, and electronic commerce have increased the efficiency and competitiveness of multinational firms, globalization has not benefitted all regions equally. Core regions and/or large metropolitan areas will undoubtedly remain as central nodes in the global economy, suggesting that location will continue to be relevant despite collapse of time and space in the cyber realm.

Developing regions are likely to reexamine some of the more traditional ways in which to speed up the modernization process. Many will turn to **agricultural-led industrialization** as an alternative. Agricultural-led industrialization refers to *the growth of industry which is directly supported by agricultural activities and exports.* This kind of development strategy may be viewed as a measure of last resort. In some cases it may be the only viable option for regions left out of global manufacturing networks.

Growth of the primary sector and food production systems in general has the potential to create many forward and backward linkages in a regional economy. Once harvested, agricultural products will be subject to processing and distribution. These activities have the dual advantage of stimulating suppliers of production inputs (that is, producers of fertilizers, machinery, tools, and energy), and encouraging the retention of value added in home-grown industries which process, package, and transport agricultural commodities. The new millennium is likely to bring about a wealth of investment in agriculture and affiliated sectors throughout the developing world.

Despite the global trend toward free-market activities and capitalism, many weak and peripheral command economies continue to be supported through trade and income transfers. Semiperipheral command economies such as Russia and China continue to show preference toward anticapitalist regimes as trading partners.

Consider the case of Cuba, a communist state of approximately 11 million people that has been largely sustained by its relationship with the republics of the former Soviet Union. The longevity of command economy, however, must be questioned. Though communist leaders such as Fidel Castro are persistent in their efforts to denounce capitalism, they are finding it increasingly difficult to justify their political ideologies when it comes to matters of economy. Cuba continues its desperate attempt to hold on to an outmoded system as the world around it undergoes economic transformation (Figure 10.10).

Command economies such as Cuba and North Korea are likely to endure continued economic hardship in the name of nationalism. Many command economy holdouts have been propped up by external assistance from the larger and more successful pseudo-command or hybrid economies. With internal economic problems, faltering state-owned industries, and a diminishing capacity to support outside economies, Russia and China will increasingly cut back their support of dependent nations, as they undergo economic transformation themselves. Many

Figure 10.10 Politics and economic development go hand in hand. Critics contend that Cuba's economy and people continue to suffer as a result of Castro's short-sighted economic reforms and Marxist orientation.
10.10: ©Bob Crandall/Stock Boston

small command economies such as Cuba may soon find themselves searching for new survival strategies. Ironically, the troubled Cuban economy continues to be sustained by remittances from Cuban-Americans (that is, money transfers from the United States to relatives back home). Income transfers to Cuba from abroad now account for more income than does Cuba's lucrative sugar trade. Cuba is also becoming more dependent upon tourism.

Tourism is likely to play a prominent role in the economic development of many peripheral areas. The growing tourist trade has been a vital source of income for many developing economies, particularly those that have been unable to gain a global competitive advantage in exports of secondary sector commodities. The tourist industry, however, represents a luxury consumption item and is geographically limited to high-amenity regions. Furthermore, tourism is considered to be an unreliable source of income as economic and business cycles can bring about recession or depression and the subsequent slowdown in tourist activities and travel. Yet, worldwide recession, concerns over political and economic instabilities throughout the SOUTH, and the recent rise of global terrorist activities have done little to diminish the overall revenue-generating potential of this high-growth industry.

Recently, "ecotourism" has grown in popularity. **Ecotourism** refers to *tourist activities and travel to distinct high-amenity regions which offer unique scenery and a one-of-a-kind setting or natural environment.* Ecotourist hotspots such as the Patagonia region of Argentina, the Brazilian rainforest, remote highland areas of Central Asia, and national parks and game reserves in Southeast Asia and East Africa offer visitors a unique visual experience and a chance to enjoy the rich and diverse plant and animal life. These highly unusual and out-of-the-way places offer both physical and cultural attractions which cannot be found elsewhere on the planet. These characteristics set these areas apart from typical and substitutable tourist destinations which offer pleasant climates and more mainstream tourist activities. With a global boom in world tourism, regions in developing economies may increasingly examine their tourist potential and market their unique geographic qualities as they attempt to compete for a share of tourist income. Developing a successful ecotourist trade is highly contingent upon several factors including a region's accessibility and reputation, the preferences and perceptions of potential tourists in core economies, the quantity, quality, and competitiveness of accommodations, the uniqueness of the tourist experience as related to the physical and cultural attractions found within that region, and the general political stability and travel climate.

It is far too early to tell whether or not agricultural-led industrialization or ecotourism will work as development strategies or remedies for peripheral economies. Many believe that there are far too many forces at work (both regionally and globally) and too many obstacles to overcome to ever ensure the success of any one strategy. The "multipolar" aspects of the world economy, the organization of nations about region-specific trading blocs, and diversification of multinational assets across interregional production networks, have made regional development policies increasingly less effective at promoting economic parity. The impacts of regional economic development policies will most likely vary both geographically and sectorally, and are not likely to undermine processes and forces at the global level. Though regions of the world economy have become more interdependent, they have also become more divided and less powerful. Even powerful and prominent nations such as the United States, find themselves less able to dictate international economic policy and influence the direction of economic change. In short, the diminished political and economic clout of individual nations has reduced their ability to single-handedly influence other regions or economies. The development agendas of individual regions or nations are likely to remain subordinate to the agendas of nations or groups of nations unified under continental trade accords. Developing regions and nations with distinct competitive global advantages in a given industry may find themselves unable to exploit their positions should external conditions or forces combine to nullify that competitive edge.

TECHNOLOGY, HAZARDS, AND HUMAN-ENVIRONMENT RELATIONS

Technology continues to redefine possibilities for human economic activity and economic change. While technology has allowed human agents to manipulate the parameters of physical and production systems, it has not necessarily guaranteed that progress will be made in the human condition. The world's physical and production systems continue to be at risk from both man-made hazards (for example, waste, toxins, emissions, and industrial pollution) and **natural hazards**—*events in nature (predictable or unpredictable) which can be harmful to people, property, land, production systems, or the environment*. The most common types of natural hazards are droughts, floods, earthquakes, typhoons, landslides, tornadoes, and severe weather. Hazards have become an ongoing threat to the sustainability of regional and global production systems and the people which make up those systems.

As discussed by Jones (1991), technology has actually increased regions' vulnerability to natural hazards. Disturbing the natural environment by disrupting nature's delicate balance has increased the likelihood of experiencing natural disasters. In many instances, increased vulnerability to hazards and environmental degradation are by-products of development and technologically

induced change. For example, human economic activity since the start of the Industrial Revolution has dramatically increased atmospheric concentrations of the **greenhouse gases**—*carbon dioxide (CO_2), chlorofluorocarbons (CFCs), nitrous oxides, and methane*. Heavy industry, transportation sectors, and urban population growth have contributed greatly to the increased emissions of greenhouse gases over the past several decades. Rising concentrations of these gases have also been blamed on slash and burn agriculture, wildfires, farm animal waste, decomposing waste in landfills, rice paddy production, and decaying organic matter. As a result, there is scientific evidence that the earth is proceeding through an accelerated warming period as attributed to the so-called "greenhouse effect." The **greenhouse effect** *is the subsequent warming of the earth's atmosphere as a direct result of increasing concentrations of greenhouse gases and their efficient radiation-absorption capabilities*. Sunlight entering the earth's atmosphere as long-wave infrared radiation and heat is naturally deflected off of the earth's surface, flowing back into space. The presence of greenhouse gases traps much of the radiation near the surface causing **global warming**—*the increase in land, lower atmosphere, and water-surface temperatures as attributed to the greenhouse effect*. Although it is difficult to predict how global warming will affect each and every region of the world, there is evidence to suggest that it will lead to:

1. major shifts in climate, greater variability in weather systems and more extreme changes in temperature, precipitation, wind speeds, etc.;
2. the inundation, erosion, and regression of coastal areas in relation to the melting of the polar ice caps and rising sea levels, not to mention destruction of coral reefs and mangrove forests, inland saltwater intrusion in low-lying areas, and loss of agricultural land;
3. changing productivity of cropland, shifting agricultural zones, and loss of biodiversity in forest systems; and
4. changes in the regional water supplies in relation to shifting climates (Rosenberg, 1988).

As a consequence of global warming, experts have estimated that the expected rise in sea level will range between four and seven feet. An extreme rise in sea level could have a devastating impact on many low-lying coastal regions. Rising sea levels are likely to result in a substantial reduction of inhabitable land area in nations such as Bangladesh, Egypt, and the United States. Specifically, the coastal and lowland regions of Louisiana, Florida, Maryland, and New Jersey are likely to be hardest hit (Cutter et al., 1991). A growing concern over global warming has helped to rekindle interest in renewable energy sources and has prompted government and industry to take a serious look at ways in which to improve energy efficiency and reduce greenhouse gas emissions.

From a hazards perspective, the combination of global warming, greater variability in climate and weather systems, and the rapid expansion of human settlements has made droughts, storm surges, hurricanes, monsoons, blizzards, avalanches, earthquakes, and storm conditions seem to occur with more frequency and severity. It seems as if natural events are becoming more dangerous to the increasing number of humans which must contend with these events. The growing vulnerability of human agents to hazards, however, is not so much an indication that nature is becoming increasingly more violent with time, but rather it stems from the fact that rapid population growth, urbanization, and development have proceeded in ways that place an increasing number of people at risk. As the world's population expands, more people are exposed to potential hazards. In addition, an increasing number of people are inadequately prepared to deal with natural hazards or underestimate the risk of the forces found within nature (Jones, 1991). As greater numbers of people reside along floodplains, fault lines, hurricane-prone regions, or hazard-sensitive areas, there will be increasing casualties and staggering economic losses. Many existing settlements have been already established in areas predisposed to hazards. Many people choose to stay in these areas despite the risk, opting to deal with hazards on an as-they-need-to basis. Many remain and repair damages in nature's aftermath, having established social and economic roots to the areas in question. People in impoverished regions face the greatest difficulties in dealing with hazards as they have little or no resources to allocate toward preventative measures or clean up.

Global warming and natural hazards are not the only big environmental concerns. There is a growing concern over the destruction of the earth's **ozone layer**—*a stratospheric layer of gaseous ozone (O_3) that is responsible for filtering out the harmful ultraviolet rays of the sun*. There is a growing "hole" in the ozone layer, predominantly over the continent of Antarctica. There is scientific evidence which links this anomaly to the increase in chlorofluorocarbon emissions worldwide (a by-product of development). Increased exposure to ultraviolet radiation has been blamed for the rising incidence of skin cancer among humans, reduced ability of human immune systems to fight infection and disease, decreasing productivity in agricultural systems from biological and genetic damage to plants, aquatic species, and animal wildlife. Increasing concentrations of air pollutants such as sulfur and nitrous oxides have also given rise to the phenomenon known as **acid rain**—*the deposition of destructive acids which fall to the earth's surface in the form of precipitation and dust particles from pollutants washed down from the atmosphere*. Acid rain continues to damage vegetation and lower soil and surface water quality. In addition, acid rain has left visible scars on the human economic landscape as it slowly wears away structures made of stone and metal.

Using technology in response to potential hazards has allowed human beings to gain a false sense of security. Consider the case of the proverbial "one-hundred-year flood"—a flood that occurs (on average) about once every 100 years. This type of flood continues to be a real danger for people living along the floodplain of the Mississippi River. The recent occurrence of a 100-year flood is a testimony to the failure of human technology to deal with the unbridled force of nature. Flood walls, levees, and sandbag barriers did little to control the mighty Mississippi once flood waters began to rage. The annihilation of entire towns was not an uncommon sight as the Mississippi washed away the homes, hopes, and dreams of those lying in its wake. Not only were human lives and material possessions lost in the flood, but the river claimed large tracts of productive farmland, washing away precious topsoil and the livelihoods of many farmers. The destructive power of floods of this magnitude has refocused attention on the importance of maintaining natural wetlands and drainage areas in river systems. It is also a reminder that human agents must inevitably deal with the consequences of manipulating the landscape. Serious questions have also been raised as to the limitations of technology and human's ability to control natural and man-made hazards. The aftermath of such events has forced many to reexamine risk and the opportunity costs of risk exposure (by place and setting).

The recent destruction of settlements in Honduras during the 1998 hurricane season is a testimony to the power of nature. Hurricane Mitch alone caused billions of dollars of property damage as excessive winds and rainfall led to floods and mud slides, not to mention thousands of deaths in one of the worst hurricane seasons on record. Numerous developing nations in Central America and the Caribbean remain at risk as tropical storms and hurricanes continue to threaten their economic landscapes. Residents in storm- and hurricane-prone regions in the tropics have learned to cope with risk as a part of their everyday lives (Figure 10.11).

The seemingly increasing exposure to natural hazards and the recent occurrences of severe and unpredictable weather patterns have been tied to global climate changes. Severe droughts, wildfires, and floods throughout Asia, Africa, and Latin America have been linked to oceanic and atmospheric changes. These events not only remind us of the commanding power of nature, but also reveal the interconnectiveness of regional climates and the cumulative impact human activity has had on the earth's physical systems.

Increasing variability in climate systems has been directly connected to global warming. Severe weather conditions in many regions have also been attributed to **El Niño**—*the periodic and somewhat unpredictable warming of ocean currents in the southeast Pacific* (in waters off the coast of Peru and Chile), as well as its complement **La Niña**—*the periodic cooling of ocean* *currents in the southeast Pacific which generally follows El Niño.* ENSO (El Niño Southern Oscillation) events have a long history of being correlated to weather patterns in geographic regions around the globe. The nations of Latin America, the Caribbean, and sub-Saharan Africa have been especially susceptible to ENSO events, experiencing periodic droughts or floods in years directly before, during, or after El Niño. The El Niño phenomenon continues to cause immense amounts of human suffering and economic losses throughout many of the poorest regions of the SOUTH. Research efforts have intensified to improve seasonal forecasting of weather patterns in ENSO-sensitive areas. Scientists continue their work to fine-tune their predictive models and formulate proactive food and economic security measures to prevent problems in areas commonly affected by ENSO events.

Technology and development have also set the stage for new types of hazards, some of which are larger in scale and potentially more devastating than those found in nature. For example, the reality of the nuclear age has brought hope and promise in terms of a reliable and efficient energy source, yet it has also brought the threat of nuclear accidents, radioactive exposure, and nuclear war. Even a limited nuclear conflict with many survivors could have a devastating impact on the earth and its inhabitants as they face the prospects of a "nuclear winter." Dust in the atmosphere could block out sunlight for an indefinite period of time, causing surface temperatures to plummet. Subsequent changes in climate would severely reduce agricultural capacity and production, and would most likely lead to widespread famine. Entire economies would collapse and world society as we know it would be thrust into chaos and disorder. Safe drinking water would be scarce in many regions, as would food and medical supplies. The sick and the unburied dead would cause a rash of disease and epidemics, wiping out large segments of the world's population. Health and environmental problems related to the subsequent destruction of the ozone layer and delayed radioactive fallout would eventually claim a staggering number of human lives and destroy most of the earth's wildlife. Fortunately, this disturbing scenario is avoidable should common sense prevail.

Nonetheless, the proliferation of chemical, biological, and nuclear weapons amongst hostile nations in the developing world poses a great threat to world peace and prosperity. Leaders in the international community continue to encourage dialogue to reduce the likelihood of interregional confrontation. Terrorist activities, however, are on the rise. The growing international terrorist element threatens the very sustainability of the entire human civilization. While our ability to resolve conflict and sidestep confrontation has been motivated, in part, by our understanding of the repercussions of engagement, many political scientists feel that we are no closer to realizing world peace today than we were 50 years ago. Govern-

Figure 10.11   Storm-ravaged Honduras in the wake of Hurricane Mitch, a glaring example of how natural disasters continue to wreak havoc on hazard-prone areas.

10.11: ©Sean Sprague/Stock Boston

ment and industry, nevertheless, are working hard to advance unification and harmony through science, technology, and policy.

Prospects for world peace are indeed good should people and regions learn to put aside their cultural, religious, political, and economic differences and work collectively toward the common good. Economic development must proceed in ways that do not compromise the ability of present and future generations in all regions to live, prosper, and make progress in the human condition. We must not limit ourselves to the notion of economic growth as a prerequisite for development, nor must we let the benefits of growth be confined to a handful of regions. The basic life-support systems which fuel the world economy are sustainable so long as human beings act, in good faith and conscience, as kind stewards of the earth's physical systems and for the betterment of humankind.

CONCLUDING REMARKS

Findings solutions to the economic development problems of the nations of the SOUTH is something that has been made exceedingly difficult in a world economy dominated by forces that are largely beyond the control of individual regions and nations. Regional economic-development policy making in general is becoming ever more complex due to the myriad of considerations and impact assessments that must be made before policy implementation. To many regional analysts, peripheralness is a long-term liability, and in and of itself constitutes a major impediment to the economic development process. As industrialized economies of the core and semiperiphery, as well as multinationals, exercise their global competitive advantages and clout, they will continue to set the tempo and direction of regional and global economic change. This may or may not be compatible with the development strategies or objectives of individual regions or nations. In fact, concentration of power and the accumulation of capital and wealth in core and semi-peripheral economies is likely to amplify existing development problems and disparities between the core and periphery. The true development challenge for many poor nations of the SOUTH may be restricted to finding new ways in which to manage their economic realities in light of the constraints, adverse economic conditions, and circumstances which, in all likelihood, will remain outside their control.

KEY WORDS AND PHRASES

acid rain   286
agricultural-led industrialization   283
circular and cumulative causation   271
depression   278
diseconomies of size and diversity   281
economies of synergy   270
ecotourism   284

El Niño   286
global warming   285
green consumerism   275
green development   274
greenhouse gases   285
greenhouse effect   285
infrastructure-led development   281
Kondratiev wave   276
Kuznets cycles   277

La Niña   286
natural hazards   285
overaccumulation crisis   278
ozone layer   286
prosperity   277
recession   278
recovery/growth   277
stagflation   278
technocratic ideology   270
world view   274

REFERENCES

Adams, W.M. *Green Development*. London, U.K.: Routledge (1990).

Berry, B.J.L. *Long-Wave Rhythms in Economic Development and Political Change.*" Baltimore, MD: Johns Hopkins University Press (1991).

Berry, B.J.L., E.C. Conkling, and D.M. Ray. *The Global Economy*. Englewood Cliffs, NJ: Prentice Hall (1993), Chapter 11, pp. 312–338.

Boulding, K.E. *The Principles of Economic Policy*. London, U.K.: Staples Press (1959).

Business Week. "Growth vs. Environment: A Push for Sustainable Development." by E. Smith with bureau reports (May 11, 1992), pp. 66–75.

Caring for the Earth: A Strategy for Sustainable Living. Gland, Switzerland: IUCN/UNEP/WWF (October, 1991).

Clawson, D.L., and J.S. Fisher. *World Regional Geography: A Development Approach* (6th edition). Upper Saddle River, NJ: Prentice Hall (1998).

Cutter, S.L., H.L. Renwick, and W.H. Renwick. *Exploitation, Conservation, Preservation: A Geographical Perspective on Natural Resource Use* (2nd edition). New York: J. Wiley & Sons (1991).

Daly, H.E. *Steady-State Economics* (2nd edition). Covelo, CA: Island Press (1991).

Freeman, C. (ed.). *Technological Innovation and Long Waves in World Economic Development*. London, U.K.: Frances Pinter (1984).

Gainesville Sun. "Cuba's Tough Choices." by D. McFarlin, Gainesville, Florida (December 6, 1998), pp. 1G,3G.

Global Environment Outlook. United Nations Environment Program (UNEP), New York: Oxford University Press (1997).

Gorman, S.P. "Network Analysis of the Internet and Its Provider Networks." *MA Thesis* (Department of Geography), University of Florida, Spring 1999.

Johnson, R. (ed.). *Ending Hunger: An Idea Whose Time Has Come*. New York: Praeger (1985).

Jones, D.K.C. "Environmental Hazards." In *Global Change and Challenges: Geography for the 1990s*. R. Bennett and R. Estall (eds.). London: Routledge (1991), pp. 27–56.

Knox, P., and J. Agnew. *The Geography of the World Economy* (2nd edition). London: Edward Arnold (1994), Chapter 1, pp. 3–15.

Kondratiev, N.D. "The Long Waves in Economic Life." *Review of Economic Statistics* 17 (1935), pp. 105–115.

Kuznets, S. *Secular Movements in Production and Prices*. Boston, MA: Houghton Mifflin (1930).

Kuznets, S. *Economic Change*. New York: W.W. Norton (1954).

Kuznets, S. *Economic Growth of Nations*. Cambridge, MA: Harvard University Press (1965).

Levacic, R. *Economic Policy-Making*. Totowa, NJ: Barnes & Noble (1987).

May, B. *The Third World Calamity.* London, U.K.: Routledge (1981).

Meadows, D.H., D.L. Meadows, and J. Randers. *Beyond the Limits: Confronting Global Collapse, Envisioning a Sustainable Future.* Post Mills, VT: Chelsea Green (1992).

Myrdal, G. *Economic Theory and Underdeveloped Regions.* London: Duckworth (1957).

Nijkamp, P., and S.A. Rienstra. "Private Sector Involvement in Financing and Operating Transport Infrastructure." *The Annals of Regional Science* 29 (1995), pp. 221–236.

Pred, A. *The Spatial Dynamics of U.S. Urban-Industrial Growth, 1800–1914.* Cambridge, MA: The MIT Press (1966).

Rosenberg, N.J. "Greenhouse Warming: Causes, Effects and Control." *Renewable Resources Journal* 6 (Autumn 1988), pp. 4–8.

Seitz, J.L. *Global Issues: An Introduction.* Oxford: Blackwell (1995).

Smith, N. *Uneven Development.* Oxford: Basil Blackwell (1984).

Smyser, W.R. *Refugees: Extended Exile.* (published with the Center for Strategic and International Studies in Washington, D.C.), New York: Praeger (1987).

Stutz, F.P., and A.R deSouza. *The World Economy* (3rd edition). Upper Saddle River, NJ: Prentice Hall (1998).

Timberlake, L. *Only One Earth: Living for the Future.* London, U.K.: Earthscan (1987).

van Duijn, J.J. *The Long Wave in Economic Life.* London, U.K.: Allen and Unwin (1983).

Wallerstein, M.B. *Food for War—Food for Peace: United States Food Aid in a Global Context.* Cambridge, MA: The MIT Press (1980).

World Resources 1998–99: A Guide to the Global Environment. New York: Oxford University Press (1998).

INDEX